Workbook to Accompany Anatomy & Physiology | REVEALED®

Robert B. Broyles, Jr.
Butler County Community College

Boston   Burr Ridge, IL   Dubuque, IA   New York   San Francisco   St. Louis
Bangkok   Bogotá   Caracas   Kuala Lumpur   Lisbon   London   Madrid   Mexico City
Milan   Montreal   New Delhi   Santiago   Seoul   Singapore   Sydney   Taipei   Toronto

Higher Education

WORKBOOK TO ACCOMPANY ANATOMY & PHYSIOLOGY REVEALED

Published by McGraw-Hill, a business unit of The McGraw-Hill Companies, Inc., 1221 Avenue of the Americas, New York, NY 10020. Copyright © 2008 by The McGraw-Hill Companies, Inc. All rights reserved. No part of this publication may be reproduced or distributed in any form or by any means, or stored in a database or retrieval system, without the prior written consent of The McGraw-Hill Companies, Inc., including, but not limited to, in any network or other electronic storage or transmission, or broadcast for distance learning.

Some ancillaries, including electronic and print components, may not be available to customers outside the United States.

 This book is printed on recycled, acid-free paper containing 10% postconsumer waste.

2 3 4 5 6 7 8 9 0 QPD/QPD 0 9 8 7

ISBN 978–0–07–340354–0
MHID 0–07–340354–7

Publisher: *Michelle Watnick*
Senior Sponsoring Editor: *James F. Connely*
Senior Developmental Editor: *Kathleen R. Loewenberg*
Marketing Manager: *Lynn M. Breithaupt*
Lead Project Manager: *Mary E. Powers*
Lead Production Supervisor: *Sandy Ludovissy*
Lead Media Producer: *John J. Theobald*
Designer: *John Joran*
Compositor: *Carlisle Publishing Services*
Typeface: *10/12 Stone Serif*
Printer: *Quebecor World Dubuque, IA*

www.mhhe.com

P R E F A C E

When I first became aware of Anatomy & Physiology | REVEALED®, it immediately became obvious to me that this was the ultimate hands-on learning tool for any student studying the human body. Through the use of layer-by-layer dissection photographs of actual cadavers, students can gain insight into the intricate design of the human body. Radiological images and animations offer their unique perspectives as well. Not every student has access to a cadaver, and even those that do, cannot take it home for review. Now, however, students can take a cadaver home with them, through this accurate and detailed study of the human body.

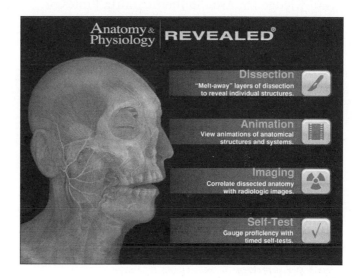

Audience

This workbook is designed to complement the Anatomy & Physiology | REVEALED® software, a powerful learning tool. Regardless of their computer skills, students will be up and running in a short time as they learn how to master the concepts within the user-friendly interface. Developed to support one-semester or two-semester anatomy and physiology courses, human anatomy and human physiology courses, and a variety of health-related career programs, this workbook, combined with Anatomy & Physiology | REVEALED®, will ensure students the very best opportunity for getting the most out of their classes. For many schools, the combination of the two will take the place of their laboratory manual.

Key Features

- The inside front cover offers key screen captures as reminders for basic navigation steps.
- The first chapter walks students through an introduction to the format and various sections of Anatomy & Physiology | REVEALED®.

- Exercises follow the arrangement of Anatomy & Physiology | REVEALED® requiring students to complete steps and answer follow-up questions.
- Perforated pages allow exercises to be easily removed and handed in to instructors.
- "Heads-Up!" sections offer tips and reminders about the program.
- "Check-Point" questions serve as brief self-tests to check comprehension.
- Useful tables include bonus questions on related topics.
- To help students master terminology, the inside back cover lists key parts of anatomical and physiological terms, their meanings and example words.

Acknowledgements

I offer my heartfelt thanks to colleagues who've offered suggestions and/or class-tested parts of this workbook. Their contributions are much appreciated. I'd also like to recognize the team at McGraw-Hill Higher Education: Publisher, Michelle Watnick; Sponsoring Editor, Jim Connely; Marketing Manager, Lynn Breithaupt; and Project Manager, Mary Powers. I would like to give a special thank you to my Editor, Kathy Loewenberg. It is through her wise council and firm guidance that this project has come to fruition. Susan Vorwald, our campus MHHE Sales Rep was instrumental in bringing us all together to begin this project in the first place. Thank you, Susan. I would also like to acknowledge Dr. Ann Stalheim Smith, Professor Emeritus of Human Body at Kansas State University, who taught me how to teach while I was an undergrad in her classroom so many years ago. Finally, this project could not have been possible without the loving support of my Beloved Bride – Patricia, who has sacrificed a lot along the way. Thank you.

Reviewers

Valerie A. Bennett
Clarion University of Pennsylvania

Deanna Ferguson
Gloucester County College

Cynthia R. Hartzog
Catawba Valley Community College

Mark F. Hoover
Penn State Altoona

Julie Huggins
Arkansas State University

Elisabeth C. Martin
College of Lake County

Elizabeth M. Meyer
College of Lake County

Mark Paternostro
Pennsylvania College of Technology

Dawn E. Roberts
University of Massachusetts Amherst

Eric L. Sun
Macon State College

Judith Moon Wolff
Delaware Technical and Community College

A Special Note to Students

Anatomy & Physiology | REVEALED® is not a learning tool that is exhausted with a single use, nor is it any less valuable the more times you use it. The high quality cadaver photos in Anatomy & Physiology | REVEALED® are designed to provide you a unique opportunity to match visuals of the human body with what you are learning from your instructor. With each use, you will glean more information to add to your knowledge base, and as you do, you will also gain the confidence that you need at exam time. With this in mind, use Anatomy & Physiology | REVEALED® on a daily basis to hone your skills, refresh your memory, and stay sharp.

Be sure to take advantage of the Self-Test aspect of Anatomy & Physiology | REVEALED®. It allows you to determine how well you understand the information presented, and will reveal areas that you may need to revisit.

This workbook is written for YOU, the student. If you have any comments on how it may be improved, please let me know – I am always open to suggestions! My desire is that this workbook, in combination with Anatomy & Physiology | REVEALED®, will help you to excel in your study of the human body.

Wishing you the best...

Bob Broyles
Biology Department
Butler County Community College
901 S. Haverhill Rd.
El Dorado, Kansas 67042
bbroyles@butlercc.edu

ABOUT THE AUTHOR

Bob embarked on his teaching career while in the third grade, bringing critters from his rural home to the kindergarten classroom. While in graduate school, he began teaching undergraduate anatomy and physiology labs. This set the stage for his current position at Butler County Community College, where he teaches anatomy and physiology lecture and lab, and cadaver dissection.

Bob has a passion for teaching; desiring to guide others to understand the complexities of the natural world as experienced in their daily lives. His other passions include organic gardening (he is a Master Gardener), landscaping for wildlife (primarily birds), prairie restoration, photography, and bird studies.

Bob's previous projects with McGraw-Hill Publishing include the Instructor's PowerPoint Presentations for Saladin's *Human Anatomy* text, and multiple textbook reviews.

Bob resides with his wife Patricia on a small farm, where they grow prairie wildflowers, as well as trees and shrubs that benefit birds, bees, and butterflies.

CONTENTS

APR VOLUME THREE— CARDIOVASCULAR, LYMPHATIC, AND RESPIRATORY

Chapter 5 The Cardiovascular System   161

CONTENTS

APR VOLUME FOUR— DIGESTIVE, URINARY, REPRODUCTIVE, AND ENDOCRINE

Chapter 8 The Digestive System   281

Chapter 9 The Urinary System   313

Chapter 10 The Reproductive System   329

CHAPTER 1

Introduction: Becoming Familiar with Anatomy & Physiology | Revealed®

Overview: Getting Started

We will use our first anatomical system, the skeletal system, to walk through the many facets of this powerful learning tool called **Anatomy & Physiology | Revealed®**. After you have mastered this chapter, you will be able to apply these skills to the remaining systems of the body. Let's get started!

Insert **Anatomy & Physiology | Revealed® Volume 1 – Skeletal and Muscular Systems** into your computer's CD drive and the following image will appear.

This is the **Home screen**. In the larger **IMAGE AREA** to the right you will find all of the information for the current CD: the volume - *Anatomy & Physiology | Revealed® Volume 1;* and the subject matter – **Skeletal System** and **Muscular System.**

Let's move counterclockwise through the other windows of the **Home screen** to familiarize ourselves with the navigational tools provided. We will identify each window, and the tools for each, and then we will go through a detailed application of these tools.

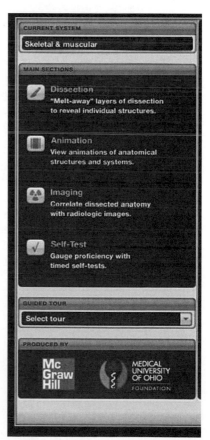

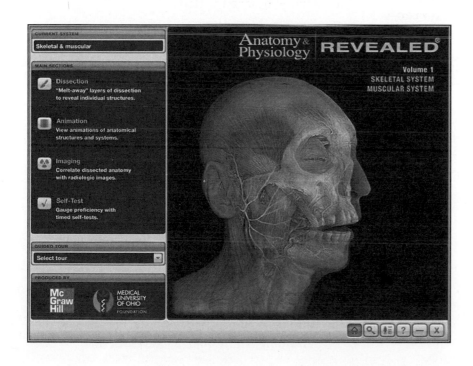

The left window is divided into several smaller windows.

The **CURRENT SYSTEM** window.

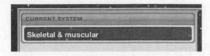

The **MAIN SECTIONS** window.

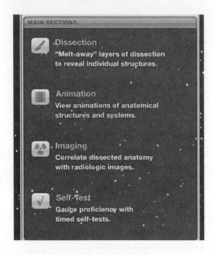

And the **GUIDED TOUR** window.

The drop-down menu within the GUIDED TOUR window will walk you through the major aspects of *Anatomy & Physiology | Revealed®*.

At the bottom right of the **Home screen** is a series of **SUB NAVIGATION** buttons.

As we will see shortly, these buttons allow you to quickly move between screens.

The **CURRENT SYSTEM** window tells you which of the body systems the current CD covers.

There are four **Section Buttons** located in the **MAIN SECTIONS** window:

- The **Dissection** button.
- The **Animation** button.
- The **Imaging** button.
- The **Self-Test** button.

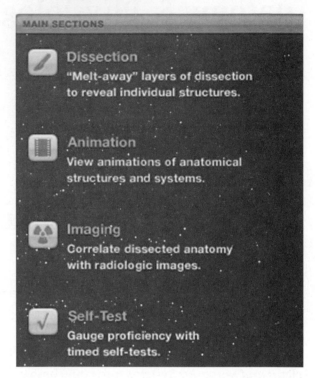

We will look at these **MAIN SECTIONS** one at a time following this opportunity for you to evaluate what you have learned.

CHECK POINT:

Becoming Familiar with *Anatomy & Physiology | Revealed®*

1. What is the name for the first screen you see when opening *Anatomy & Physiology | Revealed®*?
2. What is the name for the large window that occupies roughly two-thirds of the screen while viewing *Anatomy & Physiology | Revealed®*?
3. While working with *Anatomy & Physiology | Revealed®*, is there a location where I can find which system disc is in use? If yes, where?
4. What button would you click on for a helpful guided tour of *Anatomy & Physiology | Revealed®*?
5. What are the four **MAIN SECTIONS** in *Anatomy & Physiology | Revealed®*?

The Dissection Section

By clicking the **Dissection** button, the **SELECT A VIEW** window becomes available.

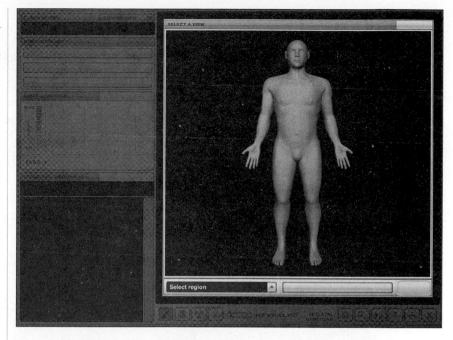

At the bottom of the **SELECT A VIEW** window are two menus. The first one is the **Select region** menu.

Click on this menu to observe the available options.

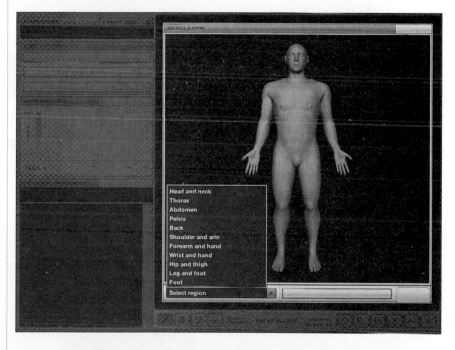

Click on the **Head and neck** option, and the **Select view** menu becomes available.

—H E A D S   U P !—————————————————

*There are two similar sounding screens in **Anatomy & Physiology |
Revealed**®, the **SELECT A VIEW** window, and the **Select view** menu. The
SELECT A VIEW window (all uppercase letters) refers to the large window
in the **IMAGE AREA** where the **Select region** menu is located. The **Select
view** menu becomes available in the **SELECT A VIEW** window after a
region has been selected from the **Select region** menu.*

Click on the **Select view** menu,
and the options at the right become
available.

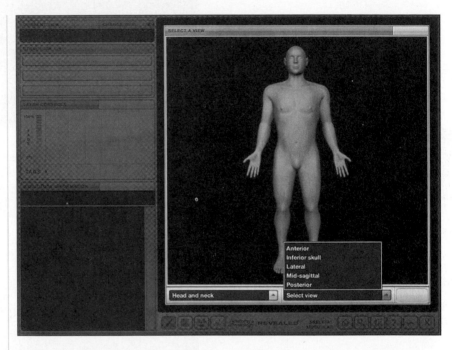

Select **Anterior** from this
menu. The anterior head and neck
of the model in the **SELECT A
VIEW** window will become highlighted, and the **GO** button will
flash green.

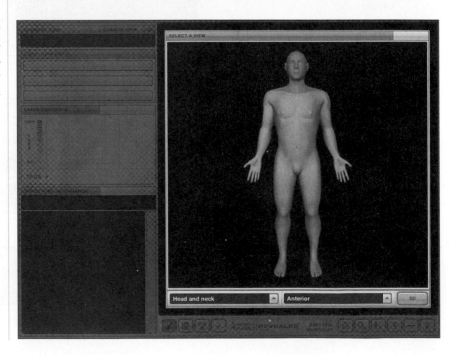

Click on the **GO** button to view the screen at right.

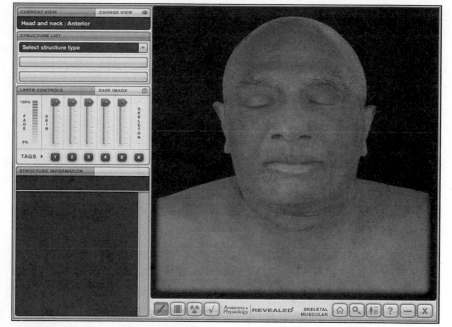

We now see a superficial view of the anterior head and neck in the **IMAGE AREA** and a new set of navigation tools in the windows on the left side of the screen. Let's look at those windows and navigation tools individually.

On the left, the top window is the **CURRENT VIEW** window.

This window informs you of which structures and views have been selected. Note also the **CHANGE VIEW** button. Click this button to change the **region** or **view** within the **Dissection** mode.

The **STRUCTURE LIST** window has multiple options.

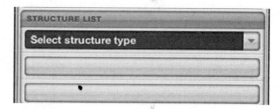

Click on the **Select structure type** menu.

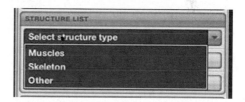

Select **Skeleton** from the list and the **Select structure group** menu becomes available.

Click on the **Select structure group** menu, and select **Ethmoid bone and features.**

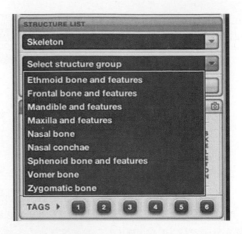

The ethmoid bone is now highlighted in the **IMAGE AREA,** and the **Perpendicular plate (ethmoid)** is listed in the **STRUCTURE LIST** window.

The **STRUCTURE INFORMATION** window lists specific information on the **Location** and **Description,** and there is a **Comment** section concerning this structure.

Often, when going through this sequence of steps, the **Select structure** menu will have more than one option. When this occurs, choose the structure of interest, and a similar screen as the one at right will appear.

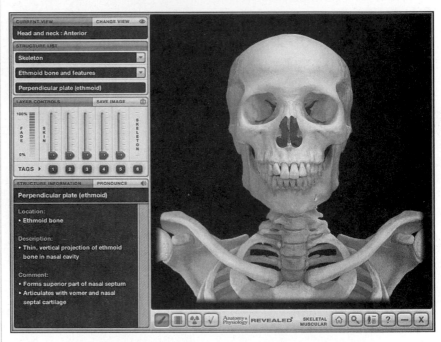

In the center of the left side of the screen is the **LAYER CONTROLS** window.

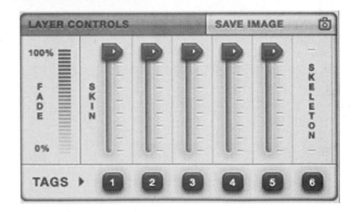

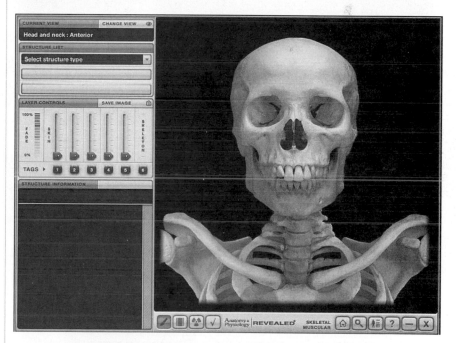

This window allows you to peel layers from the cadaver to expose the deeper structures of the body. It consists of several red sliders that, when left clicked and moved down by your mouse, reveal sequentially deeper layers. The levels range from **SKIN,** with all of the sliders in the up position, to **SKELETON,** when all of the sliders are in the down position.

At this time, spend a few moments adjusting your view by aligning the sliders in different configurations.

If at any time you wish to save an image for further study, click on the **SAVE IMAGE** button, and a **Save As** window will appear.

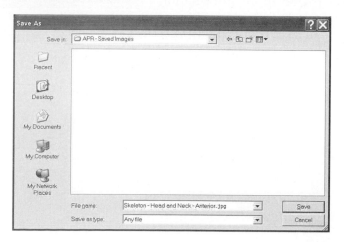

When you save an image, be sure to save it with a ".jpg" or other graphics file extension to make your images easier to find and view.

Returning to the **LAYER CON-TROLS** window, notice the row of blue buttons that are lined across the bottom labeled **TAGS.** By pressing these buttons, you bring that slider down, exposing the deepest layers for that slider in the **IMAGE AREA.** Equally importantly, it also reveals blue "pins" on important structures in that same view. If you mouse-over the pins, the label for each particular pin will show the name of that structure. For example, if you press **TAG** number **6,** the skeleton will be visible with the blue "pins". By moving your mouse over the pin in the middle of the "forehead", you will see this view, with the "frontal" bone labeled.

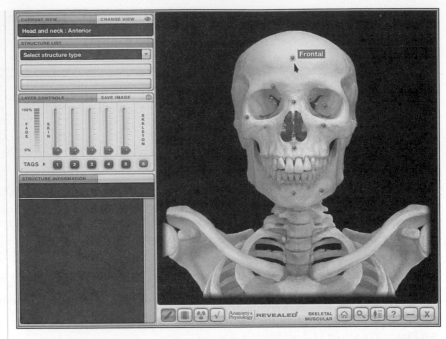

If you left click on the pin, the entire structure becomes highlighted, and the **STRUCTURE IN-FORMATION** window to the left will list information concerning that structure.

Note the **PRONOUNCE** button in the **STRUCTURE INFORMA-TION** window. By clicking this button, you can hear the correct pronunciation of that structure.

The four **Icons** at the bottom left of the **IMAGE AREA** allow you to move quickly to each of the four **MAIN SECTIONS – Dissection, Animation, Imaging,** and **Self-Test.** The **Animation** icon currently has a green background. This indicates that an animation is available to provide more information about the selected structure. To view that animation, click on the icon.

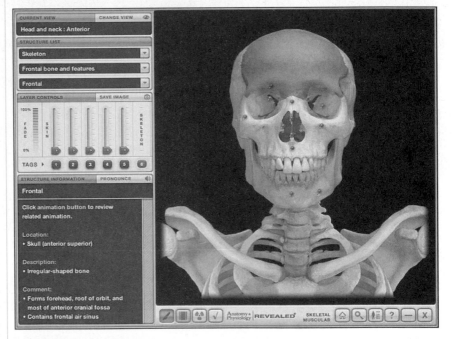

If you mouse-over each of these buttons, you will see the function of each, as in the following examples.

The remaining six **SUB NAVIGATION** buttons below the **IMAGE AREA** have the following functions:

The **HOME** button returns you to the **Home screen.**

The **SEARCH** button provides direct access to any structure found within the dissection or imaging sections.

The **ANATOMY TERMS** button provides access to definitions of common anatomy terms.

The **HELP** button provides help for each of the four **MAIN SECTIONS** specific to the one you are viewing.

The **MINIMIZE** button allows you to hide *Anatomy & Physiology | Revealed*® without having to quit the program.

Click the **EXIT** button to quit the program.

C H E C K   P O I N T :

The Dissection Section

1. What are the two different buttons you can click to enter the **Dissection** section of *Anatomy & Physiology | Revealed*®? (Note: Only one is available from the **Home screen.**)
2. What is the difference between the **SELECT A VIEW** window and the **Select view** menu in *Anatomy & Physiology | Revealed*®?
3. What is the difference between the red sliders and the blue **TAG** buttons in the **LAYER CONTROLS** window of *Anatomy & Physiology | Revealed*®?
4. When you click on a tag, what happens when you:
   a) mouse-over a pin?
   b) Left click on a pin?
5. What button do you select to change the image you are viewing in *Anatomy & Physiology | Revealed*®?

The Animation Section

Click the **HOME** button below the **IMAGE AREA** and you will return to the **Home screen.**

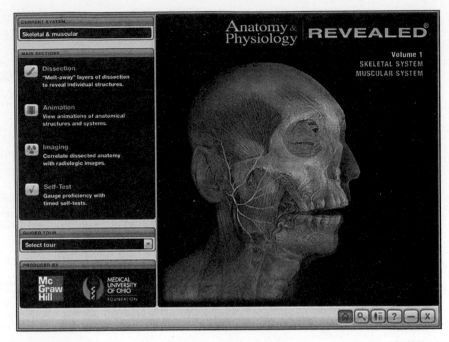

Click on the **Animation** button in the **MAIN SECTIONS** window.

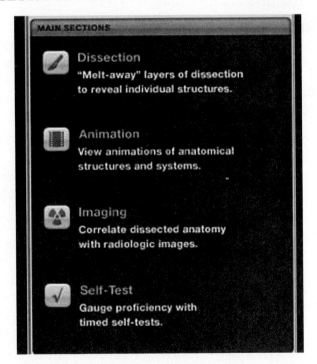

The **Select topic** menu will appear.

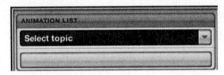

Click on the **Select topic** menu, and select **Anatomy.**

Click on the **Select animation** menu.

Select **Skull.**

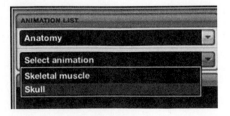

The **Animation screen** will now become available in the **IMAGE AREA**. Click the **Play** button, and the narrated animation will play.

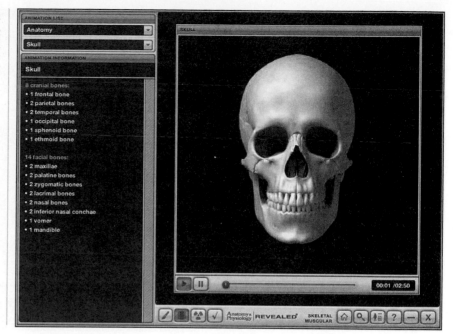

Note the list of skull bones in the **ANIMATION INFORMATION** window to the left. You can click the **Pause** button to stop the animation and the **Play** button to resume the animation.

Take time now to repeat these steps to view this and the other available animations.

CHECK POINT:

The Animation Section

1. Name the animation topics available in the **Select topic** menu.
2. Name the animation categories in the **Select animation** menu.
3. View the skeletal muscle animation, and discuss what is responsible for voluntary movement of the human body.

The Imaging Section

Return to the **Home screen.**

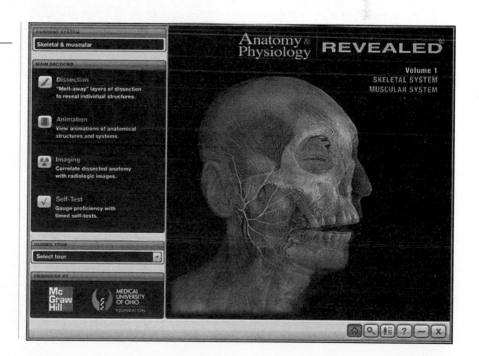

Select the **Imaging** button to access the associated radiologic images.

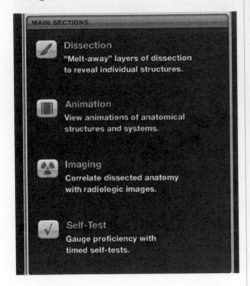

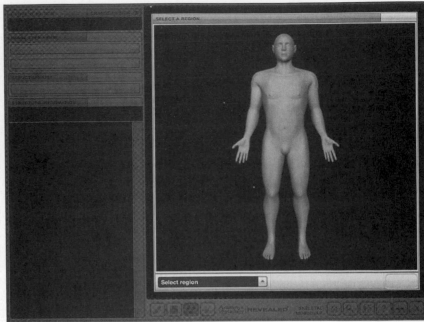

Click on the **Select region** menu.

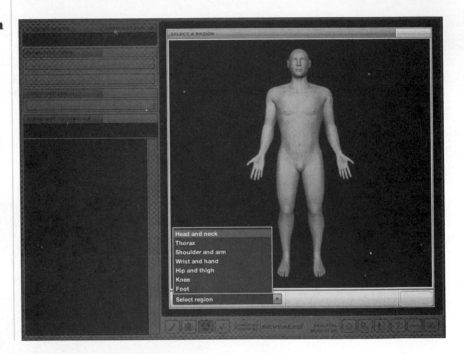

Select **Head and neck.**

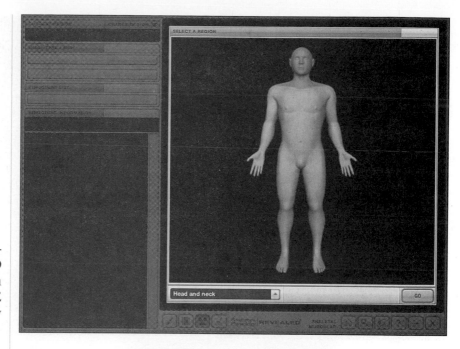

The head and neck are highlighted on the model, and the **GO** button is flashing green. Click on the **GO** button, and the **IMAGE TYPE / VIEW** window will show **X-ray.**

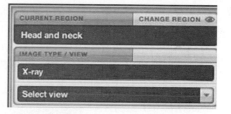

Click on the **Select view** menu.

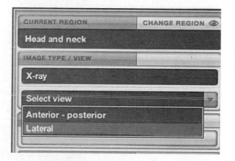

Select **Lateral.** A lateral-view X-ray of the head and neck will be in the **IMAGE AREA.**

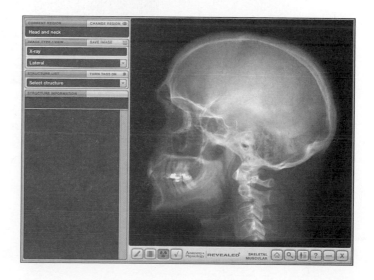

Click on the **Select structure** menu in the **STRUCTURE LIST** window to the left of the image. Select **Hyoid.**

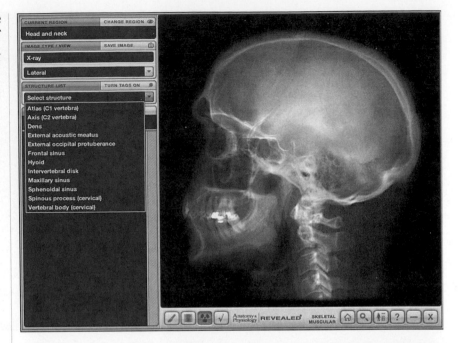

The **IMAGE AREA** now shows a highlighted hyoid bone, and the **STRUCTURE INFORMATION** window contains a **PRONOUNCE** button and a list of information concerning that bone.

Take this opportunity to repeat these steps to view all of the available images.

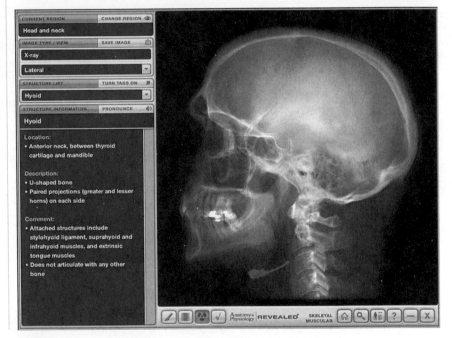

C H E C K   P O I N T:

The Imaging Section

1. Which images are available for the **Head and neck?**
2. Which images are available for the **Hip and thigh?**
3. Which images are available for the **Knee?**

The Self-Test Section

The **Self-Test** section of *Anatomy & Physiology | Revealed*® provides an opportunity for you to assess your understanding of the structures and concepts that have been presented. These self-tests are valuable for you to use in preparation for upcoming examinations or quizzes over the material. These allow you to find your strengths *and* your weaknesses, and thus you will be able to fine-tune your study time as needed. Be sure to use this section regularly before you move on to new material, so that you can solidify these structures in your mind.

Return to the **Home screen.**

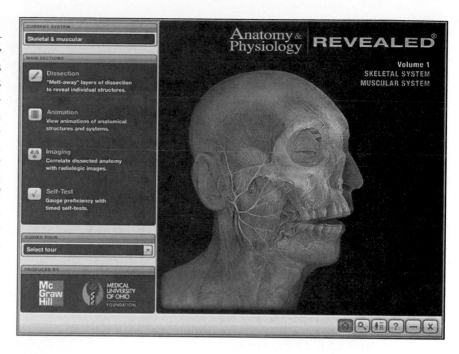

Select the **Self-Test** button.

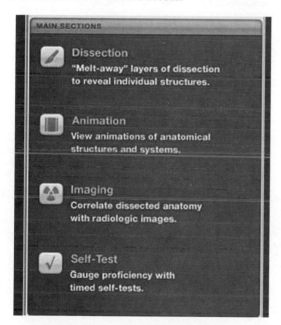

In the **TEST TYPE** window,

click on **Select test topic.**

Select **Skeleton** from the drop-down menu.

Click on the **Select region** menu.

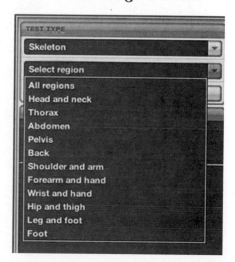

Select **Head and neck.**

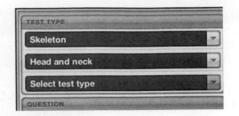

H E A D S   U P !

*It's important to distinguish between the **TEST TYPE** window and the **Select test type** menu. The **TEST TYPE** window provides multiple menus to choose from to fine-tune the test that you want to take. The **Select test type** menu then allows you to choose between a **Multiple choice** test or a **Structure identification** test.*

From the **TEST TYPE** menu,

•

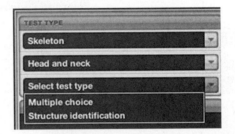

select **Multiple choice,** and a multiple choice question with five answer options appears in the **QUESTION** window, as well as an associated highlighted structure in the **IMAGE AREA.**

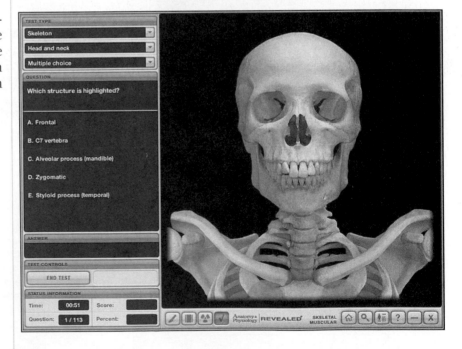

To answer the question, click on the answer that you think is correct. If your answer is correct, the **ANSWER** window will tell you **"Correct."**

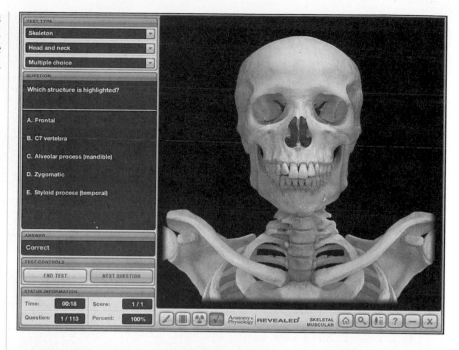

In the **TEST CONTROLS** window, you have the option to **END TEST** or answer the **NEXT QUESTION.** Note also the **STATUS INFORMATION** window, where a timer will keep track of how long it takes for you to answer the questions, your **Score** and **Percent** are recorded, and the number of **Questions** answered is tallied.

If you click on the **NEXT QUESTION** button, another question will be presented.

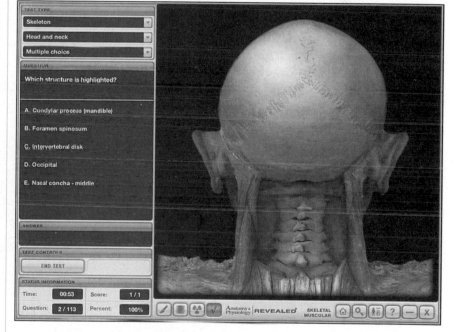

If you select an incorrect answer, the correct answer will appear in the **ANSWER** window.

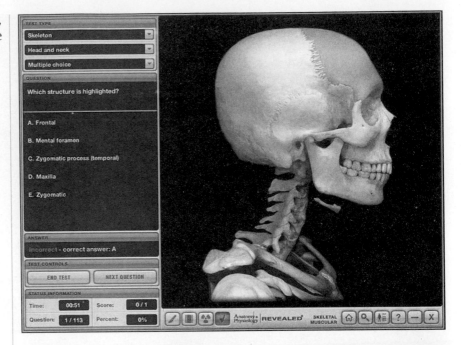

When you choose to end the test, click on the **END TEST** button, and the **RESULTS** window will display your **Score** and **Percent** correct.

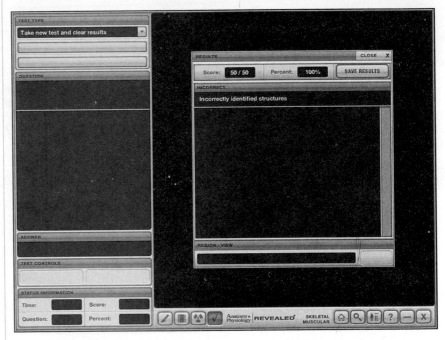

You can click on the **SAVE RESULTS** button to save your score to a file. When you click the button, a **Save As** window will appear and you can choose where to save the file.

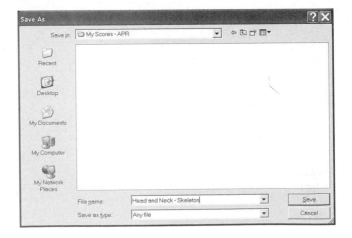

After saving your results from the **Multiple choice** test, you will return to the screen containing your results in the **IMAGE AREA.**

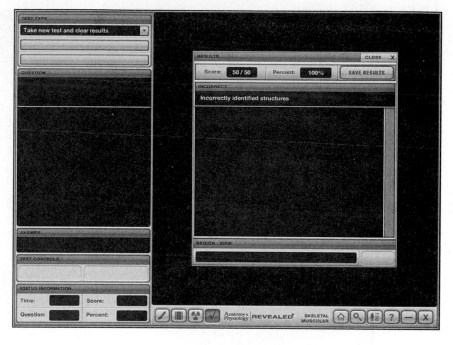

You can now complete a **Structure identification** test directly from this screen. At the top left, in the **TEST TYPE** window, is a menu entitled **Take new test and clear results.**

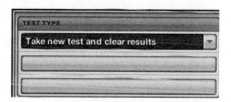

Click on this menu. Select **Skeleton** from the list of choices.

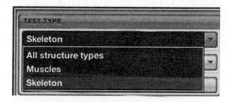

Click on the **Select region** menu.

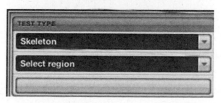

Select **Head and neck**.

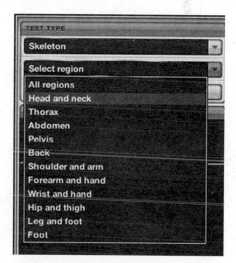

Click on the **Select test type** menu.

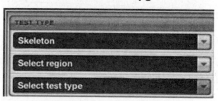

Choose **Structure identification** and the first question becomes available.

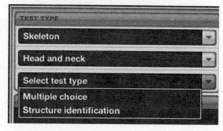

Read the question and click on that structure in the **IMAGE AREA.** Just like the **Multiple choice** tests, the **Structure identification** tests will confirm if your answer is correct. If it is wrong, it will identify the correct structure in the **IMAGE AREA.**

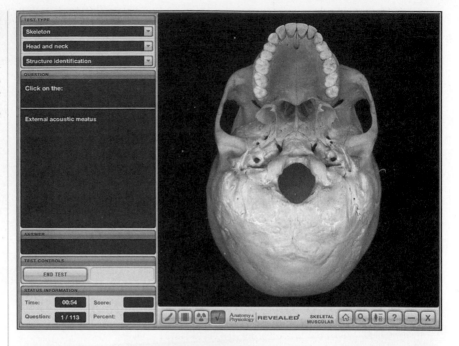

Upon ending the test, the **RESULTS** window and the saving of your results is identical to the **Multiple choice** test discussed earlier.

Your instructor may ask you to send in your test results from the **Self-Test** section. To do this, save your results as a Microsoft Word® document file (.doc) or a text file (.txt) so that it can be opened on their computer. Always follow the guidelines that your instructor presents to you.

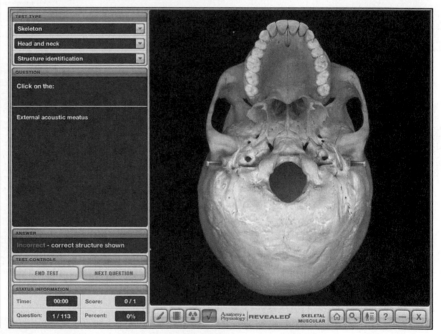

C H E C K   P O I N T :

The Self-Test Section

1. What is the difference between the **TEST TYPE** window and the **Select test type** menu?
2. What button do you click on to discontinue a **Self-Test** that you are taking?
3. Describe the process of saving your results from a **Self-Test**.

IN REVIEW

What Have I Learned?

1. Name the four **MAIN SECTIONS** in *Anatomy & Physiology | Revealed*®.

2. What is the difference between the **SELECT A VIEW** window and the **Select view** menu?

3. List the steps you would go through from inserting the *Anatomy & Physiology | Revealed*® CD into your computer to viewing a deep dissection of the posterior view of the arm and hand.

4. List the steps you would go through from inserting the *Anatomy & Physiology | Revealed*® CD into your computer to viewing the animation of **Appositional bone growth.**

5. Name the function of each of the following buttons:

   a)

   b)

   c)

   d)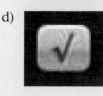

6. Name the function of each of these six buttons:

   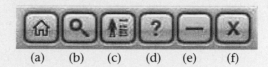
   (a)   (b)   (c)   (d)   (e)   (f)

   a)

   b)

   c)

   d)

   e)

   f)

7. What are the three choices found in the **Select structure type** menu?

8. Using *Anatomy & Physiology | Revealed*® define **Protraction.**

9. Using *Anatomy & Physiology | Revealed*® define **Oblique Plane.**

10. Using *Anatomy & Physiology | Revealed*® define **Anatomical position.**

11. Using *Anatomy & Physiology | Revealed*® define both **Proximal** and **Distal.**

12. Using *Anatomy & Physiology | Revealed*® define **Abdomen** and **pelvis.**

CHAPTER 2

The Skeletal System

Overview: Skeletal System

We are born with 270 bones in our bodies, and even more bones form during childhood. By the time we reach adulthood though, several separate bones have fused together so that the number of our bones has decreased to around 206[1], which make up the adult skeletal system. An example of this reduction occurs in each half of our pelvis, where three separate bones—the ilium, the ischium, and the pubis—fuse into one single bone called the *os coxa*.

The skeletal system is further divided into the **axial skeleton,** consisting of the bones of the skull, vertebral column, and the thoracic cage; and the **appendicular skeleton,** which consists of the bones of the upper and lower extremities along with their associated girdles (Table 2.1).

[1]Around 206—some people develop varying numbers of miscellaneous bones, either **sesamoid bones**, which form within some tendons as a response to stress (such as the patella) or **sutural bones**, which develop within the sutures of the skull.

CHECK POINT:

Skeletal System Overview

1. The average human adult has _____ bones in their body, while the average newborn has _____ bones in theirs.
2. Explain the difference.

Naming Bony Processes and Other Landmarks

Bony landmarks are various ridges, spines, depressions, pores, bumps, grooves, and articulating structures on the surface of bones. These structures allow for the passage of blood vessels and nerves; for joints between bones; and for the attachment of ligaments, muscles, and tendons. Therefore, a working knowledge of the names of these structures will be a valuable asset when considering the structure and function of these surface features.

Some of the most commonly encountered bony landmarks are listed in Table 2.2.

Table 2.1	Summary of the Bones of the Adult Skeletal System		
Axial Skeleton—80 Bones		**Appendicular Skeleton—126 Bones**	
Skull and Hyoid	23 bones	Pectoral Girdle	4 bones
Inner Ear Ossicles	6 bones	Upper Extremities	60 bones
Vertebral Column	26 bones	Pelvic Girdle	2 bones
Sternum and Ribs	25 bones	Lower Extremities	60 bones
Total – 206 Bones			

Bonus Question: Using the above information as a reference, how would this table appear if it was a list of the bones for a newborn?

Table 2.2	Common Bony Landmarks	
FEATURE	**DESCRIPTION**	**EXAMPLE**
Landmarks of Articulation:		
Condyle:	Smooth, rounded knob	occipital condyle of skull
Facet:	Smooth, flat, slightly concave or convex articular surface	articular facets of vertebrae
Head:	Prominent expanded end of a bone, sometimes rounded	head of the femur
Elevated Landmarks:		
Process:	Any bony prominence	mastoid process of the skull
Spine:	Sharp, slender, or narrow process	spine of the scapula
Crest:	Narrow ridge	iliac crest of the pelvis
Line:	Slightly raised, elongated ridge	nuchal lines of the skull
Tuberosity:	Rough surface	tibial tuberosity
Tubercle:	Small, rounded process	greater tubercle of the humerus
Trochanter:	Massive processes unique to the femur	greater trochanter of the femur
Epicondyle:	Projection superior to a condyle	medial epicondyle of the femur
Depressions or Flat Surfaces:		
Alveolus:	Pit or socket	tooth socket
Fossa:	Shallow, broad, or elongated basin	mandibular fossa
Fovea:	Small pit	fovea capitis of the femur
Sulcus:	Groove for a tendon, nerve or blood vessel	intertubercular sulcus of the Humerus
Spaces or openings:		
Foramen:	Hole through a bone, usually round	foramen magnum of the skull
Fissure:	Slit through a bone	orbital fissure behind the eye
Meatus or canal:	Tubular passage or tunnel through a bone	auditory meatus of the ear
Sinus:	Space or cavity within a bone	frontal sinus of the skull

Bonus Question: Alveolus is a common term in human anatomy. How many other examples of alveoli can you find in *APR*?

CHECK POINT:

Naming Bone Processes and Other Landmarks

1. Explain the difference between a crest and a line.
2. Explain the difference between a condyle and an epicondyle.
3. Explain the difference between a foramen and a fissure.

EXERCISE 2.1:   Bony Landmarks

Using colored pencils, color in the illustrations at right and on the next page, using different colors for each bony landmark.

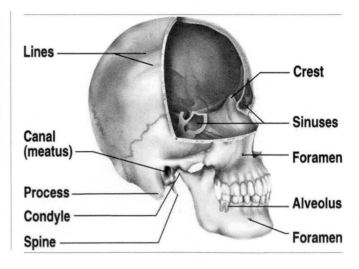

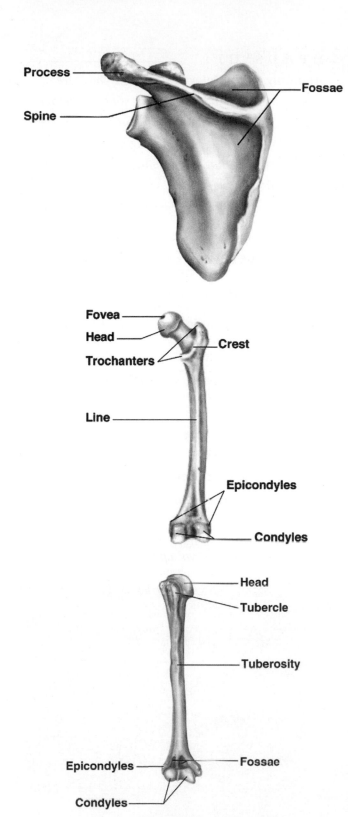

A Few Notes About Naming Processes

A process is any bony prominence—that is, a piece of bone that sticks out from the rest of the bone. When it comes to naming these processes, there are a few rules that need to be followed to minimize confusion. For example, let's consider when a process articulates with another bone, such as the zygomatic process of the temporal bone. This process is a structure on the **temporal bone** that articulates with the **zygomatic bone.** The formula for naming one of these processes is:

the *x* process of the *y* bone

where *x* = the name of bone articulated with
and   *y* = the name of the bone it is part of

So, with the ZYGOMATIC process of the TEMPORAL bone, *x* = ZYGOMATIC (the bone articulated with) and *y* = TEMPORAL (the bone it is part of). Now, how does the *zygomatic* process of the *temporal* bone compare to the *temporal* process of the *zygomatic* bone? If you are not sure, don't worry, we will cover these structures shortly.

Let's look at the styloid process of the temporal bone. *This* styloid process does *not* articulate with any other bone, but it shares its name with the styloid processes of both the ulna and radius bones of the forearm. Therefore, it must be named in reference to the bone that it is part of to prevent confusion—hence the name the styloid process of the temporal bone. What problems would you predict could occur if this distinction is not made in an emergency room scenario?

Some processes, the mastoid process for example, are unique in name and do not articulate with any other bones. These require no further clarification when naming them.

CHECK POINT:

A Few Notes About Naming Processes

1. Consider the *temporal* process of the *zygomatic* bone.
   a) This process articulates with which bone?
   b) This process is part of which bone?
2. What is the correct way to say the two styloid processes of the forearm?
3. Why is it correct to refer to the mastoid process simply as the mastoid process and not the mastoid process of the temporal bone?

Animation: The Skull

Before beginning the exercises below, view the *Anatomy & Physiology | Revealed®* animation containing an overview of the bones of the skull.

- *Insert the **Anatomy & Physiology | Revealed®: Skeletal/Muscular Systems** CD.*

- *From the **Home screen**, click on the **Animation** button.*

- *In the **Select topic** menu, select **Anatomy**.*

- *From the **Select animation** menu, select **Skull**.*

- *A list of the eight cranial bones and 14 facial bones appears on the left, and an animation screen appears in the **IMAGE AREA** on the right.*

- *Click the **Play** button and the animation will run in the window. After viewing the animation, answer the following questions:*

CHECK POINT:

The Skull

1. The bones that surround and protect the brain are referred to as the _____ bones.
2. The bones that form the underlying structure of the face are referred to as the _____ bones.
3. With one exception, the bones of the skull articulate with each other through joints known as _____. The exception is the _____.
4. There are numerous holes in the skull called _____.
5. What are the functions of these holes?

- *When you are finished with the animation, click on the **Dissection** button at the bottom of the screen to begin the following exercises, or click on the **Exit** button at the bottom right of the screen to exit **Anatomy & Physiology | Revealed®**.*

Skeletal System: Head and Neck

- *Insert the **Anatomy & Physiology | Revealed®: Skeletal/Muscular Systems** CD, or, if you are still in the **Animation** section, click the **Dissection** button at the bottom of the screen and skip the next step.*

- *In the **Home screen**, select **Dissection** in the left portion of the screen. You may click either on the **Dissection** button or on the word itself.*

- *In the **SELECT A VIEW** window that appears, click on the **Select region** menu.*

- *Choose **Head and neck** from the menu.*

- *The **Select view** menu will become available. Click here and choose the **Anterior** view.*

- *The **GO** button will flash green. Click on it, and the following image will appear:*

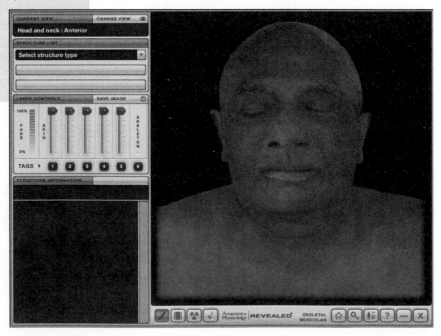

- *In the **LAYER CONTROLS** window, select **TAG 6**.*

- *The view at right now becomes available.*

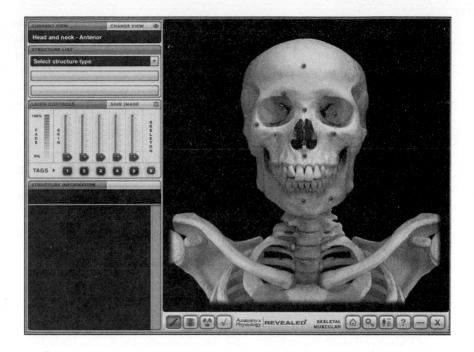

- *Mouse-over the blue pins to find the names of the bones and structures.*

- *Click on the blue pins to find information about them in the **STRUCTURE INFORMATION** window.*

— **H E A D S   U P !** ———

*The information found in the **STRUCTURE INFORMATION** window is accessed by clicking on the blue pins in the **IMAGE AREA**. This information will prove valuable for answering **Check Point** and **What Have I Learned?** questions in this **Anatomy & Physiology | Revealed®** workbook.*

- *Use this information to fill in Tables 2.3–2.6 under the anterior dissection view.*

- *Click the **CHANGE VIEW** button at the top left of the screen to select the other views of the head and neck to find the information that you need to fill in the remaining columns in the tables.*

EXERCISE 2.2   Locating Structures of the Head and Neck

There are 30 bones in the human adult head and neck, which can present a daunting task when it comes to learning each individual bone. With this in mind, the following tables are designed to allow you to discover which bones are visible in each view of *Anatomy & Physiology | Revealed®*. These tables are not meant to be tedious, but rather to help you become more familiar with each bone.

On the left of the following tables are names of structures found on the head and neck. The names of specific bones are aligned to the left, and the structures found on those bones are listed under them indented to the right. Using the *Anatomy & Physiology | Revealed®:*

Skeletal/Muscular Systems CD, open each dissection view listed across the top right of the table and put an "X" in the columns under the views where you find the structures listed in the left column. Not all bones or structures will be visible in all views.

Table 2.3	Structures of the Head and Neck—Cranial Bones				
Structures	**Dissection Views**				
	Anterior	**Inferior Skull**	**Lateral**	**Mid-sagittal**	**Posterior**
Frontal bone					
Frontal sinus					
Supraorbital notch					
Parietal bone					
Occipital bone					
External occipital protuberance					
Foramen magnum					
Occipital condyle					
Temporal bone					
Carotid canal					
External acoustic meatus					
Internal acoustic meatus					
Mastoid process					
Squamous part of temporal bone					
Styloid process of temporal bone					
Zygomatic process of temporal bone					
Sphenoid bone					
Body					
Foramen ovale					
Foramen spinosum					
Greater wing					
Sella turcica					
Sphenoidal sinus					
Ethmoid bone					
Crista galli					
Ethmoid air cells					
Nasal concha—middle					
Nasal concha—superior					
Perpendicular plate of the ethmoid bone					

Bonus Question: What is the name for the bone shaped like a "butterfly"?

Table 2.4	Structures of the Head and Neck—Facial Bones					
Structures	**Dissection Views**					
	Anterior	**Inferior Skull**	**Lateral**	**Mid-sagittal**	**Posterior**	
Maxilla						
Alveolar process of maxilla						
Infraorbital foramen						
Palatine bone						
Zygomatic bone						
Temporal process of zygomatic bone						
Nasal bone						
Vomer bone						
Mandible						
Alveolar process of mandible						
Angle of mandible						
Body of mandible						
Condylar process of mandible						
Coronoid process of mandible						
Mandibular foramen						
Mental foramen						
Ramus of mandible						

Bonus Question: What is the anatomical term that refers to the chin area? How do you suppose it received this name?

Table 2.5	Structures of the Head and Neck—Other Skull Structures					
Structures	**Dissection Views**					
	Anterior	**Inferior Skull**	**Lateral**	**Mid-sagittal**	**Posterior**	
Coronal suture						
Cranial fossa—anterior						
Cranial fossa—middle						
Cranial fossa—posterior						
Foramen lacerum						
Hard palate						
Hyoid bone						
Jugular foramen						
Lambda						
Lambdoid suture						
Nasal concha (inferior)						
Pterion						
Sagittal suture						
Septal cartilage						
Temporomandibular joint (TMJ)						
Articular disk of the TMJ						
Zygomatic arch						

Bonus Question: What is the name for the "cheekbone"?

Table 2.6	Structures of the Head and Neck—Cervical Vertebrae				
Structures	**Dissection Views**				
	Anterior	**Inferior Skull**	**Lateral**	**Mid-sagittal**	**Posterior**
Atlas (C1 vertebra)					
Axis (C2 vertebra)					
Cervical vertebrae					
Spinous process (cervical)					
Transverse process (cervical)					
Vertebral body (cervical)					
Intervertebral disk					

Bonus Question: Which bones are characterized by the presence of "transverse foramina"?

─ H E A D S   U P ! ─────

A   N O T E   A B O U T   T H E
E X E R C I S E S :
The exercises to come are designed to help you learn the individual bones and their structures as they appear in each dissection view of Anatomy & Physiology | Revealed®. On the following pages, you will be presented with the same image as you will find in the dissection view of Anatomy & Physiology | Revealed®, with the TAGS revealing the blue pins on the skeletal structures. There will be letters superimposed on each illustration associated with each of the blue pins. Following each illustration, a series of blanks will be listed in alphabetical order. Mouse-over the blue pins in the IMAGE AREA of Anatomy & Physiology | Revealed®, and you will find the name for each bone or structure. Write that name in the associated blank following the illustration.

It will be assumed that you know how to use Anatomy & Physiology | Revealed® well enough to arrive at each view of the dissection. If you are having difficulties locating the views presented, you should review Chapter 1: Introduction or the illustrations inside the front cover of this workbook to refresh your memory.

EXERCISE 2.3:   Anterior View of the Skull

- *Insert the Anatomy & Physiology | Revealed®: Skeletal/Muscular Systems CD, or, if you are still in the Dissection section, click the CHANGE VIEW button at the top of the screen and skip the next step.*

- *In the Home screen, select the Dissection menu in the left portion of the screen. You may click either on the Dissection button or on the word itself.*

- *In the SELECT A VIEW window that appears, click on the Select region menu.*

- *Choose Head and neck from the menu, if it is not already selected.*

- *The Select view menu will become available. Click here and choose the Anterior view.*

- *The GO button will flash green. Click on it.*

- *The superficial anterior image will appear.*

- *Click **TAG 6** under LAYER CONTROLS. You will see the following image:*

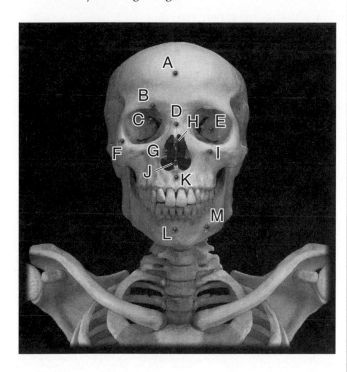

- *Mouse-over the blue pins on the screen to find the information necessary to fill in the following blanks:*

A. _____

B. _____

C. _____

D. _____

E. _____

F._____

G. _____

H._____

I._____

J._____

K. _____

L. _____

M. _____

Anterior View of the Skull

1. What two bones contain teeth?
2. What is the name for the sockets of the teeth? (You may have to revisit the earlier sections of this chapter to answer this one!)
3. The nasal septum consists of what specific bones or structures of bones?

EXERCISE 2.4:   Inferior View of the Skull

- *While you are in the **Anterior view**, click on the **CHANGE VIEW** button.*

- *In the **SELECT A VIEW** window that appears, **Head and neck** will be in the **Select region** box.*

- *In the **Select view** menu, choose the **Inferior skull** view.*

- *Click on the **GO** button and the inferior view will appear.*

- *Click **TAG 1** under LAYER CONTROLS. You will see the following image:*

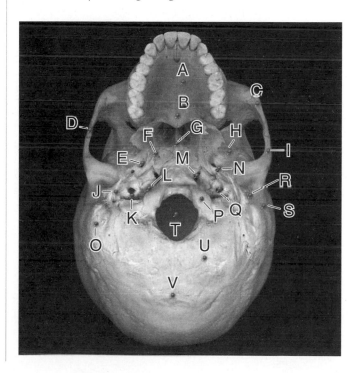

- *Mouse-over the blue pins on the screen to find the information necessary to fill in the following blanks:*

A. _____

B. _____

C. _____

D. _____

E. _____

F. _____

G. _____

H. _____

I. _____

J. _____

K. _____

L. _____

M. _____

N. _____

O. _____

P. _____

Q. _____

R. _____

S. _____

T. _____

U. _____

V. _____

EXERCISE 2.5:   Lateral View of the Skull

- *While you are in the **Inferior view**, click on the **CHANGE VIEW** button.*

- *In the **SELECT A VIEW** window that appears, **Head and neck** will be in the **Select region** box.*

- *In the **Select view** menu, choose the **Lateral** view.*

- *Click on the **GO** button and the lateral view will appear.*

- *Click **TAG 6** under LAYER CONTROLS. You will see the following image:*

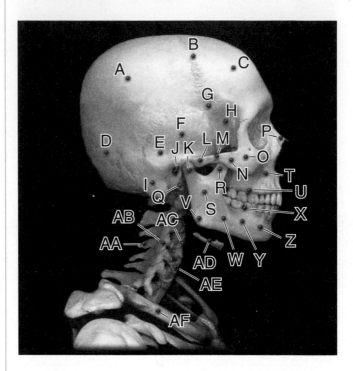

- *Mouse-over the blue pins on the screen to find the information necessary to fill in the following blanks:*

A. _____

B. _____

C. _____

D. _____

E. _____

F. _____

G. _____

H. _____

I. _____

J. _____

K. _____

L. _____

M. _____

N. _____

O._____

P._____

Q._____

R._____

S._____

T._____

U._____

V._____

W._____

X._____

Y._____

Z._____

AA._____

AB._____

AC._____

AD._____

AE._____

AF._____

CHECK POINT:

Lateral View of the Skull

1. What two bones make up most of the lateral skull (one on each side)?
2. What suture is their point of articulation?
3. What bone does not articulate with any other bone?

EXERCISE 2.6: Mid-Sagittal View of the Skull

- *While you are in the* **Lateral view**, *click on the* **CHANGE VIEW** *button.*

- *In the* **SELECT A VIEW** *window that appears,* **Head and neck** *will be in the* **Select region** *box.*

- *In the* **Select view** *menu, choose the* **Mid-sagittal** *view.*

- *Click on the* **GO** *button and the mid-sagittal view will appear.*

- *Click* **TAG 4** *under LAYER CONTROLS. You will see the following image:*

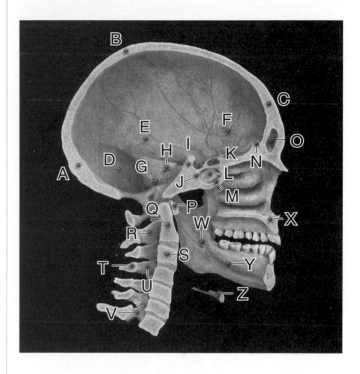

- *Mouse-over the blue pins on the screen to find the information necessary to fill in the following blanks:*

A._____

B._____

C._____

D._____

E._____

F._____

G._____

H._____

I._____

J._____

K._____

L._____

M._____

N._____

O._____

P._____

Q._____

R. _____

S. _____

T. _____

U. _____

V. _____

W. _____

X. _____

Y. _____

Z. _____

CHECK POINT:

Mid-Sagittal View of the Skull

1. When your dentist wants to numb your lower jaw by anesthetizing the nerves that serve the teeth and skin, what "hole" in what bone would be used to access those nerves?
2. Which cervical vertebra lacks a vertebral body?
3. What two bones articulate with that vertebra?

EXERCISE 2.7:   Posterior View of the Skull and Vertebral Column

- *While you are in the **Mid-sagittal view**, click on the **CHANGE VIEW** button.*

- *In the **SELECT A VIEW** window that appears, **Head and neck** will be in the **Select region** box.*

- *In the **Select view** menu, choose the **Posterior** view.*

- *Click on the **GO** button and the posterior view will appear.*

- *Click **TAG 5** under LAYER CONTROLS. You will see the following image:*

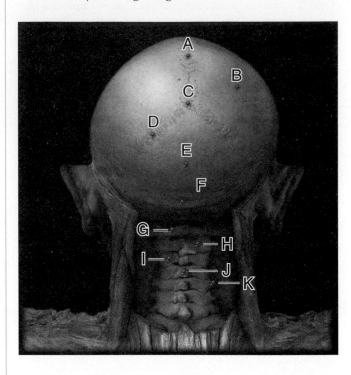

- *Mouse-over the blue pins on the screen to find the information necessary to fill in the following blanks:*

A. _____

B. _____

C. _____

D. _____

E. _____

F. _____

G. _____

H. _____

I. _____

J. _____

K. _____

CHECK POINT:

Posterior View of the Skull and Vertebral Column

1. What suture forms the joint between the parietal and occipital bones?
2. What bone forms most of the posterior skull?
3. What is the attachment site for the nuchal ligament on the skull?

EXERCISE 2.8:   Imaging

Use these *Anatomy & Physiology | Revealed®* radiological images to answer the following questions:

- *From the* **Home screen**, *or the* current screen, *select the* **Imaging** button.

- *Select* **Head and neck** *from the* **Select region** *menu.*

- *Click on the flashing* **GO** *button.*

- **X-ray** *will appear in the* **IMAGE TYPE/VIEW** *menu.*

- *From the* **Select view** *menu, select* **Anterior-posterior**.

- *From the* **Select structure** *menu, select each of the structures listed, and use the information listed in the* **STRUCTURE INFORMATION** *window to answer the following questions.*

1. What two structures increase the surface area of the nasal cavity and warm the inhaled air?
2. Name the mucous-membrane lined cavity in the body of the sphenoid bone.

3. Where does this cavity drain?
4. Name the joint formed between the atlas (C1 vertebra) and the occipital bone.
5. Through which notch or foramen pass the supraorbital nerve, artery, and vein?
6. What bone contains this notch or foramen?

- *Click on the* **CHANGE REGION** *button.*

- *From the* **Select view** *menu, select* **Lateral**.

- *From the* **Select structure** *menu, select each of the structures listed, and use the information listed in the* **STRUCTURE INFORMATION** *window to answer the following questions.*

1. Name the mucous-membrane lined cavity lateral to the nasal cavity and inferior to the orbit.
2. What structure of what bone supports your body weight?
3. List the two names for the structure unique to the axis (C2 vertebra) and that represents the body of the atlas (C1 vertebra).
4. Name the paired mucous-membrane lined cavities within the bone superior to the orbits. What is the name of that bone?

IN   REVIEW

Self-Test

- *From the opening screen or the present screen, click on the* **Self-Test** *icon.*
- *From the* **Select test topic** *menu, select* **Skeleton**.
- *From the* **Select region** *menu, select* **Head and neck**.
- *From the* **Select test type** *menu, select* **Multiple choice**.
- *Answer the questions listed, and then when finished, click on the* **END TEST** *button.*
- *You can save your results to document your progress as you learn the structures in* **Anatomy & Physiology | Revealed®** *by clicking the* **SAVE RESULTS** *button.*
- *Click* **Take new test and clear results**, *and select* **Skeleton**.
- *Again, click* **Head and neck**.
- *Click* **Structure identification**.

- *Answer the questions listed, and then when finished, click on the* **END TEST** *button.*
- *You can save your results to document your progress as you learn the structures in* **Anatomy & Physiology | Revealed®** *by clicking the* **SAVE RESULTS** *button.*

What Have I Learned?

The following questions cover the material that you just read, the introduction to the skeleton and the skeletal structures of the head and neck. Apply what you have learned in answering these questions:

1. The hard palate consists of which bones?

2. What is the name for the large foramen on the inferior side of the occipital bone?

Continued

IN REVIEW   Continued

Self-Test

3. What passes through this foramen?

4. Name the two bones that form the bridge of the nose.

5. At the junction of the temporal and occipital bones, the internal jugular vein passes through which foramen?

6. The greater and lesser wings are parts of what bone?

7. Name the bony canal of the external ear. What bone is it a part of?

8. What is the attachment site for the sternocleido-mastoid muscle on the skull?

9. What bony landmark is defined as "any bony prominence"?

10. What bony landmark is defined as "a pit or socket"?

11. What bony landmark is defined as "a hole through a bone, usually round"?

12. What bony landmark is defined as "a smooth, rounded knob"?

Skeletal System: Trunk, Shoulder Girdle, and Upper Limb

EXERCISE 2.9:   Skeletal System—Thorax, Anterior View

- *Insert **Anatomy & Physiology | Revealed®
  Skeletal/Muscular Systems** CD, or, if you
  are already in the **Dissection** section, click the
  CHANGE VIEW button at the top left of the
  screen and skip the next step.*

- *In the **Home screen**, select the **Dissection**
  menu in the left portion of the screen. You may
  click either on the **Dissection** button or on the
  word itself.*

- *In the **SELECT A VIEW** window, click on the
  Select region button.*

- *Choose **Thorax** from the menu.*

- *The **Select view** menu will show **Anterior**
  view.*

- *The **GO** button will flash green. Click on it, and
  the thorax screen will appear.*

- *Click **TAG 5**. You will see the image at right.*

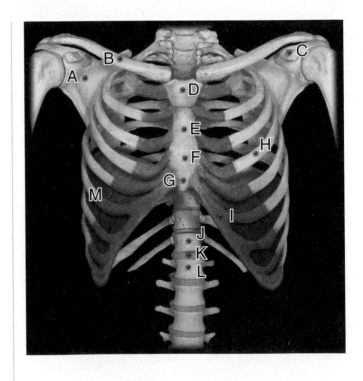

• *Mouse-over the blue pins on the screen to find the information necessary to fill in the following blanks:*

A. _____

B. _____

C. _____

D. _____

E. _____

F. _____

G. _____

H. _____

I. _____

J. _____

K. _____

L. _____

M. _____

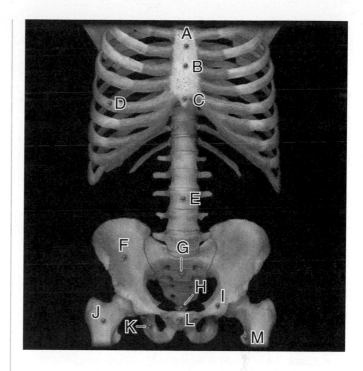

CHECK POINT:

Thorax, Anterior View

1. Name the structure that attaches the ribs to the sternum.
2. Name the three regions of the sternum.
3. On which bone would you find the glenoid cavity?

EXERCISE 2.10:   Skeletal System—Abdomen, Anterior View

• *Click on the **CHANGE VIEW** button in the top-left portion of the screen.*

• *In the **SELECT A VIEW** window, click on the **Select region** button.*

• *Choose **Abdomen** from the menu.*

• *The **Select view** menu will show **Anterior view**.*

• *The **GO** button will flash green. Click on it, and the abdomen screen will appear.*

• *Click **TAG 6**. You will see the image at top right.*

• *Mouse-over the blue pins on the screen to find the information necessary to fill in the following blanks:*

A. _____

B. _____

C. _____

D. _____

E. _____

F. _____

G. _____

H. _____

I. _____

J. _____

K. _____

L. _____

M. _____

CHECK POINT:

Abdomen, Anterior View

1. What differentiates the floating ribs from the other ribs?
2. Name the structure that consists of three to five variably fused, poorly developed vertebrae that form a small, triangular bone.

3. Name the landmark for establishing female pelvic dimensions.
4. Name the pelvic bone that has a large, wing-like superior extension. What is the name for this extension?

EXERCISE 2.11:   Skeletal System—Pelvis, Superior View

- *Click on the **CHANGE VIEW** button in the top-left portion of the screen.*

- *In the **SELECT A VIEW** window, click on the **Select region** button.*

- *Choose **Pelvis** from the menu.*

- *The **Select view** menu will show **Superior** view.*

- *The **GO** button will flash green. Click on it, and the pelvis screen will appear.*

- *Click **TAG 2**. You will see the following image:*

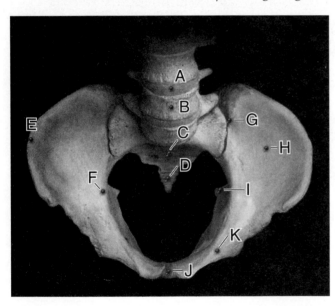

- *Mouse-over the blue pins on the screen to find the information necessary to fill in the following blanks:*

A. _____

B. _____

C. _____

D. _____

E. _____

F. _____

G. _____

H. _____

I. _____

J. _____

K. _____

CHECK POINT:

Pelvis, Superior View

1. What is another name for the pelvic inlet?
2. Which cartilage softens in late pregnancy to allow slight separation of the pubic bones? What would be the advantage of this?
3. Name the landmark for administering anesthetic (pudendal block) during childbirth.
4. Name the landmark for intramuscular injections.

EXERCISE 2.12:   Skeletal System—Back, Posterior View

- *Click on the **CHANGE VIEW** button in the top-left portion of the screen.*

- *In the **SELECT A VIEW** window, click on the **Select region** button.*

- *Choose **Back** from the menu.*

- *The **Select view** menu will show **Posterior** view.*

- *The **GO** button will flash green. Click on it, and the back screen will appear.*

- *Click **TAG 6**. You will see the following image:*

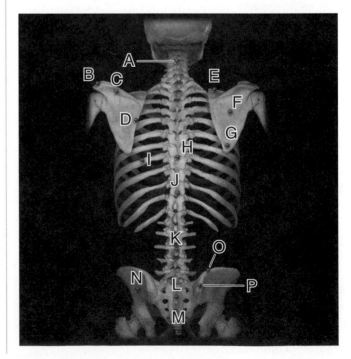

• *Mouse-over the blue pins on the screen to find the information necessary to fill in the following blanks:*

A. _____

B. _____

C. _____

D. _____

E. _____

F._____

G. _____

H._____

I._____

J._____

K. _____

L. _____

M. _____

N. _____

O._____

P. _____

CHECK POINT:

Back, Posterior View

1. Name the bones characterized by having a transverse foramen and bifid spinous processes.
2. What bony structure is marked by a shallow skin depression or "dimple" in the lower back?
3. How many vertebrae are found in the vertebral column?

EXERCISE 2.13:   Skeletal System—Shoulder and Arm, Anterior View

• *Click on the **CHANGE VIEW** button in the top-left portion of the screen.*

• *In the **SELECT A VIEW** window, click on the **Select region** button.*

• *Choose **Shoulder and arm** from the menu.*

• *From the **Select view** menu, select **Anterior** view.*

• *The **GO** button will flash green. Click on it, and the anterior view of the shoulder and arm screen will appear.*

• *Click **TAG 6**. You will see the following image:*

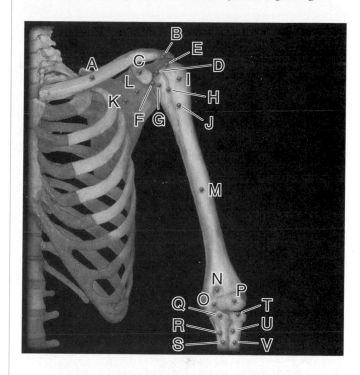

• *Mouse-over the blue pins on the screen to find the information necessary to fill in the following blanks:*

A. _____

B. _____

C. _____

D. _____

E. _____

F._____

G. _____

H. _____

I. _____

J. _____

K. _____

L. _____

M. _____

N. _____

O. _____

P. _____

Q. _____

R. _____

S. _____

T. _____

U. _____

V. _____

D. _____

E. _____

F. _____

G. _____

H. _____

I. _____

J. _____

K. _____

L. _____

EXERCISE 2.14:   Skeletal System—Shoulder and Arm, Posterior View

- *Click on the **CHANGE VIEW** button in the top-left portion of the screen.*

- **Shoulder and arm** *appears on the* **Select region** *menu.*

- *From the **Select view** menu, select **Posterior** view.*

- *The **GO** button will flash green. Click on it, and the posterior view of the shoulder and arm screen will appear.*

- *Click **TAG 5**. You will see the following image:*

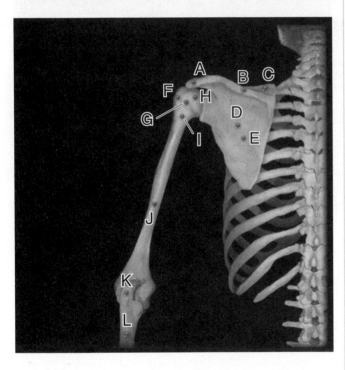

- *Mouse-over the blue pins on the screen to find the information necessary to fill in the following blanks:*

A. _____

B. _____

C. _____

CHECK POINT:

Shoulder and Arm

1. Name the part of the humerus bone that is a common site of fractures.
2. Name the subcutaneous superior point of the shoulder.
3. Name the structure of the humerus that receives the coronoid process of the ulna in full flexion of the elbow.
4. What two bones of the arm are characterized by having a styloid process? Where else have we seen a styloid process?
5. What structure of what bone forms the point of the elbow?

EXERCISE 2.15:   Skeletal System—Forearm and Hand, Anterior View

- *Click on the **CHANGE VIEW** button in the top-left portion of the screen.*

- *In the **SELECT A VIEW** window, click on the **Select region** button.*

- *Choose **Forearm and hand** from the menu.*

- *From the **Select view** menu, select **Anterior** view.*

- *The **GO** button will flash green. Click on it, and the anterior view of the forearm and hand screen will appear.*

• Click **TAG 6**. *You will see the following image:*

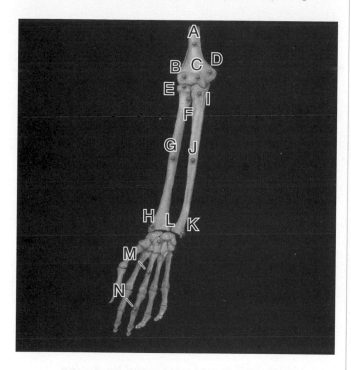

• *Mouse-over the blue pins on the screen to find the information necessary to fill in the following blanks:*

A. _____

B. _____

C. _____

D. _____

E. _____

F. _____

G. _____

H. _____

I. _____

J. _____

K. _____

L. _____

M. _____

N. _____

EXERCISE 2.16:   Skeletal System—Forearm and Hand, Posterior View

• *Click on the **CHANGE VIEW** button in the top-left portion of the screen.*

• **Forearm and hand** *appears on the **Select region** menu.*

• *From the **Select view** menu, select **Posterior** view.*

• *The **GO** button will flash green. Click on it, and the posterior view of the forearm and hand screen will appear.*

• *Click **TAG 5**. You will see the following image:*

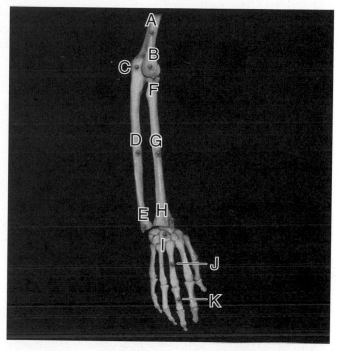

• *Mouse-over the blue pins on the screen to find the information necessary to fill in the following blanks:*

A. _____

B. _____

C. _____

D. _____

E. _____

F. _____

G. _____

II. _____

I. _____

J. _____

K. _____

CHECK POINT:

Forearm and Hand

1. What term refers to the bones of the wrist?
2. Name the particular structure of the arm referred to as the "funny bone".
3. What term refers to the bones of the palm of the hand?

EXERCISE 2.17:   Skeletal System—Wrist and Hand, Anterior View

- Click on the **CHANGE VIEW** button in the top-left portion of the screen.

- In the **SELECT A VIEW** window, click on the **Select region** button.

- Choose **Wrist and hand** from the menu.

- From the **Select view** menu, select **Anterior view**.

- The **GO** button will flash green. Click on it, and the anterior view of the wrist and hand screen will appear.

- Click **TAG 6**. You will see the following image:

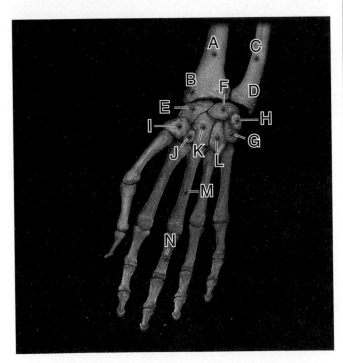

- Mouse-over the blue pins on the screen to find the information necessary to fill in the following blanks:

A. _____

B. _____

C. _____

D. _____

E. _____

F. _____

G. _____

H. _____

I. _____

J. _____

K. _____

L. _____

M. _____

N. _____

EXERCISE 2.18:   Skeletal System—Wrist and Hand, Posterior View

- Click on the **CHANGE VIEW** button in the top-left portion of the screen.

- **Wrist and hand** appears on the **Select region** menu.

- From the **Select view** menu, select **Posterior view**.

- The **GO** button will flash green. Click on it, and the posterior view of the wrist and hand screen will appear.

- Click **TAG 4**. You will see the following image:

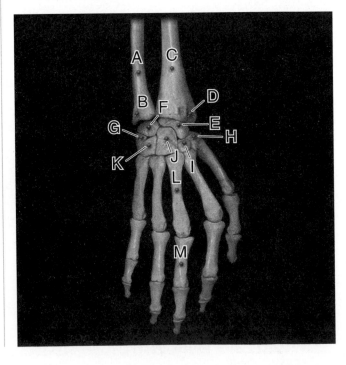

- *Mouse-over the blue pins on the screen to find the information necessary to fill in the following blanks:*

A. _____

B. _____

C. _____

D. _____

E. _____

F. _____

G. _____

H. _____

I. _____

J. _____

K. _____

L. _____

M. _____

CHECK POINT:

Wrist and Hand

1. Name the wrist bone most often fractured.
2. What term applies to the bones of the fingers?
3. What is the largest wrist bone?
4. Which wrist bone name means "three corners"?

IN REVIEW

What Have I Learned?

The following questions cover the material that you just learned: the shoulder girdle, upper limb, and trunk. Apply what you have learned in answering these questions:

1. Name the joint where the ilium and sacrum articulate.

2. Name the joint where the two pubic bones articulate.

3. Name the structure commonly called the "tailbone".

4. Name the triangular bone wedged between the hip bones and that consists of five fused vertebrae.

5. What is the lumbar curvature formed by?

6. Name the eight wrist bones.

7. Name the group of vertebrae that articulate with the ribs.

8. Name the landmark for intramuscular injections of the shoulder.

9. Name the two bone structures, and the bones that they are part of, that form the shoulder joint.

10. Name the prominent ridge on the posterior side of the scapula.

11. Name the wrist bone that has the process called the hamulus.

12. What term describes a bone embedded in a tendon? Which wrist bone is embedded in a tendon?

Skeletal System: Pelvic Girdle and Lower Limb

EXERCISE 2.19: Skeletal System—Hip and Thigh, Anterior View

- *Insert **Anatomy & Physiology | Revealed®: Skeletal/Muscular Systems** CD, or, if you are already in the **Dissection** section, click the **CHANGE VIEW** button at the top left of the screen and skip the next step.*

- *In the **Home screen**, select the **Dissection** menu in the left portion of the screen. You may click either on the **Dissection** button or on the word itself.*

- *In the **SELECT A VIEW** window, click on the **Select region** button.*

- *Choose **Hip and thigh** from the menu.*

- *From the **Select view** menu, select **Anterior** view.*

- *The **GO** button will flash green. Click on it, and the anterior view of the hip and thigh image will appear.*

- *Click **TAG 6**. You will see the following image:*

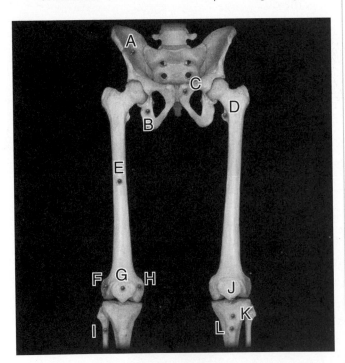

- *Mouse-over the blue pins on the screen to find the information necessary to fill in the following blanks:*

A. _____

B. _____

C. _____

D. _____

E. _____

F. _____

G. _____

H. _____

I. _____

J. _____

K. _____

L. _____

EXERCISE 2.20: Skeletal System Hip and Thigh, Posterior View

- *Click on the **CHANGE VIEW** button in the top-left portion of the screen.*

- ***Hip and thigh** appears on the **Select region** menu.*

- *From the **Select view** menu, select **Posterior** view.*

- *The **GO** button will flash green. Click on it, and the posterior view of the hip and thigh screen will appear.*

- *Click **TAG 6**. You will see the following image:*

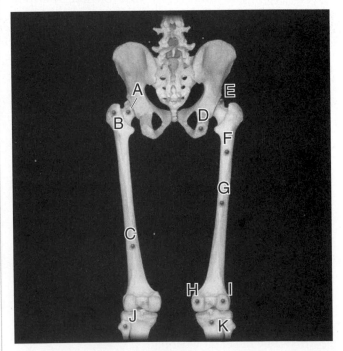

- *Mouse-over the blue pins on the screen to find the information necessary to fill in the following blanks:*

A. _____

B. _____

C. _____

D. _____

E. _____

F. _____

G. _____

H. _____

I. _____

J. _____

K. _____

CHECK POINT:

Hip and Thigh

1. Name the longest bone in the body.
2. Name a leg bone embedded in a tendon.
3. Name the socket of the hip joint. What three bones contribute to its structure?
4. Name the bone commonly referred to as the "shin."
5. Where on the femur is the most common site of fractures, especially in the elderly?
6. Name the leg bone that does not contribute to the knee joint.

EXERCISE 2.21:   Skeletal System—Leg and Foot, Anterior View

- *Click on the* **CHANGE VIEW** *button in the top-left portion of the screen.*

- *In the* **SELECT A VIEW** *window, click on the* **Select region** *button.*

- *Choose* **Leg and foot** *from the menu.*

- *From the* **Select view** *menu, select* **Anterior view.**

- *The* **GO** *button will flash green. Click on it, and the anterior view of the leg and foot screen will appear.*

- *Click* **TAG 4**. *You will see the following image:*

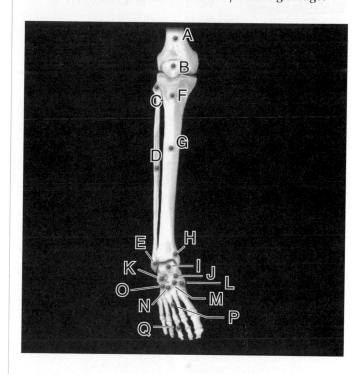

- *Mouse-over the blue pins on the screen to find the information necessary to fill in the following blanks:*

A. _____

B. _____

C. _____

D. _____

E. _____

F. _____

G. _____

H. _____

I. _____

J. _____

K. _____

L. _____

M. _____

N. _____

O. _____

P. _____

Q. _____

EXERCISE 2.22:   Skeletal System—Leg and Foot, Posterior View

- *Click on the **CHANGE VIEW** button in the top-left portion of the screen.*

- **Leg and foot** *appears on the **Select region** menu.*

- *From the **Select view** menu, select **Posterior** view.*

- *The **GO** button will flash green. Click on it, and the posterior view of the leg and foot screen will appear.*

- *Click **TAG 6**. You will see the following image:*

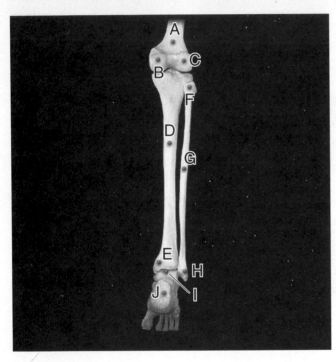

- *Mouse-over the blue pins on the screen to find the information necessary to fill in the following blanks:*

A. _____

B. _____

C. _____

D. _____

E. _____

F._____

G. _____

H. _____

I._____

J._____

EXERCISE 2.23:   Skeletal System—Foot, Plantar View

- *Click on the **CHANGE VIEW** button in the top-left portion of the screen.*

- *In the **SELECT A VIEW** window, click on the **Select region** button.*

- *Choose **Foot** from the menu.*

- **Plantar** *view will appear in the **Select view** menu.*

- *The **GO** button will flash green. Click on it, and the plantar view of the foot screen will appear.*

- *Click **TAG 6**. You will see the following image:*

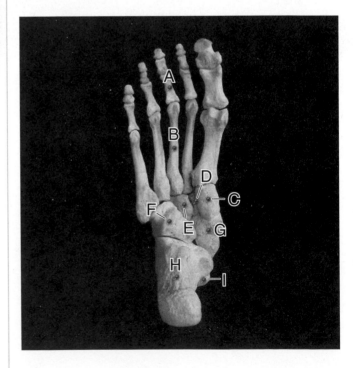

• Mouse-over the blue pins on the screen to find the information necessary to fill in the following blanks:

A. _____

B. _____

C. _____

D. _____

E. _____

F._____

G. _____

H. _____

I._____

CHECK POINT:

Foot, Plantar View

1. Name the five small long bones of the foot. What specific structure of these bones are the surface contact points on the plantar foot?
2. The intermediate cuneiform is the smallest of which group of bones? Which bone is the largest of this group?
3. Inversion and eversion of the foot occur at the joint between which two bones of the foot?

IN REVIEW

What Have I Learned?

The following questions cover the material that you have just learned: the pelvic girdle and lower limb. Apply what you have learned in answering these questions:

1. Where on the femur does the lateral collateral ligament attach?

2. How many tarsal bones are there on each foot? Name them.

3. The head of which bone is commonly referred to as the "ball of the foot"?

4. Name the small bones of the toes. How many of these bones make up each toe?

5. Name the two enlargements of the distal femur. What structures do they articulate with?

CHAPTER 3

The Muscular System

Overview: Muscular System

Without the skeletal muscular system, the bones that we learned in the previous unit would be unable to move. Our bodies are equipped with some 600 skeletal muscles to not only put those 206 bones into motion, but also to generate as much as 85% of our body heat, maintain our posture, control the openings involved with the entrance and exit of materials, and to express our emotions and thoughts through movements of our facial muscles.

Three important structural terms to understand as you begin your study of the skeletal muscular system are a muscle's **origin**, **insertion,** and **belly**. Most muscles are attached to different bones at each end. This assures that each muscle or its tendon will span at least one joint. When a muscle contracts, it causes movement where one bone remains relatively stationary and the other bone will move. The end of the muscle attached to the relatively stationary bone is called the **origin**, while the end of the muscle attached to the moving bone is called the **insertion**. One way to remember the difference is to think of your birthplace, your **origin**. No matter where you may move throughout your life, your **origin** remains the same – it *doesn't move!* You may **insert** yourself at several locations throughout your life—away to college, a job in a different town, and so on. These require *moving!* Also, many muscles are narrow at each end, their origin and insertion, and thick in the middle. This thicker middle region is called the **belly**.

Animation: Skeletal Muscle

Before beginning the following exercises, view the *Anatomy & Physiology | Revealed®* animation covering the anatomy of skeletal muscle.

- *Insert* Anatomy & Physiology | Revealed® **Skeletal / Muscular** CD.
- *Click on the* **ANIMATIONS** *button.*
- *In the* **Select topic** *menu, select* **Anatomy**.
- *From the* **Select animation** *menu, select* **Skeletal muscle**.
- *Click the* **Play** *button and the animation will run in the window.*
- *When you are finished viewing the animation, click on the* **Dissection** *button at the bottom of the screen to begin the next exercises, or click on the* **EXIT** *button at the bottom right of the screen to exit* Anatomy & Physiology | Revealed®.

Anatomy & Physiology | Revealed® has several animations available to aid your study of different systems. Watch for the Animation button at the bottom of the screen to be highlighted, which indicates that an animation is available for the specific structure(s) you are viewing.

Muscular System: Head and Neck

EXERCISE 3.1: Skeletal Muscle—Head and Neck, Anterior View

- *Insert* Anatomy & Physiology | Revealed® *Skeletal / Muscular CD, or, if you are still in the* **Animation** *section, click the* **Dissection** *button at the bottom of the screen, and skip the next step.*

- *In the **Home screen**, select the **Dissection** button in the left portion of the screen. You may click either on the **Dissection** button or on the word itself.*

- *On the **SELECT A VIEW** window that appears, click on the **Select region** button.*

- *Choose **Head and neck** from the menu.*

- *The **Select view** menu will now become available. Click here and choose the **Anterior** view.*

- *The **GO** button will now flash green. Click on it, and the screen at right will appear.*

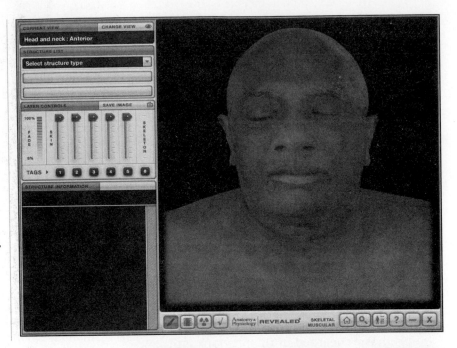

- *Click on **TAG 1**, and the following image will appear:*

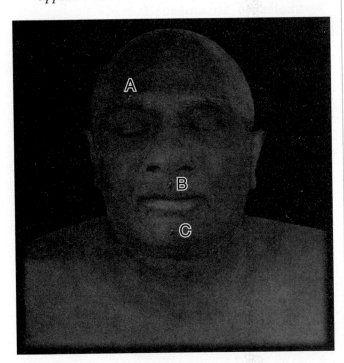

- *Mouse-over the blue pins on the screen to find the information necessary to fill in the following blanks:*

    A. _____

    B. _____

    C. _____

CHECK POINT:

Head and Neck, Anterior View

1. Name the ridge superior to each orbit on the anterior side.
2. What bone is that ridge part of?
3. What is the name of the shallow midline groove of the upper lip?

- *Now click on **TAG 2** and the following image will appear:*

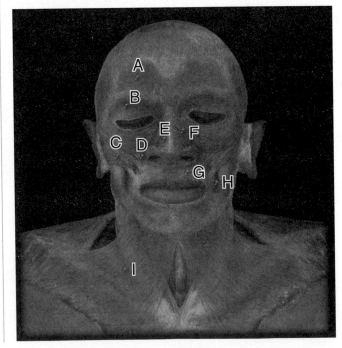

- *Mouse-over the blue pins on the screen to find the information necessary to fill in the following blanks:*

    A. _____

    B. _____

    C. _____

    D. _____

    E. _____

    F. _____

    G. _____

    H. _____

    I. _____

CHECK POINT:

Head and Neck, Anterior View, cont'd

4. Name the muscle that closes the eye when winking or blinking.
5. Name the muscle responsible for compression of the cheek as in inflating a balloon or playing a wind instrument.
6. Name the muscle that closes and protrudes the lips.

- *Now click on* **TAG 3,** and the following image will appear:

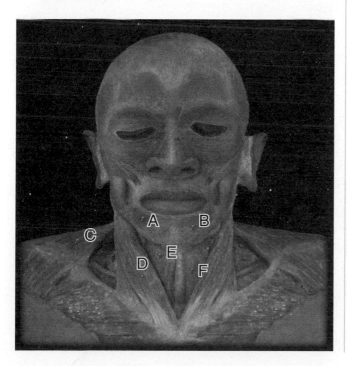

- *Mouse-over the blue pins on the screen to find the information necessary to fill in the following blanks:*

    A. _____

    B. _____

    C. _____

    D. _____

    E. _____

    F. _____

CHECK POINT:

Head and Neck, Anterior View, cont'd

7. Name two muscles whose insertion is the hyoid bone.
8. Name the muscle responsible for depression of the angle of the mouth to grimace.
9. Name the muscle responsible for depression of the lower lip while pouting.

- *Now click on* **TAG 4,** and the following image will appear:

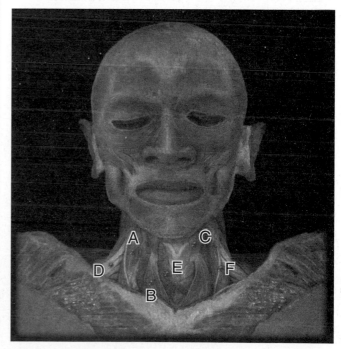

• *Mouse-over the blue pins on the screen to find the information necessary to fill in the following blanks:*

A. _____

B. _____

C. _____

D. _____

E. _____

F. _____

CHECK POINT:

Head and Neck, Anterior View, cont'd

10. The roots of the brachial plexus are located between the anterior and posterior _____ muscles.
11. Name a muscle responsible for elevation of the larynx and depression of the hyoid bone.
12. Name the four infrahyoid muscles.

Animations: Frontalis, Levator Labii Superioris Alaeque Nasi, Orbicularis Oculi, Orbicularis Oris, and Sternocleidomastoid

• *Click on the **ANIMATIONS** button at the bottom of the screen.*

• *In the **Select topic** menu, select **Muscle actions**.*

• *In the **Select animation** menu, select and view the following animations:*

– **Frontalis**
– **Levator labii superioris alaeque nasi**
– **Orbicularis oculi**
– **Orbicularis oris**
– **Sternocleidomastoid**

EXERCISE 3.2: Skeletal Muscle—Head and Neck, Lateral View

• *Insert* Anatomy & Physiology | Revealed® **Skeletal / Muscular** CD, or, if you are still in the **Animation** section, click the **Dissection button** at the bottom of the screen, and skip the next step.

• *In the **Home screen**, select the **Dissection** button in the left portion of the screen. You may click either on the **Dissection button** or on the word itself.*

• *On the **SELECT A VIEW** window that appears, click on the **Select region** button.*

• *Choose **Head and neck** from the menu.*

• *The **Select view** menu will now become available. Click here and choose the **Lateral** view.*

• *The **GO** button will now flash green. Click on it, and the head and neck lateral view screen will appear.*

• *Click on **TAG 1** , and the following image will appear:*

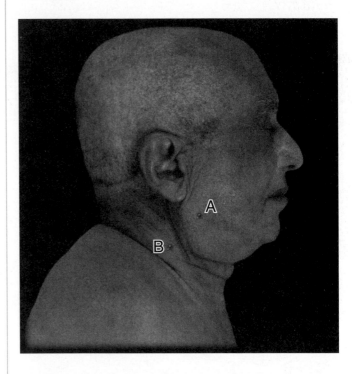

• *Mouse-over the blue pins on the screen to find the information necessary to fill in the following blanks:*

A. _____

B. _____

CHECK POINT:

Head and Neck, Lateral View

1. What location on the mandible provides an attachment site for the masseter muscle?
2. What other muscle attaches at this point?

• *Click on* **TAG 2**, *and the following image will appear:*

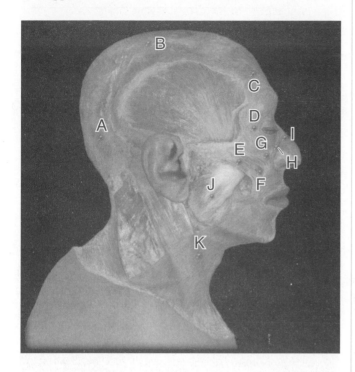

• *Mouse-over the blue pins on the screen to find the information necessary to fill in the following blanks:*

A. _____

B. _____

C. _____

D. _____

E. _____

F. _____

G. _____

H. _____

I. _____

J. _____

K. _____

Head and Neck, Lateral View, cont'd

3. Name the muscle responsible for elevation of the upper lip in a sneer.
4. Name the two muscles responsible for elevation of the upper lip in a smile.
5. Name the muscle that elevates and creases the skin of the neck as well as depresses the lower lip and the angle of the mouth.

• *Click on* **TAG 3**, *and the following screen will appear:*

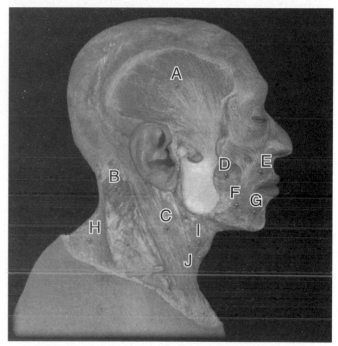

• *Mouse-over the blue pins on the screen to find the information necessary to fill in the following blanks:*

A. _____

B. _____

C. _____

D. _____

E. _____

F. _____

G. _____

H. _____

I. _____

J. _____

Head and Neck, Lateral View, cont'd

6. Name a muscle with two bellies (superior and inferior) joined by an intermediate tendon.
7. What is a raphe?
8. What is the "kissing muscle"?

- *Click on* **TAG 4**, *and the following image will appear:*

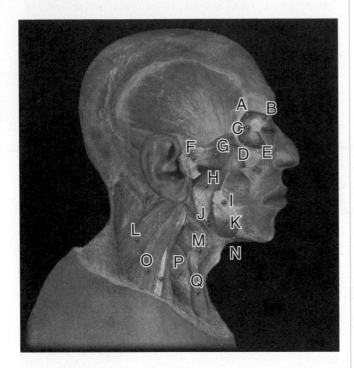

- *Mouse-over the blue pins on the screen to find the information necessary to fill in the following blanks:*

A. _____

B. _____

C. _____

D. _____

E. _____

F. _____

G. _____

H. _____

I. _____

J. _____

K. _____

L. _____

M. _____

N. _____

O. _____

P. _____

Q. _____

Head and Neck, Lateral View, cont'd

9. Name a muscle responsible for the protrusion of the mandible.
10. Name a muscle responsible for the elevation of the scapula, as in shrugging the shoulders.
11. Name the muscle involved in abduction of the eyeball.

- *Click on* **TAG 5**, *and the following image will appear:*

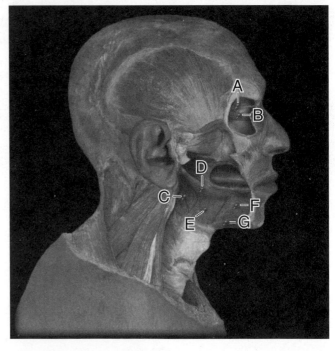

- *Mouse-over the blue pins on the screen to find the information necessary to fill in the following blanks:*

A. _____

B. _____

C. _____

D. _____

E. _____

F. _____

G. _____

Head and Neck, Lateral View, cont'd

12. Name the muscle involved with adduction of the eyeball.
13. Name the muscle whose tendon passes through a trochlea.
14. Which muscle allows you to stick out your tongue?

Animations: Frontalis, Levator Labii Superioris Alaeque Nasi, Orbicularis Oculi, Orbicularis Oris, and Sternocleidomastoid

- *Click on the* **ANIMATIONS** *button at the bottom of the screen.*

- *In the* **Select topic** *menu, select* **Muscle actions**.

- *In the* **Select animation** *menu, select and view the following animations:*
  - **Frontalis**
  - **Levator labii superioris alaeque nasi**
  - **Orbicularis oculi**
  - **Orbicularis oris**
  - **Sternocleidomastoid**

EXERCISE 3.3: Skeletal Muscles—Head and Neck, mid-Sagittal view

- *Insert* Anatomy & Physiology | Revealed® **Skeletal / Muscular** *CD, or, if you are still in the* **Animation** *section, click the* **DISSECTION** *button at the bottom of the screen, and skip the next step.*

- *In the* **Home screen**, *select the* **Dissection** *button in the left portion of the screen. You may click either on the* **Dissection** *button or on the word itself.*

- *On the* **SELECT A VIEW** *window that appears, click on the* **Select region** *button.*

- *Choose* **Head and neck** *from the menu.*

- *The* **Select view** *menu will now become available. Click here and choose the* **Mid-sagittal** *view.*

- *The* **GO** *button will now flash green. Click on it, and the mid-sagittal view will appear.*

- *Click on* **TAG 2**, *and the following image will appear:*

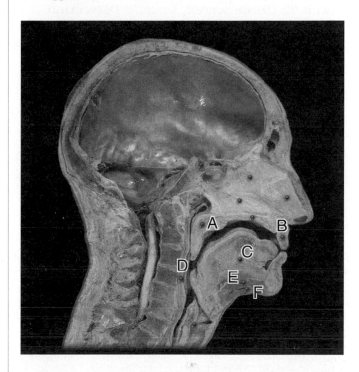

- *Mouse over the blue pins on the screen to find the information necessary to fill in the following blanks:*

  A. _____

  B. _____

  C. _____

  D. _____

  E. _____

  F. _____

Mid-Sagittal View

1. Name the muscular structure that separates the oropharynx from the nasopharynx.
2. Name a muscle that blends with the musculature of the tongue.

EXERCISE 3.4:   Skeletal Muscle—Head and Neck, Posterior View

- *Insert* Anatomy & Physiology | Revealed® *Skeletal / Muscular CD, or, if you are already in the* **Dissection** *section, click the* **CHANGE VIEW** *at the top of the screen, and skip the next step.*

- *In the* **Home screen**, *select the* **Dissection** *button in the left portion of the screen. You may click either on the* **Dissection** *button or on the word itself.*

- *In the* **SELECT A VIEW** *window that appears, click on the* **Select region** *button.*

- *Choose* **Head and neck** *from the menu, if it's not already selected.*

- *The* **Select view** *menu will now become available. Click here and choose the* **Posterior** *view.*

- *The* **GO** *button will now flash green. Click on it, and the posterior view will appear.*

- *Click on* **TAG 1**, *and the following image will appear:*

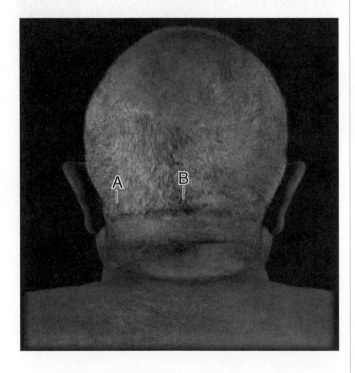

- *Mouse-over the blue pins on the screen to find the information necessary to fill in the following blanks:*

    A. _____

    B. _____

CHECK POINT:

Head and Neck, Posterior View

1. Name the three muscles that attach to the mastoid process.
2. What is the attachment point for the nuchal ligament to the skull?

- *Click on* **TAG 2**, *and the following image will appear:*

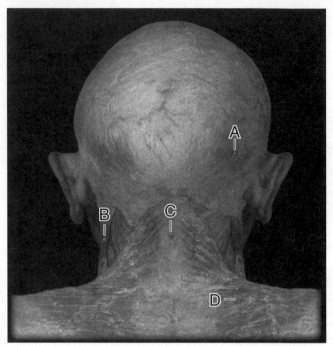

- *Mouse-over the blue pins on the screen to find the information necessary to fill in the following blanks:*

    A. _____

    B. _____

    C. _____

    D. _____

CHECK POINT:

Head and Neck, Posterior View, cont'd

3. Name two muscles attached to the nuchal ligament.
4. Name the two origins and the one insertion for the sternocleidomastoid muscle.
5. Name the large superficial muscle located from the posterior neck to the shoulders and the posterior midline.

- *Click on* **TAG 3**, *and the following image will appear:*

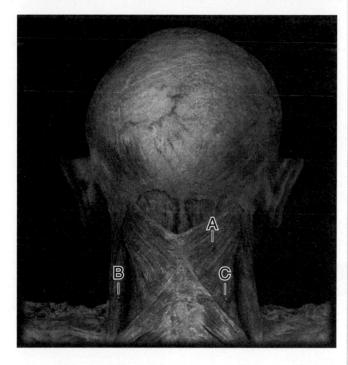

- *Mouse-over the blue pins on the screen to find the information necessary to fill in the following blanks:*

    A. _____

    B. _____

    C. _____

- *Click on* **TAG 4**, *and the following image will appear:*

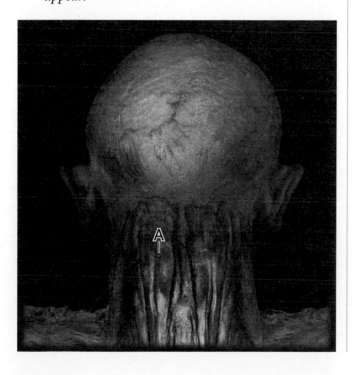

- *Mouse-over the blue pin on the screen to find the information necessary to fill in the following blank:*

    A. _____

- *Click on* **TAG 6**, *and the following image will appear:*

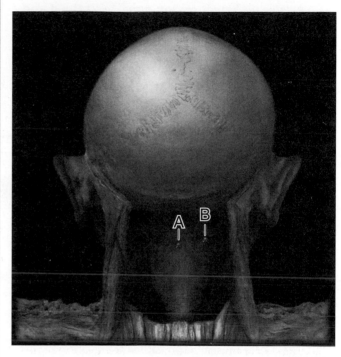

- *Mouse-over the blue pins on the screen to find the information necessary to fill in the following blanks:*

    A. _____

    B. _____

CHECK POINT:

Head and Neck, Posterior View, cont'd

6. Name a muscle responsible for elevation of the pharynx during swallowing.

Self-Test

- *From the opening screen or the present screen, click on the* **SELF-TEST** *button.*
- *From the* **Select test topic** *menu, select* **Muscles**.
- *From the* **Select region** *menu, select* **Head and neck**.
- *From the* **Select test type** *menu, select* **Multiple choice**.
- *Answer the questions listed, and when finished, click on the* **END TEST** *button.*
- *You can save your results to document your progress as you learn the structures in* Anatomy & Physiol-

ogy | Revealed® *by clicking the* **SAVE RESULTS** *button.*
- *Now click* **Take new test and clear results** , *and select* **Muscles**.
- *Again, click* **Head and neck**.
- *Click* **Structure identification**.
- *Answer the questions listed, and when finished, click on the* **END TEST** *button.*
- *Again, you can save your results to document your progress as you learn the structures in* Anatomy & Physiology | Revealed® *by clicking the* **SAVE RESULTS** *button.*

What Have I Learned?

The following questions cover the material that you have just learned: the muscles of the head and neck. Use the **STRUCTURE INFORMATION** to answer these questions:

1. Name a muscle responsible for elevation of the larynx.

2. Name the muscle that flares the nostrils.

3. The scapula is elevated by which muscles?

4. When this muscle contracts, the head rotates so that the face turns downward and to the opposite side.

5. Name three muscles responsible for closing the mouth.

6. Name three muscles responsible for depression of the hyoid bone.

7. What muscle is responsible for flexion of the head to look downward?

8. Name the group of muscles responsible for the peristaltic waves of swallowing.

9. Name three muscles involved in moving the tongue.

10. Name the muscle involved in elevating the eyebrow and creasing the skin of the forehead.

11. Name a muscle responsible for depression of the larynx.

12. There is a muscle complex that lies deep to the scalp from the forehead to the posterior skull. What is the name of that complex, and the two muscles that it is made of?

13. List all of the muscles involved with eye movement, and describe the movement involved with each muscle.

14. Name the anatomical structure commonly called the "chin."

Muscular System: Trunk, Shoulder Girdle, and Upper Limb

EXERCISE 3.5:   Skeletal Muscle—Thorax, Anterior View

- *Insert* Anatomy & Physiology | Revealed® *Skeletal / Muscular CD, or, if you are still in the* **Dissection** *section, click the* **CHANGE VIEW** button *at the top of the screen, and skip the next step.*

- *In the* **Home screen***, select the* **Dissection** *button in the left portion of the screen. You may click either on the* **Dissection** *button or on the word itself.*

- *In the* **SELECT A VIEW** *window that appears, click on the* **Select region** *button.*

- *Choose* **Thorax** *from the menu.*

- *The* **Select view** *menu will now show* **Anterior** *view.*

- *The* **GO** *button will flash green. Click on it, and the screen at right will appear.*

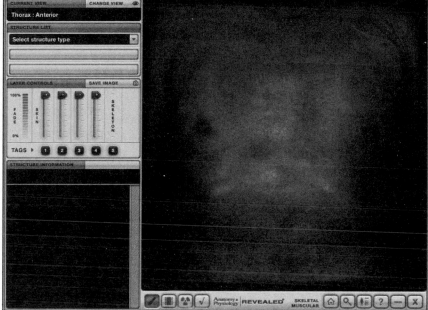

- *Click on* **TAG 1** *and the following image will appear:*

- *Mouse-over the blue pins on the screen to find the information necessary to fill in the following blanks:*

  A._____

  B._____

  C. _____

  D. _____

CHECK POINT:

Thorax, Anterior View

1. What is the name for the inferior border of costal cartilages 7–10?
2. What structures attach to this location?
3. What are the two names for the shallow notch in the superior border of the manubrium, visible superficially?

• *Click on* **TAG 2** *and the following image will appear:*

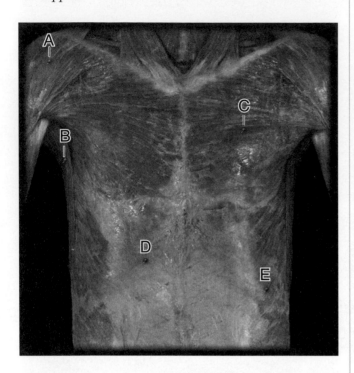

• *Mouse-over the blue pins on the screen to find the information necessary to fill in the following blanks:*

    A. _____

    B. _____

    C. _____

    D. _____

    E. _____

CHECK POINT:

Thorax, Anterior View, cont'd

4. Name the muscle involved with adduction, extension, and medial rotation of the humerus.
5. Name the muscle involved with abduction, flexion, extension, lateral, and medial rotation of the humerus.
6. What is the name for the fibrous compartment enclosing the rectus abdominis muscle?

• *Click on* **TAG 3**, *and the following image will appear:*

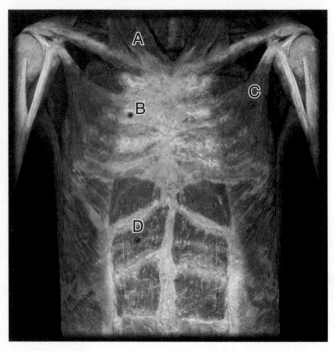

• *Mouse-over the blue pins on the screen to find the information necessary to fill in the following blanks:*

    A. _____

    B. _____

    C. _____

    D. _____

CHECK POINT:

Thorax, Anterior View, cont'd

7. Name the muscle that consists of three to four bellies, separated by tendinous intersections.
8. Name the muscle with its origin at the medial clavicle and the manubrium of the sternum and its insertion at the mastoid process.
9. Name the muscle that stabilizes the scapula and is involved in its lateral rotation.

IN REVIEW

What Have I Learned?

The following questions cover the material that you have just learned—the muscles of the thorax. Use the **STRUCTURE INFORMATION** for the muscles you have learned in answering these questions:

1. Name the structure formed by the tendons of three abdominal muscles.

2. Name the three primary muscles of respiration.

3. Name the muscle responsible for the adduction, extension, and medial rotation of the humerus.

4. Name the two muscles that stabilize the scapula.

5. Name the muscle that is the site of intramuscular injections of the arm.

• *Click on* **TAG 4**, *and the following screen will appear:*

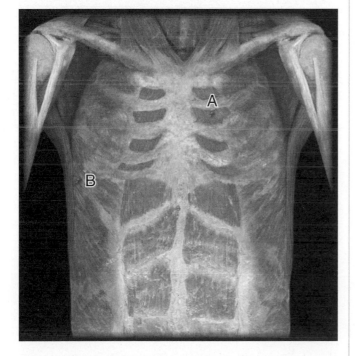

• *Mouse-over the blue pins on the screen to find the information necessary to fill in the following blanks:*

    A. _____

    B. _____

Animations: Deltoid, Rectus Abdominis, and Sternocleidomastoid

• *Click on the* **ANIMATIONS** *button at the bottom of the screen.*

• *In the* **Select topic** *menu, select* **Muscle actions**.

• *In the* **Select animation** *menu, select and view the following animations:*
  – **Deltoid**
  – **Rectus abdominis**
  – **Sternocleidomastoid**

• *Click on the* **Dissection** *button at the bottom of the* **IMAGE AREA** *to return to the* **Dissection** *section.*

EXERCISE 3.6:   Skeletal Muscle—Abdomen, Anterior View

- *Insert* Anatomy & Physiology | Revealed® **Skeletal / Muscular** CD, *or, if you are still in the* **Dissection** *section, click the* **CHANGE VIEW** *button at the top of the screen, and skip the next step.*

- *In the* **Home screen**, *select the* **Dissection** *button in the left portion of the screen. You may click either on the* **Dissection** *button or on the word itself.*

- *In the* **SELECT A VIEW** *window that appears, click on the* **Select region** *button.*

- *Choose* **Abdomen** *from the menu.*

- *The* **Select view** *menu will now show* **Anterior** *view.*

- *The* **GO** *button will now flash green. Click on it.*

- *When the* superficial abdomen *screen appears, click on* **TAG 2** *and the following image will appear:*

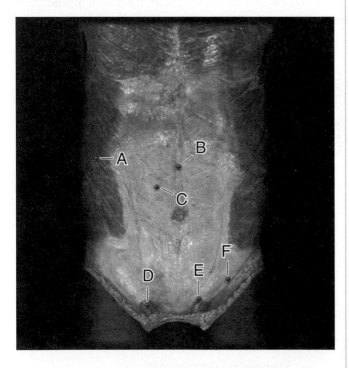

- *Mouse-over the blue pins on the screen to find the information necessary to fill in the following blanks:*

    A. _____

    B. _____

    C. _____

    D. _____

E. _____

F._____

CHECK POINT:

Abdomen, Anterior View

1. Name the common site for male inguinal hernias.
2. Opening the abdominal wall by incision through the _____ avoids cutting muscle fibers.
3. What abdominal muscle has its fibers running at right angles to the internal abdominal oblique?

- *Click on* **TAG 3**, *and the following image will appear:*

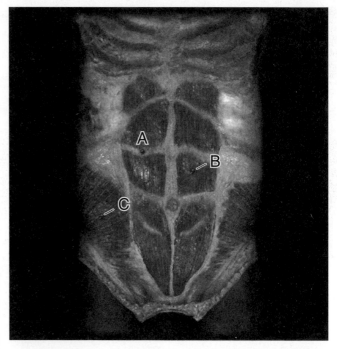

- *Mouse-over the blue pins on the screen to find the information necessary to fill in the following blanks:*

    A. _____

    B. _____

    C. _____

• Click on **TAG 5**, and the following image will appear:

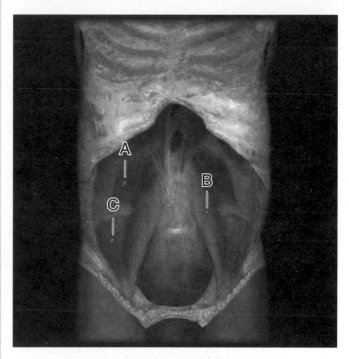

Abdomen, Anterior View, cont'd

4. Name the structures that subdivide the rectus abdominis muscle into three to four bellies.
5. What abdominal muscle has its fibers running at right angles to the external abdominal oblique?
6. Name the abdominal muscles in this view important in straining and abdominal breathing.

• Click on **TAG 4**, and the following image will appear:

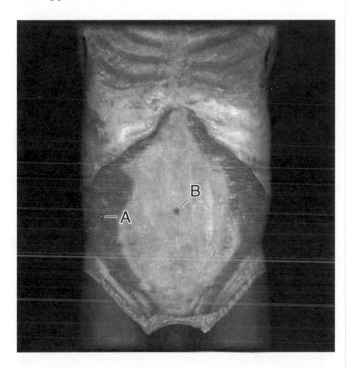

• Mouse-over the blue pins on the screen to find the information necessary to fill in the following blanks:

A. _____

B. _____

C. _____

Abdomen, Anterior View, cont'd

10. Name a muscle of the posterior abdominal wall involved in respiration.

• Mouse-over the blue pins on the screen to find the information necessary to fill in the following blanks:

A. _____

B. _____

Abdomen, Anterior View, cont'd

7. Name the abdominal muscle whose fibers run in a transverse plane.
8. What is the anatomical term for "flat tendons"?
9. What two structures come together to form the linea alba?

IN REVIEW

What Have I Learned?

The following questions cover the material that you have just learned, the muscles of the thorax. Use the **STRUCTURE INFORMATION** to answer these questions:

1. Name the abdominal wall muscles responsible for abdominal breathing.

2. What is the term for a "seam" where two structures meet?

3. Two individual muscles of the abdomen unite to form a single muscle, the most powerful flexor of the hip. Name those two individual muscles and the muscle they unite to form.

4. Two abdominal wall muscles have their structures running at right angles to each other. What are those two muscles?

5. Name the abdominal wall muscles important in straining, such as while lifting.

Animation: Rectus Abdominis

- *Click on the* **ANIMATIONS** *button at the bottom of the screen.*

- *In the* **Select topic** *menu, select* **Muscle actions**.

- *In the* **Select animation** *menu, select and view the following animation:*
  - **Rectus abdominis**

EXERCISE 3.7:   Skeletal Muscle—Pelvis, Superior View

- *Insert* Anatomy & Physiology | Revealed® **Skeletal / Muscular** *CD, or, if you are already in the* **Dissection** *section, click the* **CHANGE VIEW** *button at the top of the screen, and skip the next step.*

- *In the* **Home screen**, *select the* **Dissection** *button in the left portion of the screen. You may click either on the* **Dissection** *button or on the word itself.*

- *In the* **SELECT A VIEW** *window that appears, click on the* **Select region** *button.*

- *Choose* **Pelvis** *from the menu.*

- *The* **Select view** *menu will show* **Superior** *view.*

- *The* **GO** *button will now flash green. Click on it.*

- *When the anterior abdomen screen appears, click on* **TAG 1**, *and the following image will appear:*

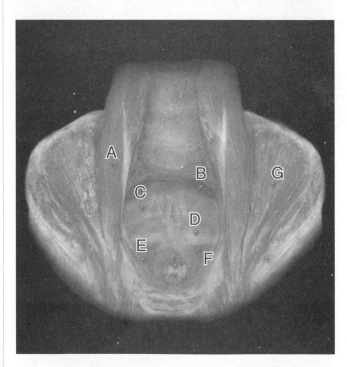

- *Mouse-over the blue pins on the screen to find the information necessary to fill in the following blanks:*

   A._____

   B._____

   C. _____

IN REVIEW

What Have I Learned?

The following questions cover the material that you have just learned: the muscles of the pelvis. Use the **STRUCTURE INFORMATION** for the muscles you have learned to answer these questions:

1. Name the pelvic muscle involved with lateral rotation of the femur and that exits the pelvis through the greater sciatic foramen.

2. Name the two muscles that make up the pelvic diaphragm. What are their functions?

3. Name the structure that serves as the origin for part of the levator ani muscle.

D. _____

E. _____

F. _____

G. _____

EXERCISE 3.8:   Skeletal Muscle—Back, Posterior View

- *Insert* Anatomy & Physiology | Revealed® **Skeletal / Muscular** *CD, or if you are already in the* **Dissection** *section, click the* **CHANGE VIEW** *button at the top of the screen, and skip the next step.*

- *In the* **Home screen***, select the* **Dissection** *button in the left portion of the screen. You may click either on the* **Dissection** *button or on the word itself.*

- *In the* **SELECT A VIEW** *window that appears, click on the* **Select region** *button.*

- *Choose* **Back** *from the menu.*

- *The* **Select view** *menu will now show* **Posterior** *view.*

- *The* **GO** *button will flash green. Click on it.*

- *When the back screen appears, click on* **TAG 1***, and the image at right will appear.*

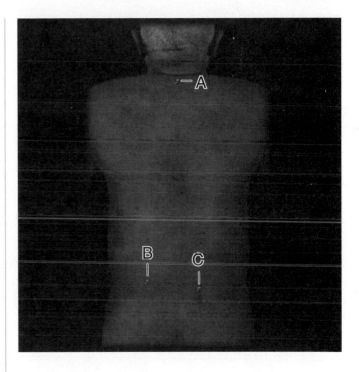

- *Mouse-over the blue pins on the screen to find the information necessary to fill in the following blanks:*

A. _____

B. _____

C. _____

Back, Posterior View

1. Name the landmark for intramuscular injections of the hip.
2. The shallow skin depression (dimple) in the lower back marks what point?
3. What is the name for the prominent surface projection produced by the spinous process of vertebra C7?

- *Click on* **TAG 2**, *and the following image will appear:*

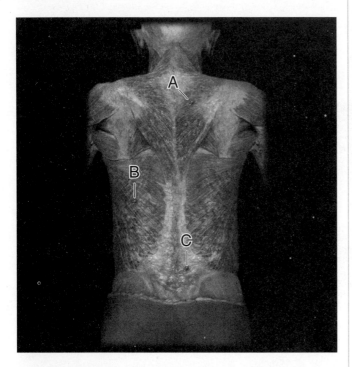

- *Mouse-over the blue pins on the screen to find the information necessary to fill in the following blanks:*

  A. _____

  B. _____

  C. _____

Back, Posterior View, cont'd

4. Name the superficial "kite-shaped" muscle of the back that spans from the nuchal line of the occipital bone to vertebra T12.

5. Name the deep fascia whose attached structures include the latissimus dorsi and erector spinae muscles.
6. Name the superficial muscle whose name describes its location as it spans from the back to the side of the body.

- *Click on* **TAG 3**, *and the following image will appear:*

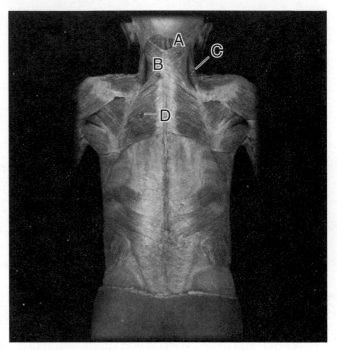

- *Mouse-over the blue pins on the screen to find the information necessary to fill in the following blanks:*

  A. _____

  B. _____

  C. _____

  D. _____

Back, Posterior View, cont'd

7. Name two muscles involved in the retraction and elevation of the scapula.
8. Name a muscle that allows the shrugging of the shoulders.

- *Click on **TAG 4**, and the following image will appear:*

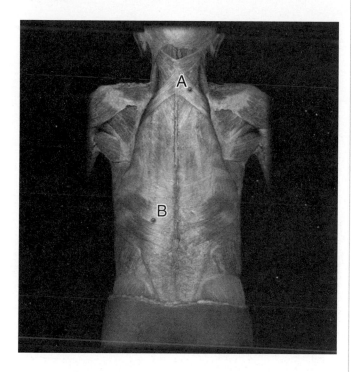

- *Mouse-over the blue pins on the screen to find the information necessary to fill in the following blanks:*

    A. _____

    B. _____

- *Click on **TAG 5**, and the following image will appear:*

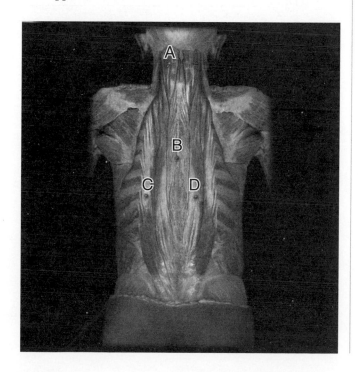

- *Mouse-over the blue pins on the screen to find the information necessary to fill in the following blanks:*

    A. _____

    B. _____

    C. _____

    D. _____

CHECK POINT:

Back, Posterior View, cont'd

9. Name the muscle known as the "antigravity muscle."
10. The muscle in Check Point consists of three separate muscles. What are they?

Animation: Rectus Abdominis

- *Click on the **ANIMATIONS** button at the bottom of the screen.*

- *In the **Select topic** menu, select **Muscle actions**.*

- *In the **Select animation** menu, select and view the following animation:*

    – **Rectus abdominis**

EXERCISE 3.9:   Skeletal Muscle—Shoulder and Arm, Anterior View

- *Insert Anatomy & Physiology | Revealed® **Skeletal / Muscular** CD, or, if you are already in the **Animation** section, click the **Dissection** button at the bottom of the screen, and skip the next step.*

- *In the **Home screen**, select the **Dissection** button in the left portion of the screen. You may click either on the **Dissection** button or on the word itself.*

- *In the **SELECT A VIEW** window that appears, click on the **Select region** button.*

- *Choose **Shoulder and arm** from the menu.*

- *From the **Select view** menu, select **Anterior view**.*

- *The **GO** button will flash green. Click on it.*

• *When the* **Shoulder and arm** *screen appears, click on* **TAG 1**, *and the following image will appear:*

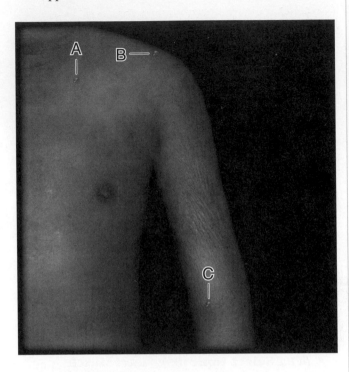

• *Mouse-over the blue pins on the screen to find the information necessary to fill in the following blanks:*

A. _____

B. _____

C. _____

Shoulder and Arm, Anterior View

1. Name the structure referred to as the collar bone.
2. Name the structure that is the flattened, lateral part of the scapular spine.
3. What is the name for the triangular concavity of the anterior elbow?

• *Click on* **TAG 2**, *and the following image will appear:*

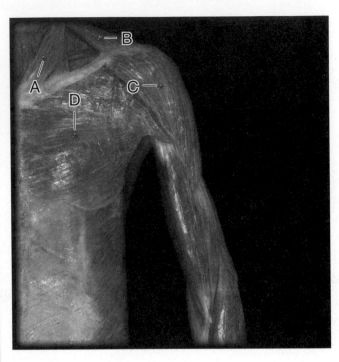

• *Mouse-over the blue pins on the screen to find the information necessary to fill in the following blanks:*

A._____

B._____

C. _____

D. _____

Shoulder and Arm, Anterior View, cont'd

4. Name the superficial muscle of the chest.
5. Name the muscle that contributes to the roundness of the shoulder.
6. Name the mostly posterior muscle that has its insertion at the clavicle and scapula.

• *Click on* **TAG 3**, *and the following image will appear:*

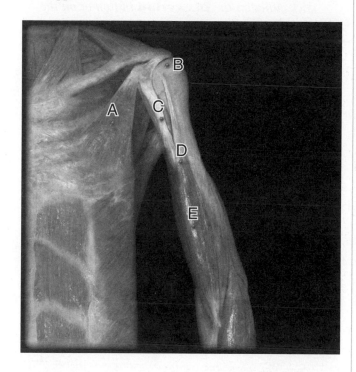

• *Mouse-over the blue pins on the screen to find the information necessary to fill in the following blanks:*

A. _____

B. _____

C. _____

D. _____

E. _____

CHECK POINT:

Shoulder and Arm, Anterior View, cont'd

7. Name the muscle of the arm that has two heads.
8. Name the tough fibrous envelope that surrounds the joint where the arm attaches to the pectoral girdle.
9. Name the two muscles referred to as the "pecs."

• *Click on* **TAG 4**, *and the following image will appear:*

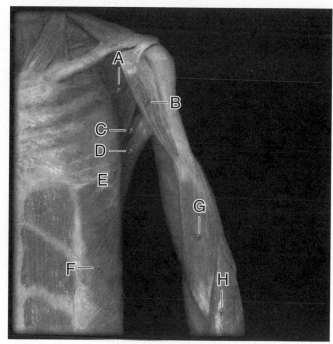

• *Mouse-over the blue pins on the screen to find the information necessary to fill in the following blanks:*

A. _____

B. _____

C. _____

D. _____

E. _____

F. _____

G. _____

H. _____

CHECK POINT:

Shoulder and Arm, Anterior View, cont'd

10. Name the four rotator cuff muscles.
11. What is the function of the rotator cuff muscles?
12. Name the muscle that is deep to the biceps brachii.

• *Click on* **TAG 5**, *and the following screen will appear:*

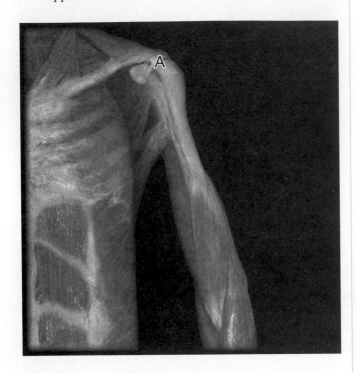

• *Mouse-over the blue pin on the screen to find the information necessary to fill in the following blank:*

A. _____

Animations: Biceps Brachii and Deltoid

• *Click on the* **ANIMATIONS** *button at the bottom of the screen.*

• *In the* **Select topic** *menu, select* **Muscle actions**.

• *In the* **Select animation** *menu, select and view the following animations:*

  – **Biceps brachii**
  – **Deltoid**

EXERCISE 3.10:   Skeletal Muscle—Shoulder and Arm, Posterior View

• *Insert* Anatomy & Physiology | Revealed® **Skeletal / Muscular** *CD, or, if you are already in the* **Animation** *section, click the* **DISSECTION** *button at the bottom of the screen, and skip the next step.*

• *In the* **Home screen**, *select the* **Dissection** *button in the left portion of the screen. You may click either on the* **Dissection** *button or on the word itself.*

• *In the* **SELECT A VIEW** *window that appears, click on the* **Select region** *button.*

• *Choose* **Shoulder and arm** *from the menu.*

• *From the* **Select view** *menu, select* **Posterior view**.

• *The* **GO** *button will flash green. Click on it.*

• *When the* **Shoulder and arm** *screen appears, click on* **TAG 1**, *and the following image will appear:*

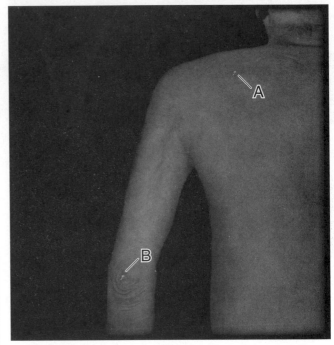

• *Mouse-over the blue pins on the screen to find the information necessary to fill in the following blanks:*

A. _____

B. _____

CHECK POINT:

Shoulder and Arm, Posterior View

1. What is the name for the point of the elbow?
2. What specific structure of what bone constitutes the point of the elbow?
3. Name the prominent ridge on the posterior surface of the scapula.

- *Click on **TAG 2**, and the following image will appear:*

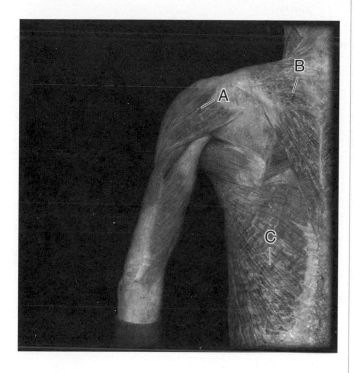

- *Mouse-over the blue pins on the screen to find the information necessary to fill in the following blanks:*

    A. _____
    B. _____
    C. _____

Shoulder and Arm, Posterior View, cont'd

4. Name the triangle-shaped muscle of the shoulder.
5. Name the large lateral muscle responsible for adduction, extension, and medial rotation of the humerus.
6. Name the muscle responsible for the elevation, medial rotation, adduction, and depression of the scapula.

- *Click on **TAG 3**, and the image at top right will appear.*

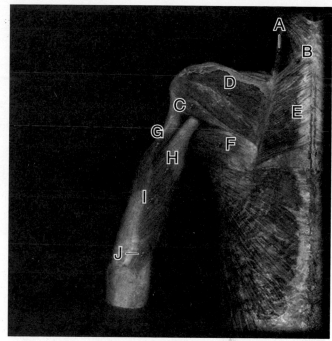

- *Mouse-over the blue pins on the screen to find the information necessary to fill in the following blanks:*

    A. _____
    B. _____
    C. _____
    D. _____
    E. _____
    F. _____
    G. _____
    H. _____
    I. _____
    J. _____

Shoulder and Arm, Posterior View, cont'd

7. Name the muscle of the arm with three heads.
8. Name the muscle found in the infraspinous fossa of the scapula.
9. Name two muscles with their insertions on the medial border of the scapula.

- *Click on* **TAG 4**, *and the following image will appear:*

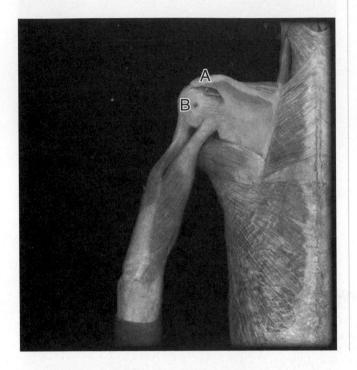

- *Mouse-over the blue pins on the screen to find the information necessary to fill in the following blanks:*

    A._____

    B._____

CHECK POINT:

Shoulder and Arm, Posterior View, cont'd

10. Name the muscle located in the supraspinous fossa of the scapula.
11. Name a muscle that holds the head of the humerus in the glenoid cavity.

IN REVIEW

What Have I Learned?

The following questions cover the material that you have just learned—the muscles of the shoulder and arm. Use the **STRUCTURE INFORMATION** for the muscles you have learned to answer these questions:

1. Name the landmark for intramuscular injections of the hip.

2. Name the superficial "kite-shaped" muscle of the back that spans from the nuchal line of the occipital bone to vertebra T12.

3. Name the superficial muscle whose name describes its location as it spans from the back to the side of the body.

4. Name two muscles involved in the retraction and elevation of the scapula.

5. Name a muscle that allows the shrugging of the shoulders.

6. Name the muscle known as the "antigravity muscle". What three muscles combine to form this muscle?

7. Name the superficial muscle of the chest.

8. Name the muscle of the arm that has two heads.

9. Name the muscle of the arm that has three heads.

10. Name the four rotator cuff muscles. What is the function of the these muscles?

IN REVIEW   Continued

11. Name the large lateral muscle responsible for adduction, extension and medial rotation of the humerus.

12. Name the muscle responsible for the elevation, medial rotation, adduction and depression of the scapula.

13. Name a muscle that holds the head of the humerus in the glenoid cavity.

EXERCISE 3.11:   Skeletal Muscles—Forearm and Hand, Anterior View

- *Insert Anatomy & Physiology | Revealed® **Skeletal / Muscular** CD, or, if you are still in the **Animation** section, click the **Dissection** button at the bottom of the screen, and skip the next step.*

- *In the **Home screen**, select the **Dissection** button in the left portion of the screen. You may click either on the **Dissection** button or on the word itself.*

- *In the **SELECT A VIEW** window that appears, click on the **Select region** button.*

- *Choose **Forearm and hand** from the menu.*

- *From the **Select view** menu, select **Anterior** view.*

- *The **GO** button will flash green. Click on it.*

- *When the **Forearm and hand** screen appears, click on **TAG 1** and the image at right will appear.*

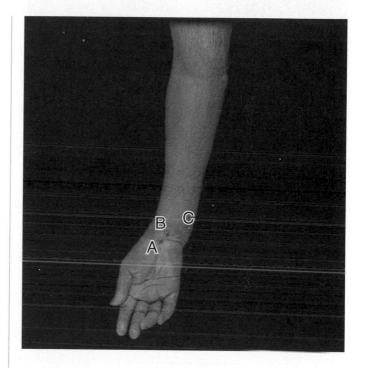

- *Mouse-over the blue pins on the screen to find the information necessary to fill in the following blanks:*

    A. _____

    B. _____

    C. _____

• *Click on* **TAG 2**, *and the following image will appear:*

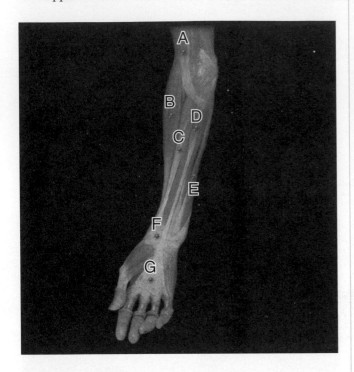

• *Mouse-over the blue pins on the screen to find the information necessary to fill in the following blanks:*

A. _____

B. _____

C. _____

D. _____

E. _____

F._____

G._____

CHECK POINT:

Forearm and Hand, Anterior View

1. Name three muscles that flex the wrist.
2. Name two muscles that flex the elbow.
3. Name the structure that forms the carpal tunnel.

• *Click on* **TAG 3**, *and the following image will appear:*

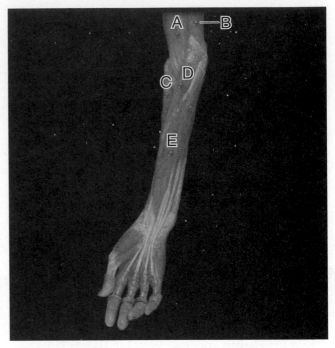

• *Mouse-over the blue pins on the screen to find the information necessary to fill in the following blanks:*

A. _____

B. _____

C. _____

D. _____

E. _____

CHECK POINT:

Forearm and Hand, Anterior View, cont'd

4. Name a muscle involved in the pronation of the forearm.
5. Name a muscle involved in the supination of the forearm.
6. Name a muscle involved with the extension of the elbow.

• *Click on* **TAG 4**, *and the following image will appear:*

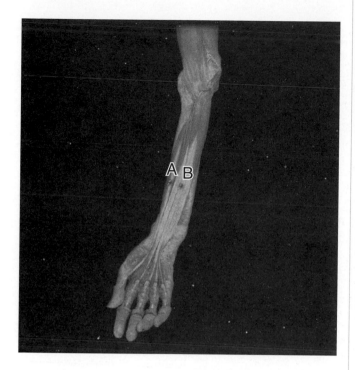

• *Mouse-over the blue pins on the screen to find the information necessary to fill in the following blanks:*

A. _____

B. _____

CHECK POINT:

Forearm and Hand, Anterior View, cont'd

7. Name the only muscle that flexes the distal interphalangeal joint.
8. Name the only muscle that flexes the distal phalanx of the thumb.
9. What is the anatomical name for the thumb?

• *Click on* **TAG 5**, *and the following image will appear:*

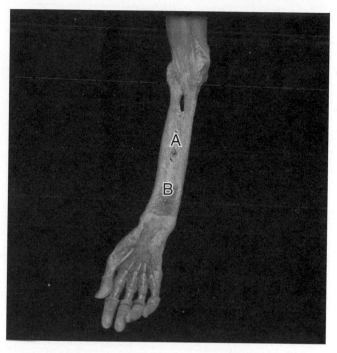

• *Mouse-over the blue pins on the screen to find the information necessary to fill in the following blanks:*

A. _____

B. _____

CHECK POINT:

Forearm and Hand, Anterior View, cont'd

10. Name the thick sheet of connective tissue between the ulna and the radius.
11. What is its function?

EXERCISE 3.12:   Skeletal Muscles—Forearm and Hand, Posterior View

• *Insert Anatomy & Physiology | Revealed®* **Skeletal / Muscular** *CD, or, if you are already in the* **Dissection** *section, click the* **CHANGE VIEW** *button at the top of the screen, and skip the next step.*

• *In the* **Home screen**, *select the* **Dissection** *button in the left portion of the screen. You may click either on the* **Dissection** *button or on the word itself.*

- *In the **SELECT A VIEW** window that appears, click on the **Select region** button.*

- *Choose **Forearm and hand** from the menu.*

- *From the **Select view** menu, select **Posterior** view.*

- *The **GO** button will flash green. Click on it.*

- *When the **Forearm and hand** screen appears, click on **TAG 1**, and the following image will appear:*

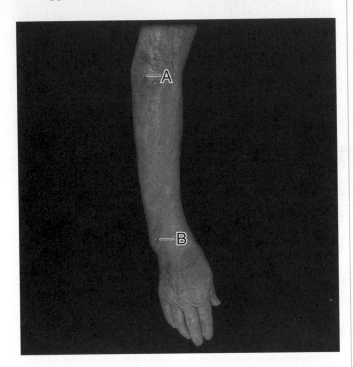

- *Mouse-over the blue pins on the screen to find the information necessary to fill in the following blanks:*

    A. _____

    B. _____

- *Click on **TAG 2**, and the following image will appear:*

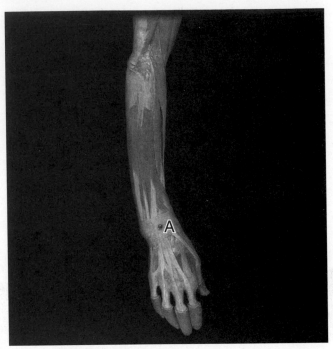

- *Mouse-over the blue pin on the screen to find the information necessary to fill in the following blank:*

    A. _____

CHECK POINT:

Forearm and Hand, Posterior View

1. What is the relationship between the retinaculum and the extensor tendons of the forearm?

- *Click on* **TAG 3**, *and the following image will appear:*

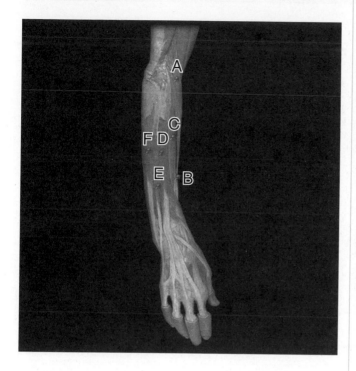

- *Mouse-over the blue pins on the screen to find the information necessary to fill in the following blanks:*

A. _____

B. _____

C. _____

D. _____

E. _____

F. _____

CHECK POINT:

Forearm and Hand, Posterior View, cont'd

2. Name three muscles that extend the wrist.
3. Name a muscle that assists in both pronation and supination of the forearm.
4. Name a muscle that extends the fifth finger.

- *Click on* **TAG 4**, *and the following image will appear:*

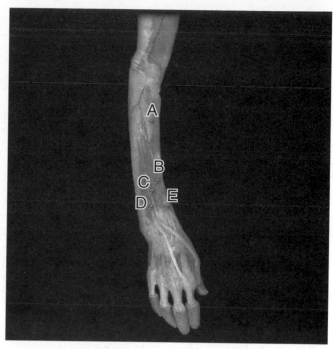

- *Mouse-over the blue pins on the screen to find the information necessary to fill in the following blanks:*

A. _____

B. _____

C. _____

D. _____

E. _____

CHECK POINT:

Forearm and Hand, Posterior View, cont'd

5. Name a muscle that both extends and abducts the thumb.
6. Name two muscles that extend the thumb.
7. Name a muscle that extends the second finger.

IN REVIEW

What Have I Learned?

The following questions cover the material that you have just learned—the muscles of the forearm and hand. Use the **STRUCTURE INFORMATION** for the muscles you have learned to answer these questions:

1. Name three muscles that flex the wrist.

2. Name two muscles that flex the elbow.

3. Name the structure that forms the carpal tunnel.

4. Name a muscle involved in the pronation of the forearm.

5. Name a muscle involved in the supination of the forearm.

6. What is the anatomical name for the thumb?

7. Name the thick sheet of connective tissue between the ulna and the radius. What is its function?

8. Name three muscles that extend the wrist.

9. Name a muscle that both extends and abducts the thumb.

EXERCISE 3.13: Skeletal Muscles - Wrist and Hand, Anterior View

- *Insert* Anatomy & Physiology | Revealed® **Skeletal / Muscular** *CD, or, if you are already in the* **Dissection** *section, click the* **CHANGE VIEW** *button at the top of the screen, and skip the next step.*

- *In the* **Home screen**, *select the* **Dissection** *button in the left portion of the screen. You may click either on the* **Dissection** *button or on the word itself.*

- *In the* **SELECT A VIEW** *window that appears, click on the* **Select region** *button.*

- *Choose* **Wrist and hand** *from the menu.*

- *From the* **Select view** *menu, select* **Anterior view.**

- *The* **GO** *button will flash green. Click on it.*

- *When the* **Wrist and hand** *screen appears, click on* **TAG 1** *and the screen at right will appear.*

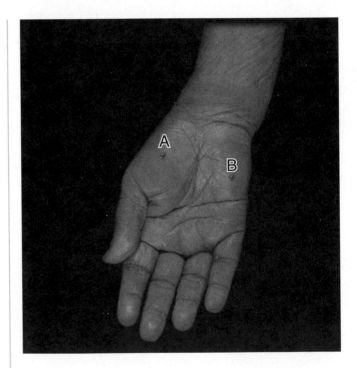

- *Mouse-over the blue pins on the screen to find the information necessary to fill in the following blanks:*

    A. _____

    B. _____

CHECK POINT:

Wrist and Hand, Anterior View

1. Name the thick, fleshy eminence at the base of the first digit.
2. Name the thick, fleshy eminence at the base of the fifth digit.

- *Click on* **TAG 2**, *and the following image will appear:*

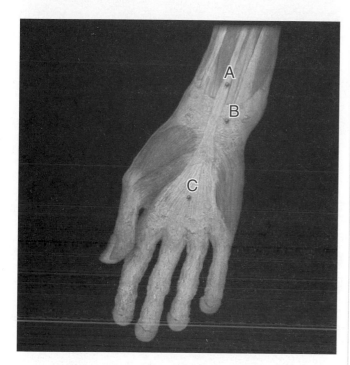

- *Mouse-over the blue pins on the screen to find the information necessary to fill in the following blanks:*

    A. _____

    B. _____

    C. _____

CHECK POINT:

Wrist and Hand, Anterior View, cont'd

3. Name a muscle often missing on one or both forearms.

- *Click on* **TAG 3**, *and the following image will appear:*

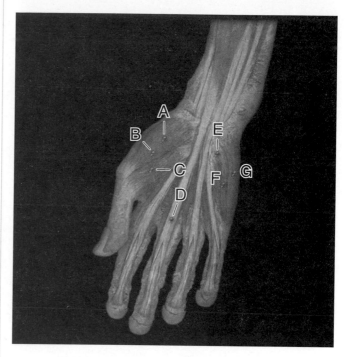

- *Mouse-over the blue pins on the screen to find the information necessary to fill in the following blanks:*

    A. _____

    B. _____

    C. _____

    D. _____

    E. _____

    F._____

    G. _____

CHECK POINT:

Wrist and Hand, Anterior View, cont'd

4. Name the three thenar muscles.
5. Name the three hypothenar muscles.
6. Which digits do each of the above muscles act upon?

- *Click on **TAG 4**, and the following image will appear:*

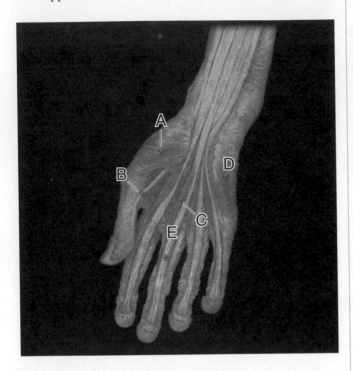

- *Mouse-over the blue pins on the screen to find the information necessary to fill in the following blanks:*

    A. _____

    B. _____

    C. _____

    D. _____

    E. _____

CHECK POINT:

Wrist and Hand, Anterior View, cont'd

7. Name the muscles that both flex the metacarpophalangeal joint and extend the interphalangeal joints.
8. Name the muscle that allows the fifth finger to touch the tip of the first finger.
9. Name the muscle that allows the tip of the first finger to touch the tips of the other fingers.

- *Click on **TAG 5**, and the following image will appear:*

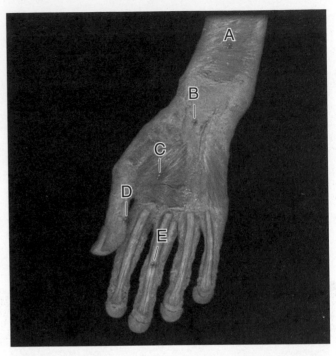

- *Mouse-over the blue pins on the screen to find the information necessary to fill in the following blanks:*

    A. _____

    B. _____

    C. _____

    D. _____

    E. _____

CHECK POINT:

Wrist and Hand, Anterior View, cont'd

10. Name the only muscle that flexes the distal phalanx of the first digit.
11. Name the only muscle that flexes the distal interphalangeal joint of digits 2–5.
12. Name the distal pronator of the forearm.

EXERCISE 3.14:   Skeletal Muscles—Wrist and Hand, Posterior View

- *Insert Anatomy & Physiology | Revealed®* **Skeletal/Muscular** *CD, or, if you are already in the* **Dissection** *section, click the* **CHANGE VIEW** *button at the top of the screen, and skip the next step.*

- *In the* **Home screen***, select the* **Dissection** *button in the left portion of the screen. You may click either on the* **Dissection** *button or on the word itself.*

- *In the* **SELECT A VIEW** *window that appears, click on the* **Select region** *button.*

- *Choose* **Wrist and hand** *from the menu.*

- *From the* **Select view** *menu, select* **Posterior view.**

- *The* **GO** *button will flash green. Click on it.*

- *When the* **Wrist and hand** *screen appears, click on* **TAG 1***, and the following image will appear:*

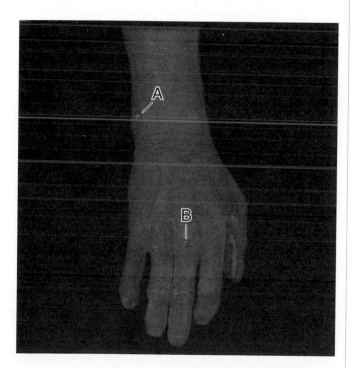

- *Mouse-over the blue pins on the screen to find the information necessary to fill in the following blanks:*

    A. _____

    B. _____

CHECK POINT:

Wrist and Hand, Posterior View

1. Flexion of which joint makes the knuckles prominent?
2. What structures are visible as the knuckles?

- *Click on* **TAG 2***, and the following image will appear:*

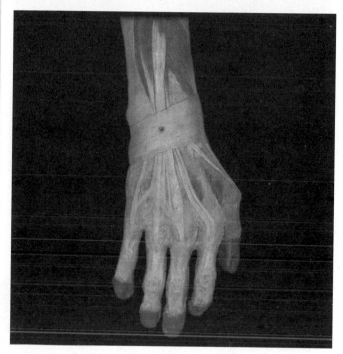

- *Mouse-over the blue pin on the screen to find the information necessary to fill in the following blank:*

    A. _____

CHECK POINT:

Wrist and Hand, Posterior View, cont'd

3. Name a muscle often missing on one or both forearms.

• *Click on* **TAG 3**, *and the following image will appear:*

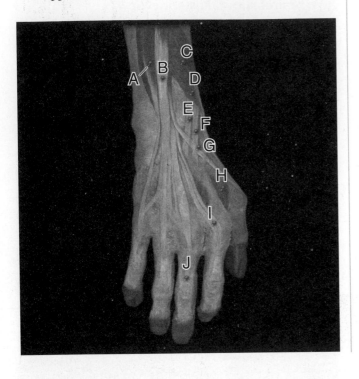

• *Mouse-over the blue pins on the screen to find the information necessary to fill in the following blanks:*

A. _____

B. _____

C. _____

D. _____

E. _____

F. _____

G. _____

H. _____

I. _____

J. _____

I N   R E V I E W

What Have I Learned?

The following questions cover the material that you have just learned—the muscles of the wrist and hand. Use the **STRUCTURE INFORMATION** for the muscles you have learned to answer these questions:

1. Name the thick, fleshy eminence at the base of the first digit.

2. Name the thick, fleshy eminence at the base of the fifth digit.

3. Name a muscle often missing on one or both forearms.

4. Name the only muscle that flexes the distal phalanx of the first digit.

5. Flexion of which joint makes the knuckles prominent?

6. What structures are visible as the knuckles?

EXERCISE 3.15:   Skeletal Muscles—Hip and Thigh, Anterior View

- *Insert* Anatomy & Physiology | Revealed® **Skeletal / Muscular** *CD, or, if you are already in the* **Dissection** *section, click the* **CHANGE VIEW** *button at the top of the screen, and skip the next step.*

- *In the* **Home screen**, *select the* **Dissection** *button in the left portion of the screen. You may click either on the* **Dissection** *button or on the word itself.*

- *In the* **SELECT A VIEW** *window that appears, click on the* **Select region** *button.*

- *Choose* **Hip and thigh** *from the menu.*

- *From the* **Select view** *menu, select* **Anterior view**.

- *The* **GO** *button will flash green. Click on it.*

- *When the* **Hip and thigh** *screen appears, click on* **TAG 1** *and the following screen will appear:*

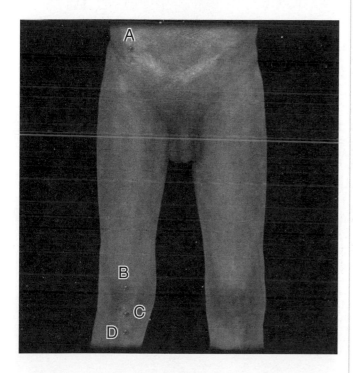

- *Mouse-over the blue pins on the screen to find the information necessary to fill in the following blanks:*

    A._____

    B._____

    C. _____

    D. _____

C H E C K   P O I N T :

Hip and Thigh, Anterior View

1. Name the superficially visible anterior subcutaneous end of the iliac crest.
2. Name the point of attachment for the quadriceps femoris muscles by way of the patellar ligament.
3. Name the ligament that connects the patella to the tuberosity of the tibia.

- *Click on* **TAG 2**, *and the following screen will appear:*

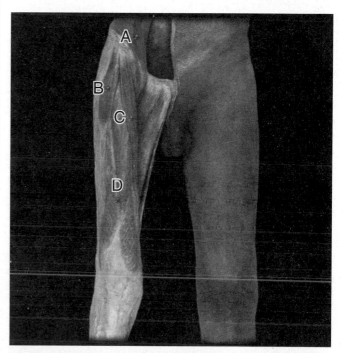

- *Mouse-over the blue pins on the screen to find the information necessary to fill in the following blanks:*

    A._____

    B._____

    C. _____

    D. _____

CHECK POINT:

Hip and Thigh, Anterior View, cont'd

4. Name the muscle whose origin is the anterior superior iliac spine of the ilium and whose insertion is the proximal medial shaft of the tibia.
5. Name the four muscles of the quadriceps femoris.
6. Name the most powerful flexor of the hip joint.

- *Click on* **TAG 3**, *and the following image will appear:*

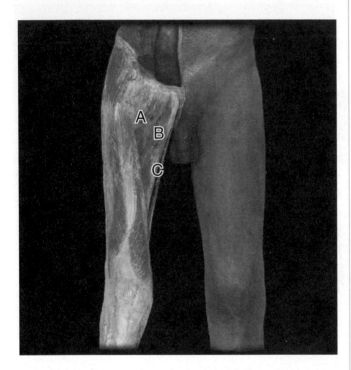

- *Mouse-over the blue pins on the screen to find the information necessary to fill in the following blanks:*

    A. _____

    B. _____

    C. _____

CHECK POINT:

Hip and Thigh, Anterior View, cont'd

7. Name the muscle of the thigh that is weak in humans and used in muscle transplants.
8. Name the muscle often involved in a "pulled groin."

- *Click on* **TAG 4**, *and the following image will appear:*

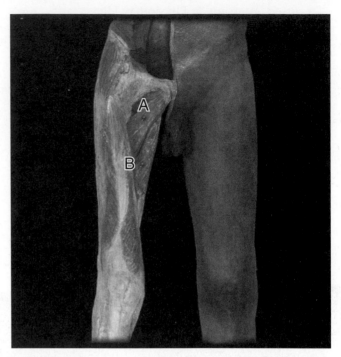

- *Mouse-over the blue pins on the screen to find the information necessary to fill in the following blanks:*

    A. _____

    B. _____

- *Click on* **TAG 5**, *and the following image will appear:*

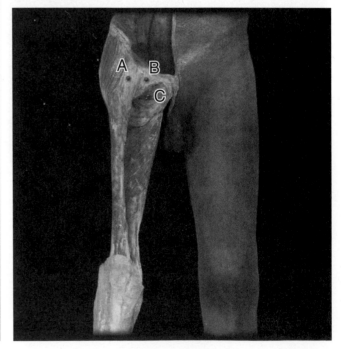

- *Mouse-over the blue pins on the screen to find the information necessary to fill in the following blanks:*

  A._____

  B._____

  C. _____

Hip and Thigh, Anterior View, cont'd

9. Name the strongest ligament around the hip joint.
10. Name the ligament that resists excessive abduction of the hip.
11. Name the ligament that resists hyperextension of the hip joint.

Animations: Quadriceps Femoris

- *Click on the **ANIMATIONS** button at the bottom of the screen.*

- *In the **Select topic** menu, select **Muscle actions**.*

- *In the **Select animation** menu, select and view the following animation:*

  – **Quadriceps femoris**

EXERCISE 3.16:   Skeletal Muscles—Hip and Thigh, Posterior View

- *Insert Anatomy & Physiology | Revealed® **Skeletal/Muscular** CD, or, If you are still in the **Animation** section, click the **Dissection** button at the bottom of the screen, click change view and skip the next step.*

- *In the **Home screen**, select the **Dissection** button in the left portion of the screen. You may click either on the **Dissection** button or on the word itself.*

- *In the **SELECT A VIEW** window that appears, click on the **Select region** button.*

- *Choose **Hip and thigh** from the menu.*

- *From the **Select view** menu, select **Posterior** view.*

- *The **GO** button will flash green. Click on it.*

- *When the **Hip and thigh** screen appears, click on **TAG 1**, and the following image will appear:*

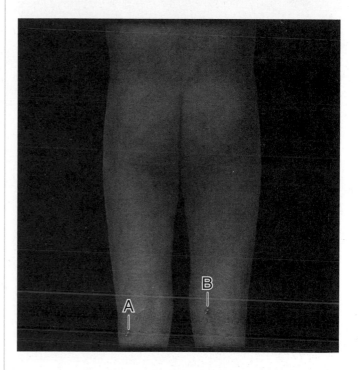

- *Mouse-over the blue pins on the screen to find the information necessary to fill in the following blanks:*

  A. _____

  B. _____

Hip and Thigh, Posterior View

1. Name the muscle whose tendon is the lateral hamstring.
2. Name the muscles whose tendons are the medial hamstring.
3. Name the structure that provides attachment for the fibular collateral ligament of the knee and the biceps femoris muscle.

• *Click on* **TAG 2**, *and the following image will appear:*

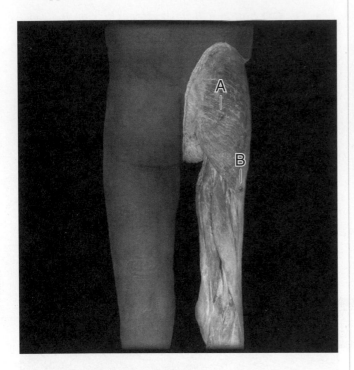

• *Mouse-over the blue pins on the screen to find the information necessary to fill in the following blanks:*

A. _____

B. _____

CHECK POINT:

Hip and Thigh, Posterior View, cont'd

4. Name a muscle of the posterior thigh that is not important in walking.
5. Name a muscle of the posterior thigh that is important for powerful extension of the femur as in running, climbing stairs, and rising from the seated position.
6. Name the structure that provides attachment for the tensor fascia lata and gluteus maximus muscles.

• *Click on* **TAG 3**, *and the following image will appear:*

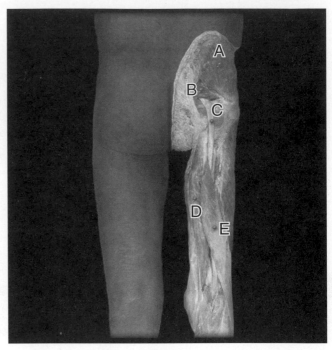

• *Mouse-over the blue pins on the screen to find the information necessary to fill in the following blanks:*

A. _____

B. _____

C. _____

D. _____

E. _____

CHECK POINT:

Hip and Thigh, Posterior View, cont'd

7. Name the two muscles that allow the non-weight-bearing limb to swing forward during walking.
8. Name the two heads of the biceps femoris.
9. Name the largest nerve in the body.

- *Click on **TAG 4**, and the following image will appear:*

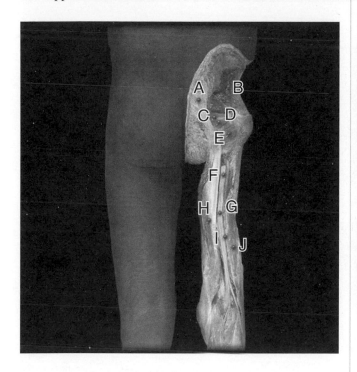

- *Mouse-over the blue pins on the screen to find the information necessary to fill in the following blanks:*

  A. _____

  B. _____

  C. _____

  D. _____

  E. _____

  F._____

  G. _____

  H._____

  I._____

  J._____

- *Click on **TAG 5**, and the following image will appear:*

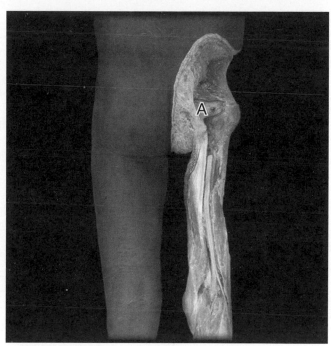

- *Mouse-over the blue pin on the screen to find the information necessary to fill in the following blank:*

  A. _____

IN REVIEW

What Have I Learned?

The following questions cover the material that you have just learned—the muscles of the hip and thigh. Use the **STRUCTURE INFORMATION** for the muscles you have learned to answer these questions:

1. Name the four muscles of the quadriceps femoris.

2. Name the most powerful flexor of the hip joint.

3. Name the muscle of the thigh that is weak in humans and used in muscle transplants.

4. Name the muscle often involved in a "pulled groin".

5. Name the strongest ligament around the hip joint.

6. Name a muscle of the posterior thigh that is important for powerful extension of the femur as in running, climbing stairs, and rising from the seated position.

7. Name the two muscles that allow the non weight-bearing limb to swing forward during walking.

8. Name the two heads of the biceps femoris.

9. Name the thick fibrous band fused to the posterior surface of the hip joint capsule.

EXERCISE 3.17: Skeletal Muscles—Leg and Foot, Anterior View

- *Insert* Anatomy & Physiology | Revealed® **Skeletal / Muscular** *CD, or, if you are still in the* **Dissection** *section, click the* **CHANGE VIEW** *button at the top of the screen, and skip the next step.*

- *In the* **Home screen**, *select the* **Dissection** *button in the left portion of the screen. You may click either on the* **Dissection** *button or on the word itself.*

- *In the* **SELECT A VIEW** *window that appears, click on the* **Select region** *button.*

- *Choose* **Leg and foot** *from the menu.*

- *From the* **Select view** *menu, select* **Anterior view.**

- *The* **GO** *button will flash green. Click on it.*

- *When the* **Leg and foot** *screen appears, click on* **TAG 1**, *and the screen at right will appear.*

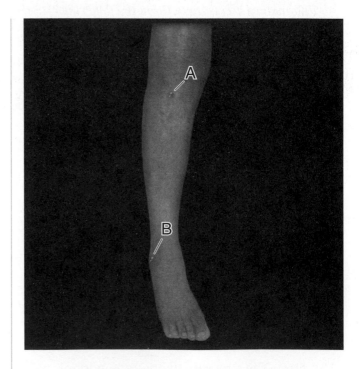

- *Mouse-over the blue pins on the screen to find the information necessary to fill in the following blanks:*

  A. _____

  B. _____

CHECK POINT:

Leg and Foot, Anterior View

1. Name the bony elevation of the anterior proximal tibia.
2. Name the lateral subcutaneous projection that contributes to the ankle joint.

- *Click on **TAG 2**, and the following image will appear:*

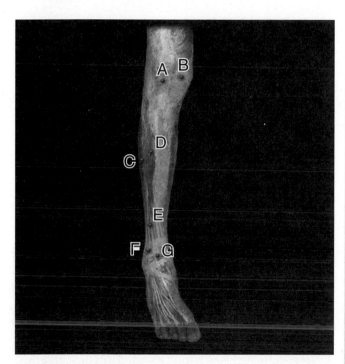

- *Mouse-over the blue pins on the screen to find the information necessary to fill in the following blanks:*

  A._____

  B._____

  C. _____

  D. _____

  E._____

  F. _____

  G. _____

CHECK POINT:

Leg and Foot, Anterior View, cont'd

3. What is the anatomical term for the first toe?
4. The tendons of which muscles are subcutaneous on the dorsum of the foot?
5. Name the structure that serves to bind in place the tendons from the anterior compartment of the leg as they cross the ankle joint.

- *Click on **TAG 3**, and the following image will appear:*

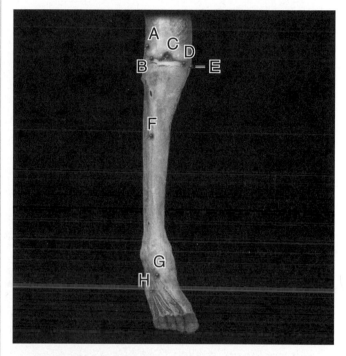

- *Mouse-over the blue pins on the screen to find the information necessary to fill in the following blanks:*

  A._____

  B._____

  C. _____

  D. _____

  E._____

  F. _____

  G. _____

  H. _____

CHECK POINT:

Leg and Foot, Anterior View, cont'd

6. Name the thick sheet of connective tissue between the tibia and fibula.
7. What is the function of this sheet of connective tissue?
8. Name the structures referred to as the "unhappy triad."

EXERCISE 3.18: Skeletal Muscles—Leg and Foot, Posterior View

- *Insert* Anatomy & Physiology | Revealed® **Skeletal / Muscular** *CD, or, if you are still in the* **Dissection** *section, click the* **CHANGE VIEW** *button at the top of the screen, and skip the next step.*

- *In the* **Home screen***, select the* **Dissection** *button in the left portion of the screen. You may click either on the* **Dissection** *button or on the word itself.*

- *In the* **SELECT A VIEW** *window that appears, click on the* **Select region** *button.*

- *Choose* **Leg and foot** *from the menu.*

- *From the* **Select view** *menu, select* **Posterior view***.*

- *The* **GO** *button will flash green. Click on it.*

- *When the* **Leg and foot** *screen appears, click on* **TAG 1** *and the following image will appear:*

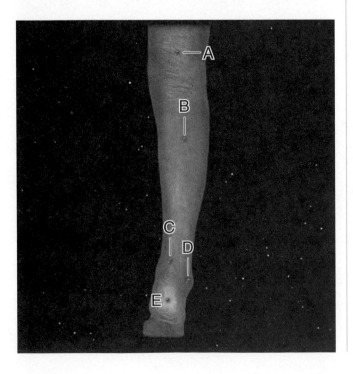

- *Mouse-over the blue pins on the screen to find the information necessary to fill in the following blanks:*

    A._____

    B._____

    C._____

    D._____

    E._____

CHECK POINT:

Leg and Foot, Posterior View

1. Name the strongest tendon in the body.
2. Give an example of this tendon's strength from the **STRUCTURE INFORMATION** window.
3. Name the tendon also known as the "Achilles" tendon.

- *Click on* **TAG 2***, and the following image will appear:*

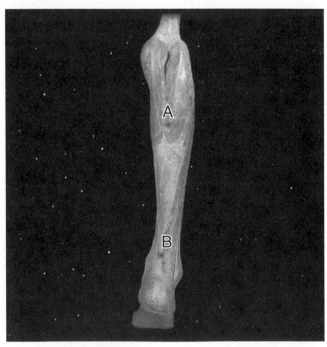

- *Mouse-over the blue pins on the screen to find the information necessary to fill in the following blanks:*

    A._____

    B._____

Leg and Foot, Posterior View, cont'd

4. The tendons of which two muscles contribute to the calcaneal tendon?
5. Name the calf muscle that consists of a medial and a lateral belly.
6. Name the superficial calf muscle.

• *Click on* **TAG 3**, *and the following image will appear:*

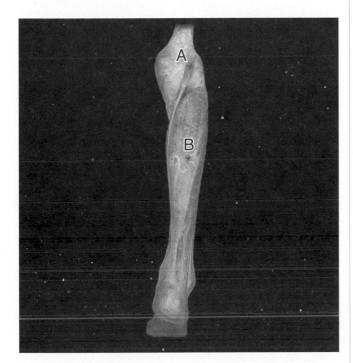

• *Mouse-over the blue pins on the screen to find the information necessary to fill in the following blanks:*

    A. _____

    B. _____

Leg and Foot, Posterior View, cont'd

7. Name the calf muscle deep to the gastrocnemius.
8. Name the long thin tendon that is a common source for tendon transplants.

• *Click on* **TAG 4**, *and the following image will appear:*

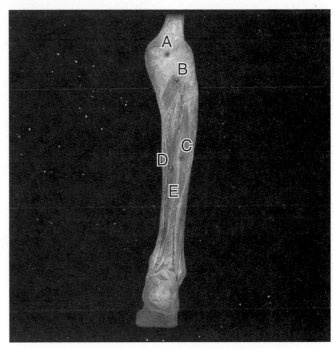

• *Mouse-over the blue pins on the screen to find the information necessary to fill in the following blanks:*

    A. _____

    B. _____

    C. _____

    D. _____

    E. _____

Leg and Foot, Posterior View, cont'd

9. Name the muscle that helps "unlock" the knee joint from full extension.
10. Name the two structures that maintain the position of the femur on the tibia in full knee flexion such as squatting.
11. Name the powerful muscle for "push-off" of the foot during walking or running.

- *Click on* **TAG 5**, *and the following image will appear:*

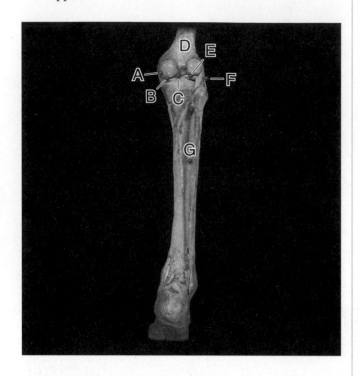

- *Mouse-over the blue pins on the screen to find the information necessary to fill in the following blanks:*

A. _____

B. _____

C. _____

D. _____

E. _____

F. _____

G. _____

CHECK POINT:

Leg and Foot, Posterior View, cont'd

12. Name the thinner and weaker of the cruciate ligaments.
13. Name the cartilaginous structure on the tibia that articulates with the medial condyle of the femur.
14. Name the structure that limits rotation between the femur and the tibia.

Animation: Gastrocnemius

- *Click on the* **ANIMATIONS** *button at the bottom of the screen.*
- *In the* **Select topic** *menu, select* **Muscle actions**.
- *In the* **Select animation** *menu, select and view the following animation:*
  - **Gastrocnemius**

IN REVIEW

What Have I Learned?

The following questions cover the material that you have just learned—the muscles of the leg and foot. Use the **STRUCTURE INFORMATION** for the muscles you have learned to answer these questions:

1. What is the anatomical term for the first toe?

2. Name the thick sheet of connective tissue between the tibia and fibula. 2. What is its function?

3. Name the structure that serves to bind the tendons from the anterior compartment of the leg in place as they cross the ankle joint.

4. Name the strongest tendon in the body.

5. The tendons of which two muscles contribute to the calcaneal tendon?

6. Name the long thin tendon that is a common source for tendon transplants.

7. Name the powerful muscle for "push-off" of foot during walking or running.

8. Name the cruciate ligaments. Which of the two is thinner and weaker?

EXERCISE 3.19:   Skeletal Muscles—Foot, Plantar View

- *Insert* Anatomy & Physiology | Revealed® **Skeletal / Muscular** *CD, or, if you are still in the* **Dissection** *section, click the* **CHANGE VIEW** *button at the top of the screen, and skip the next step.*

- *In the* **Home screen**, *select the* **Dissection** *button in the left portion of the screen. You may click either on the* **Dissection** *button or on the word itself.*

- *In the* **SELECT A VIEW** *window that appears, click on the* **Select region** *button.*

- *Choose* **Foot** *from the menu.*

- **Plantar** *view will then appear in the* **Select view** *menu.*

- *The* **GO** *button will flash green. Click on it.*

- *When the* **Foot** *screen appears, click on* **TAG 1**, *and the following image will appear:*

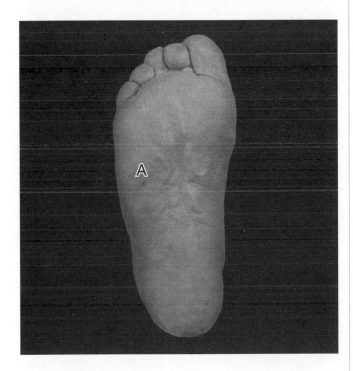

- *Mouse-over the blue pin on the screen to find the information necessary to fill in the following blank:*

    A. _____

Foot, Plantar View

1. Name the structures that serve as contact points of the foot for weight bearing.

- *Click on* **TAG 2**, *and the following image will appear:*

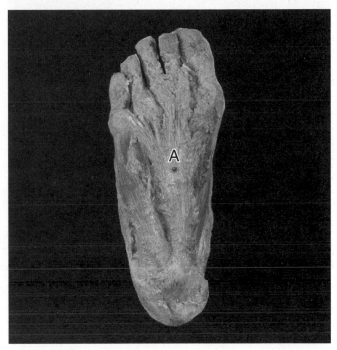

- *Mouse-over the blue pin on the screen to find the information necessary to fill in the following blank:*

    A. _____

Foot, Plantar View, cont'd

2. Name the structure that protects the muscles, vessels, and nerves of plantar foot.

- *Click on* **TAG 3,** *and the following image will appear:*

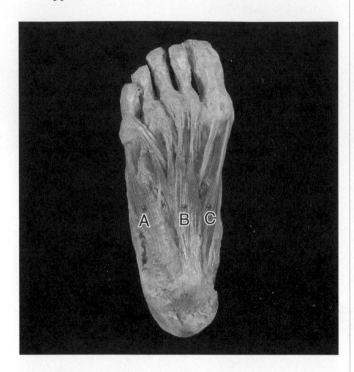

- *Mouse-over the blue pins on the screen to find the information necessary to fill in the following blanks:*

  A. _____

  B. _____

  C. _____

CHECK POINT:

Foot, Plantar View, cont'd

3. What is the anatomical term for the first toe?
4. Name the muscle responsible for flexion of toes 2–5.
5. Name the muscle that supports the medial longitudinal arch of the foot during weight bearing.

- *Click on* **TAG 4,** *and the following image will appear:*

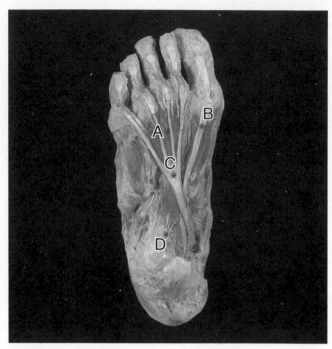

- *Mouse-over the blue pins on the screen to find the information necessary to fill in the following blanks:*

  A._____

  B._____

  C. _____

  D. _____

CHECK POINT:

Foot, Plantar View, cont'd

6. Name two muscles located on the posterior leg whose tendons run along the plantar foot.
7. Name a muscle of the foot that uses the tendons of another muscle to produce toe flexion.

• Click on **TAG 5**, *and the following image will appear:*

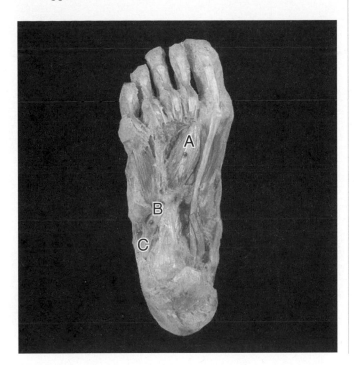

• *Mouse-over the blue pins on the screen to find the information necessary to fill in the following blanks:*

A. _____

B. _____

C. _____

Foot, Plantar View, cont'd

8. Name a muscle that resists separation ("spreading") of the metatarsals during weight-bearing.
9. Name two muscles of the lateral leg whose muscles run along the plantar foot.
10. Name a muscle responsible for adduction of the first toe.

IN REVIEW

What Have I Learned?

The following questions cover the material that you have just learned—the muscles of the plantar foot. Use the **STRUCTURE INFORMATION** for the muscles you have learned to answer these questions:

1. Name the structures that serve as contact points of the foot for weight-bearing.

2. Name the structure that protects the muscles, vessels and nerves of plantar foot.

3. Name the muscle that supports the medial longitudinal arch of the foot during weight-bearing.

4. Name a muscle that resists separation ("spreading") of the metatarsals during weight-bearing.

CHAPTER 4

The Nervous System

Overview: Nervous System

The nervous system is the master controlling and communicating system of your body. It is divided into two major divisions. The **central nervous system** (CNS) consists of the brain and spinal cord, while the **peripheral nervous system** (PNS) is comprised of nerve pathways leading to and from the CNS.

The brain in the adult is one of our largest organs. The average weight for the adult human brain is 1.4 kg, or 3 pounds. It consists of roughly 100 billion neurons and 900 billion glial cells. This means that there are one-fourth as many neurons and 2½ times as many total cells in your brain as there are stars in our Milky Way galaxy! Think about that the next time you are out at night, looking at the stars!

We will look at the central nervous system first, and then continue with the peripheral nervous system and the special senses.

HEADS UP!

*Be sure to click all **TAG** buttons in the **Dissection** section of Anatomy & Physiology | Revealed®, even if it is not necessary for completion of the assignments. This will allow you to have a better feel for the location of all of the structures that you do need to know.*

Animation: Divisions of the Brain

Before beginning the following exercises, view the *Anatomy & Physiology | Revealed*® animation covering the divisions of the brain.

- *Insert* Anatomy & Physiology | Revealed® **Nervous System** CD.
- *Click on the **ANIMATIONS** button.*
- *In the **Select topic** menu, select **Anatomy**.*
- *From the **Select animation** menu, select **Divisions of brain**.*

HEADS UP!

*Many of the animations for the nervous system have an **ANIMATION INFORMATION** window that contains an outline of the animation you are viewing.*

- *Click the **Play** button, and the animation will run in the **IMAGE AREA**.*
- *After viewing the animation, answer the following questions:*

1. The surface of the cerebrum is distinguished by thick folds and grooves. What are the terms that refer to these thick folds and grooves?

2. Name the structure of the brain responsible for regulation of body temperature, food and water intake, sleep and circadian rhythms, emotional responses, and memory.

3. Name the structure of the brain responsible for the regulation of respiration, blood pressure, and heartbeat.

- *When you are finished viewing the animation, click on the* **Dissection button** *at the bottom of the screen to begin the exercises, or click on the* **Exit** *button at the bottom right of the screen to exit* Anatomy & Physiology | Revealed®.

The Brain

EXERCISE 4.1: Nervous System—Brain, Coronal View

- *Insert* Anatomy & Physiology | Revealed® **Nervous System** *CD, or, if you are still in the* **Animation** *section, click the* **Dissection** *button at the bottom of the screen, and skip the next step.*

- *In the* **Home screen**, *select the* **Dissection** *button in the left portion of the screen. You may click either on the* **Dissection** *button or on the word itself.*

- *In the* **SELECT A VIEW** *window that appears, click on the* **Select topic** *button.*

HEADS UP!

In the **Dissection** *section of the* **Nervous System** *CD, the* **Select region** *menu does not appear in the* **SELECT A VIEW** *window. It is replaced by the* **Select topic** *menu.*

- *Choose* **Brain** *from the menu.*
- *The* **Select view** *menu will now become available. Click here and choose the* **Coronal** *view.*
- *The* **GO** *button will now flash green. Click on it. The screen at right will appear.*

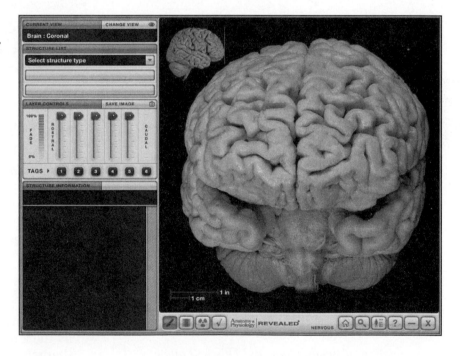

• *Click on* **TAG 1**, *and the following image will appear:*

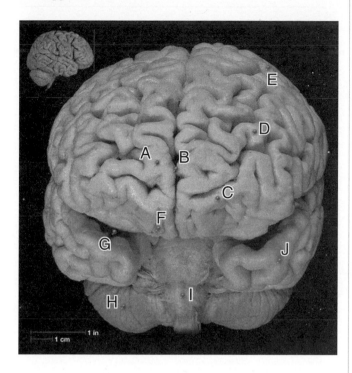

• *Mouse-over the blue pins on the screen to find the information necessary to fill in the following blanks:*

A. _____

B. _____

C. _____

D. _____

E. _____

F. _____

G. _____

H. _____

I. _____

J. _____

CHECK POINT:

Brain, Coronal View

1. Name the deep grooves that separate the temporal lobe from the frontal and parietal lobes.
2. Name the deep groove that separates the right and left cerebral hemispheres.
3. Name site of synapse for the olfactory neurons after they pass through the cribriform plate.

• *Click on* **TAG 2**, *and the following image will appear:*

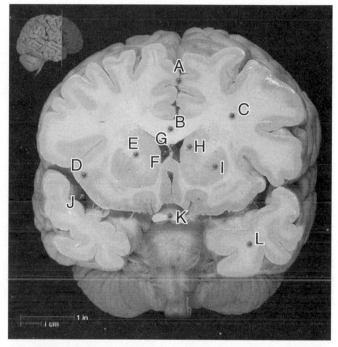

• *Mouse-over the blue pins on the screen to find the information necessary to fill in the following blanks:*

A. _____

B. _____

C. _____

D. _____

E. _____

F. _____

G. _____

H. _____

I. _____

J. _____

K. _____

L. _____

I. _____

J. _____

K. _____

L. _____

M. _____

N. _____

CHECK POINT:

Brain, Coronal View, cont'd

4. What is the term for unmyelinated nervous tissue?
5. What is the term for the collection of myelinated axons in the brain?
6. What is the large myelinated fiber tract that connects the right and left cerebral hemispheres?

- *Click on* **TAG 3**, *and the following image will appear:*

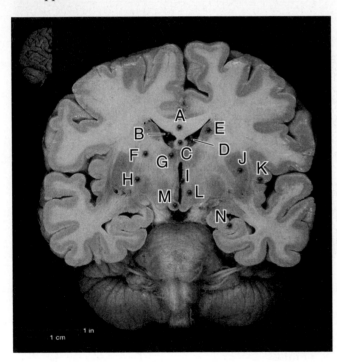

- *Mouse-over the blue pins on the screen to find the information necessary to fill in the following blanks:*

A. _____

B. _____

C. _____

D. _____

E. _____

F. _____

G. _____

H. _____

CHECK POINT:

Brain, Coronal View, cont'd

7. Name the paired, rounded projections involved in regulation of autonomic functions and memory.
8. Name the structure located in the cerebral ventricles that is the site of the production of cerebrospinal fluid (CSF).
9. Name the structure that is primarily for the relay of sensory information to the cerebral cortex.

Animation: CSF flow

Before continuing with the following exercises, view the *Anatomy & Physiology | Revealed*® animation covering the CSF flow.

- *Click on the* **ANIMATION** *button at the bottom of the screen.*
- *In the* **Select topic** *menu, select* **Anatomy**.
- *From the* **Select animation** *menu, select* **CSF flow**.
- *Click the* **Play** *button, and the animation will run in the* **IMAGE AREA**.
- *After viewing the animation, answer the following questions:*

1. In what brain structures would you expect to find (CSF)?

2. Where is this CSF produced?

3. What structure produces the CSF?

4. Beginning in the lateral ventricles, trace the flow of CSF.

5. What are arachnoid granulations?

- *Click on the **Dissection** button at the bottom of the screen to return to the dissection view.*

- *Click on **TAG 4**, and the following image will appear:*

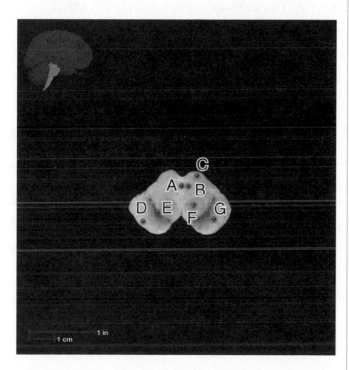

- *Mouse-over the blue pins on the screen to find the information necessary to fill in the following blanks:*

A. _____

B. _____

C. _____

D. _____

E. _____

F. _____

G. _____

Brain, Coronal View, cont'd

10. Name the structure that coordinates orienting movements of the eyes and head.
11. Name the narrow midline channel between the third and fourth ventricles.
12. Name the structure involved with suppression and modulation of pain.

- *Click on **TAG 5**, and the following image will appear:*

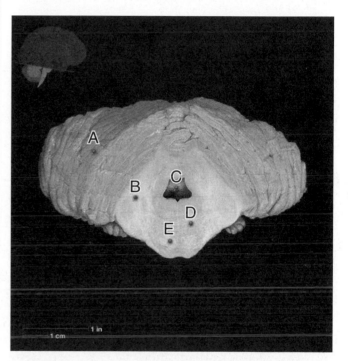

- *Mouse-over the blue pins on the screen to find the information necessary to fill in the following blanks:*

A. _____

B. _____

C. _____

D. _____

E. _____

CHECK POINT:

Brain, Coronal View, cont'd

13. Name the structure that coordinates voluntary movement.
14. Name the major afferent pathway for information from the motor cortex to the cerebellum.
15. Name the cerebrospinal fluid-filled triangular cavity that is continuous with the cerebral aqueduct and the central canal of the spinal cord.

- *Click on* **TAG 6**, *and the following image will appear:*

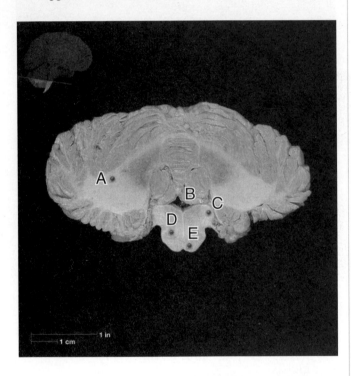

- *Mouse-over the blue pins on the screen to find the information necessary to fill in the following blanks:*

A. _____

B. _____

C. _____

D. _____

E. _____

CHECK POINT:

Brain, Coronal View, cont'd

16. Name the structure that processes and sends information to the cerebellum from many CNS nuclei and skeletal muscle proprioceptors.
17. Name the structure of the medulla oblongata that controls voluntary movement.
18. Name the structure that carries information about muscle performance from the spinal cord to the cerebellum.

EXERCISE 4.2:   Nervous System—Brain, Lateral View

- *Insert* Anatomy & Physiology | Revealed® **Nervous System** *CD, or, if you are already in the* **Dissection** *section, click the* **CHANGE VIEW** *button at the top of the screen, and skip the next step.*

- *In the* **Home screen** *, select the* **Dissection** *button in the left portion of the screen. You may click either on the* **Dissection** *button or on the word itself.*

- *In the* **SELECT A VIEW** *window that appears, click on the* **Select topic** *button.*

- *Choose* **Brain** *from the menu, if it is not already selected.*

- *The* **Select view** *menu will now become available. Click here and choose the* **Lateral** *view.*

- *The* **GO** *button will flash green. Click on it.*

- *Click on **TAG 1**, and the following image will appear:*

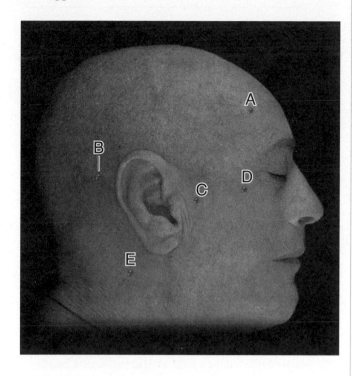

- *Mouse-over the blue pins on the screen to find the information necessary to fill in the following blanks:*

A. _____

B. _____

C. _____

D. _____

E. _____

- *Click on **TAG 3**, and the image at top right will appear.*

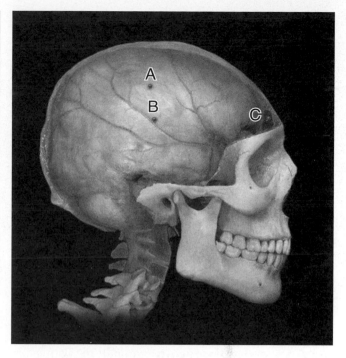

- *Mouse-over the blue pins on the screen to find the information necessary to fill in the following blanks:*

A. _____

B. _____

C. _____

Animation: Meninges

Before continuing with the follwing exercises, view the *Anatomy & Physiology | Revealed®* animation covering the meninges of the brain and spinal cord.

- *Click on the **ANIMATIONS** button at the bottom of the screen.*

- *In the **Select topic** menu, select **Anatomy**.*

- *From the **Select animation** menu, select **Meninges**.*

- *Click the **Play** button, and the animation will run in the **IMAGE AREA**.*

- *After viewing the animation, answer the following questions:*

1. List the three meninges in order from superficial to deep.

2. Name the two layers of the most superficial of the meninges.

3. Name the structures formed where these two layers split.

4. Name the space located between the middle and deepest meninges. What fills this space?

- *Click on the* **Dissection** *button at the bottom of the screen to return to the dissection view.*

- *Click on* **TAG 4**, *and the following image will appear:*

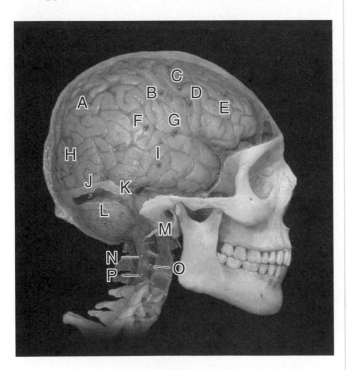

- *Mouse-over the blue pins on the screen to find the information necessary to fill in the following blanks:*

A. _____

B. _____

C. _____

D. _____

E. _____

F. _____

G. _____

H. _____

I. _____

J. _____

K. _____

L. _____

M. _____

N. _____

O. _____

P. _____

CHECK POINT:

Brain, Lateral View, cont'd

7. Name the distinct fold at the posterior border of the frontal lobe that controls voluntary movement.
8. Name the distinct fold at the anterior border of the parietal lobe that receives somatosensory information from the body.
9. Name the groove that forms the boundary between the frontal and parietal lobes.

- *Click on* **TAG 5**, *and the following image will appear:*

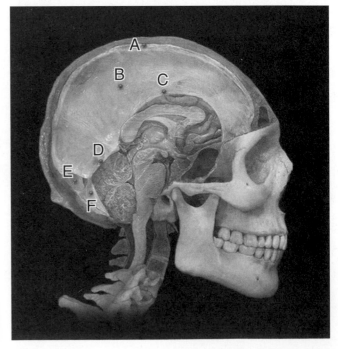

- *Mouse-over the blue pins on the screen to find the information necessary to fill in the following blanks:*

A. _____

B. _____

C. _____

D. _____

E. _____

F. _____

C H E C K   P O I N T :

Brain, Lateral View, cont'd

10. Name the large, crescent-shaped fold of the dura mater that separates the two cerebral hemispheres.
11. Name the structure that contains arachnoid granulations. What is the function of these granulations?
12. The confluence of the sinuses is the meeting point for four different sinuses. What are they?

Animation: Dural sinus blood flow

Before continuing with the following exercises, view the *Anatomy & Physiology | Revealed*® animation covering the dural sinus blood flow.

- *Click on the **ANIMATIONS** button at the bottom of the screen.*
- *In the **Select topic** menu, select **Anatomy**.*
- *From the **Select animation** menu, select **Dural sinus blood flow**.*
- *Click the **Play** button and the animation will run in the **IMAGE AREA**.*
- *After viewing the animation, answer the following questions:*

1. What are the dural venous sinuses?

2. Where are they located?

3. Name the two dural sinuses located along the midline.

4. Name the three sinuses that unite at the confluence of sinuses.

5. What vessels do the sigmoid sinuses become?

- *Click on the **Dissection** button at the bottom of the screen to return to the dissection view.*
- *Click on **TAG 6**, and the following image will appear:*

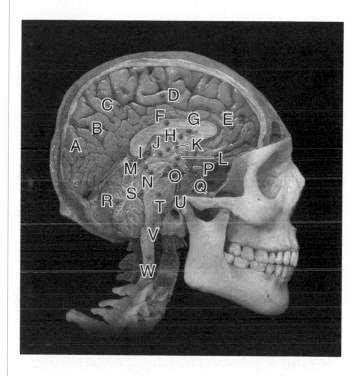

- *Mouse-over the blue pins on the screen to find the information necessary to fill in the following blanks:*

A. _____

B. _____

C. _____

D. _____

E. _____

F. _____

G. _____

H. _____

I._____

J._____

K. _____

L. _____

M. _____

N._____

O._____

P. _____

Q._____

R. _____

S. _____

T. _____

U._____

V. _____

W. _____

EXERCISE 4.3:   Nervous System—Brain, Superior View

- *Insert* Anatomy & Physiology | Revealed® **Nervous System** *CD, or, if you are already in the* **Dissection** *section, click the* **CHANGE VIEW** *button at the top of the screen, and skip the next step.*

- *In the* **Home screen**, *select the* **Dissection** *button in the left portion of the screen. You may click either on the* **Dissection** *button or on the word itself.*

- *In the* **SELECT A VIEW** *window that appears, click on the* **Select topic** *button.*

- *Choose* **Brain** *from the menu, if it is not already selected.*

- *The* **Select view** *menu will now become available. Click here and choose the* **Superior** *view.*

- *The* **GO** *button will flash green. Click on it.*

- *Click on* **TAG 1**, *and the following image will appear:*

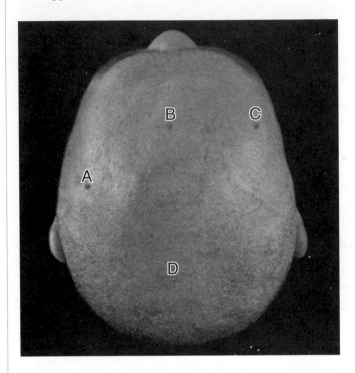

- *Mouse-over the blue pins on the screen to find the information necessary to fill in the following blanks:*

A. _____

B. _____

C. _____

D. _____

- *Click on **TAG 3**, and the following image will appear:*

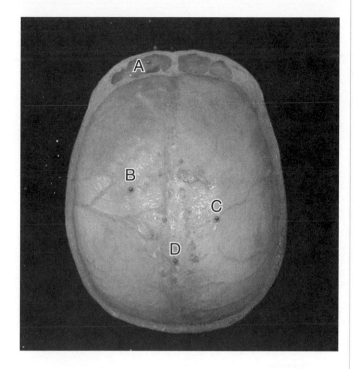

- *Mouse-over the blue pins on the screen to find the information necessary to fill in the following blanks:*

A. _____

B. _____

C. _____

D. _____

CHECK POINT:

Brain, Superior View, cont'd

4. Name the structures that allow the return of cerebrospinal fluid to the venous circulation.
5. Name an artery that courses between the dura mater and the cranium.

- *Click on **TAG 4**, and the following image will appear:*

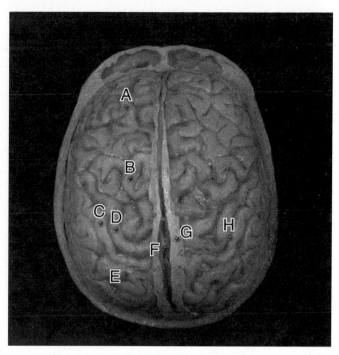

- *Mouse-over the blue pins on the screen to find the information necessary to fill in the following blanks:*

A. _____

B. _____

C. _____

D. _____

E. _____

F. _____

G. _____

H. _____

CHECK POINT:

Brain, Superior View, cont'd

6. Name an unpaired dural venous sinus that ends at the confluence of sinuses.
7. Name the structure that is also called the primary motor cortex.
8. Name the structure that is also called the primary somatosensory cortex.

- *Click on **TAG 5**, and the following image will appear:*

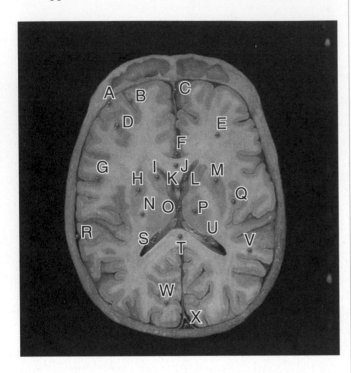

- *Mouse-over the blue pins on the screen to find the information necessary to fill in the following blanks:*

A. _____

B. _____

C. _____

D. _____

E. _____

F. _____

G. _____

H. _____

I. _____

J. _____

K. _____

L. _____

M. _____

N. _____

O. _____

P. _____

Q. _____

R. _____

S. _____

T. _____

U. _____

V. _____

W. _____

X. _____

CHECK POINT:

Brain, Superior View, cont'd

9. The cerebral ventricles are lined with tufts of capillaries covered by specialized ependymal cells. What are these cells called?
10. Both the superior sagittal sinus and the inferior sagittal sinus are located in the margins of the _____.
11. Name the slitlike cavity that separates the right and left halves of the diencephalon.

- *Click on **TAG 6**, and the following image will appear:*

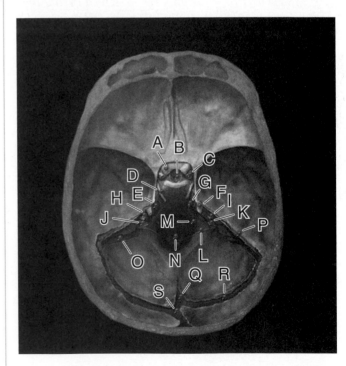

- *Mouse-over the blue pins on the screen to find the information necessary to fill in the following blanks:*

A. _____

B. _____

C. _____

D. _____

E. _____

F. _____

G. _____

H. _____

I. _____

J. _____

K. _____

L. _____

M. _____

N. _____

O. _____

P. _____

Q. _____

R. _____

S. _____

CHECK POINT:

Brain, Superior View, cont'd

12. Name the point of attachment of the pituitary gland to the hypothalamus.
13. Name the S-shaped groove on the inner aspect of the temporal bone.
14. Name the cranial nerve that controls the superior oblique muscle. What is unique about this nerve's origin?

EXERCISE 4.4:   Nervous System—Brain, Inferior View

- *Insert* Anatomy & Physiology | Revealed® **Nervous System** *CD, or, if you are already in the* **Dissection** *section, click the* **CHANGE VIEW** *button at the top of the screen, and skip the next step.*

- *In the* **Home screen***, select the* **Dissection** *button in the left portion of the screen. You may click either on the* **Dissection** *button or on the word itself.*

- *In the* **SELECT A VIEW** *window that appears, click on the* **Select topic** *button.*

- *Choose* **Brain** *from the menu, if it is not already selected.*

- *The* **Select view** *menu will now become available. Click here and choose the* **Inferior** *view.*

- *The* **GO** *button will flash green. Click on it.*

- *Click on* **TAG 1***, and the following image will appear:*

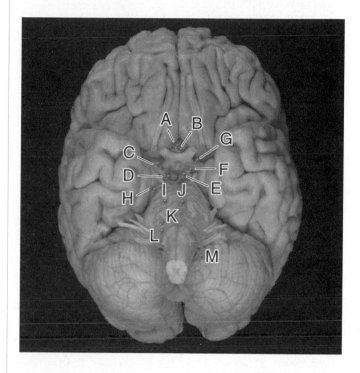

- *Mouse-over the blue pins on the screen to find the information necessary to fill in the following blanks:*

A. _____

B. _____

C. _____

D. _____

E. _____

F. _____

G. _____

H. _____

I. _____

J. _____

K. _____

L. _____

M. _____

CHECK POINT:

Brain, Inferior View

1. Name the circular anastomosis on the ventral surface of the brain also referred to as the "Circle of Willis."
2. Name the artery that passes through the transverse foramina of the cervical vertebrae.
3. Name the unpaired midline artery that ascends on the anterior surface of the pons.

• *Click on* **TAG 2**, *and the following image will appear:*

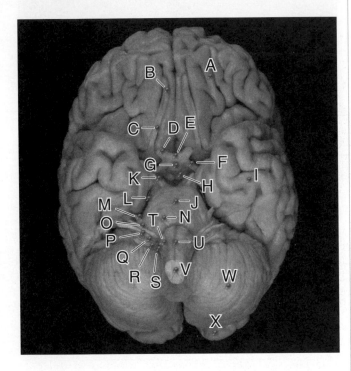

• *Mouse-over the blue pins on the screen to find the information necessary to fill in the following blanks:*

A. _____

B. _____

C. _____

D. _____

E. _____

F. _____

G. _____

H. _____

I. _____

J. _____

K. _____

L. _____

M. _____

N. _____

O. _____

P. _____

Q. _____

R. _____

S. _____

T. _____

U. _____

V. _____

W. _____

X. _____

CHECK POINT:

Brain, Inferior View, cont'd

4. Name the crossing white-matter tract between the optic nerve and the optic tracts.
5. Name the brain structure whose name means "bridge."
6. Name the most caudal portion of the brain. What functions does it have?

EXERCISE 4.5:   Nervous System—Brain, Inferior View (close-up)

• *Insert* Anatomy & Physiology | Revealed® **Nervous System** *CD, or, if you are already in the* **Dissection** *section, click the* **CHANGE VIEW** *button at the top of the screen, and skip the next step.*

• *In the* **Home screen**, *select the* **Dissection** *button in the left portion of the screen. You may click either on the* **Dissection** *button or on the word itself.*

• *In the* **SELECT A VIEW** *window that appears, click on the* **Select topic** *button.*

• *Choose* **Brain** *from the menu, if it is not already selected.*

• *The* **Select view** *menu will now become available. Click here and choose the* **Inferior (close-up)** *view.*

• *The* **GO** *button will flash green. Click on it.*

• *Click on* **TAG 1**, *and the following image will appear:*

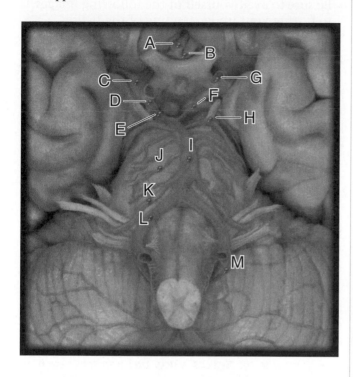

CHECK POINT:

Brain, Inferior View (close-up)

1. Name the artery whose significant branches include the ophthalmic and posterior communicating arteries.
2. Name the artery whose major branches include the pontine and superior cerebellar arteries.
3. Name the artery whose numerous branches course laterally across the surface of the pons.

• *Click on* **TAG 2**, *and the following image will appear:*

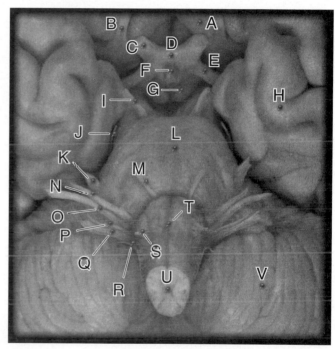

• *Mouse-over the blue pins on the screen to find the information necessary to fill in the following blanks:*

A. _____

B. _____

C. _____

D. _____

E. _____

F._____

G. _____

H._____

I._____

J._____

K. _____

L. _____

M. _____

• *Mouse-over the blue pins on the screen to find the information necessary to fill in the following blanks:*

A. _____

B. _____

C. _____

D. _____

E. _____

F._____

G. _____

H._____

I._____

J. _____

K. _____

L. _____

M. _____

N. _____

O. _____

P. _____

Q. _____

R. _____

S. _____

T. _____

U. _____

V. _____

CHECK POINT:

Brain, Inferior View (close-up), cont'd

4. Name the paired, small, rounded projections of the hypothalamus involved in the regulation of autonomic functions and memory.
5. Name the "funnel-shaped" extension of the floor of the third ventricle that continues as the pituitary stalk.
6. Name the structure that gives rise to eight pairs of spinal nerves and contains the cervical enlargement.

EXERCISE 4.6:   Imaging

Be sure to avail yourself of the following images to further your understanding of the brain:

- *Click on the **IMAGING** button at the bottom of the screen.*
- *The **SELECT A REGION** window will indicate **Brain**.*
- *Click on the flashing **GO** button.*
- *The **CURRENT REGION** window will indicate **Brain**.*
- *The **IMAGE TYPE/VIEW** window will indicate **MRI**.*
- *Click on the **Select view** menu, and select **Axial 1**.*
- *Click on the **Select structure** menu and a list of structures will appear.*
- *Select each of the structures on the list to view a highlighted view of that structure on the MRI in the **IMAGE AREA**.*
- *Return to the **Select view** menu to select each of the remaining views.*
- *Again, click on the **Select structure** menu for each of the views to see a highlighted view of those structures in the **IMAGE AREA**.*

IN REVIEW

What Have I Learned?

The following questions cover the material that you have just learned: the brain. Use the **STRUCTURE INFORMATION** to answer these questions:

1. Name the division of the brain that includes the midbrain, pons, and medulla oblongata.

2. Name the division of the brain that coordinates complex movements and smooths muscle contractions.

3. The primary hearing and smell areas are located in which lobe of the brain?

4. Memory is located in which lobe of the brain?

5. Speech perception and recognition areas are located in which hemisphere of which lobe of the brain?

6. What is the term that refers to the superficial gray matter of the cerebrum?

7. The reception of general sensory information from the body occurs in which lobe of the brain?

8. Tactile object recognition occurs in which lobe of the brain?

9. Language and verbatim repetition of terms occurs in which hemisphere of which lobe of the brain?

10. Name the structure that forms the floor of the longitudinal fissure.

11. Name the paired cavities containing cerebrospinal fluid in the brain.

12. Where is the falx cerebri located?

13. Name the crossing white-matter tracts between the optic nerves and the optic tracts.

14. Name two structures responsible for planning and execution of movement, muscle tone, and posture.

15. Name the slitlike, fluid-filled cavity that separates the right and left halves of the diencephalon.

16. Name the site of learning and memory processing.

17. Name the site of emotional behavior.

18. Name the structure that regulates body temperature, eating, and drinking.

19. Name the structure that controls the autonomic nervous system.

20. Name the lobe that is the primary visual area.

21. Name the lobe that controls voluntary motor activity.

22. Name the lobe that is the site of higher mental processing.

23. Name the structure continuous with the brain at the foramen magnum.

Cranial Nerves

Exercise 4.7: Nervous System—Cranial Nerves, Inferior Brain (CNI-XII)

- *Insert* Anatomy & Physiology | Revealed® **Nervous System** *CD, or, if you are already in the* **Dissection** *section, click the* **CHANGE VIEW** *button at the top of the screen, and skip the next step.*

- *In the* **Home screen**, *select the* **Dissection** *button in the left portion of the screen. You may*

click either on the **Dissection** *button or on the word itself.*

- *In the* **SELECT A VIEW** *window that appears, click on the* **Select topic** *button.*

- *Choose* **Cranial nerves** *from the menu.*

- *Click the* **Select view** *menu, and choose* **Inferior** *view.*

- *The* **GO** *button will flash green. Click on it.*

• *Click on* **TAG 1**, *and the following image will appear:*

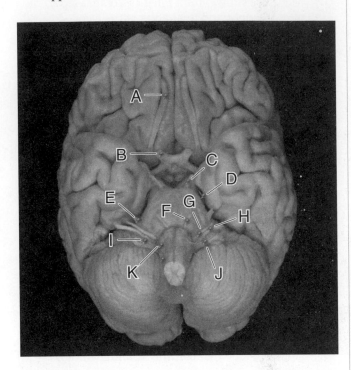

• *Mouse-over the blue pins on the screen to find the information necessary to fill in the following blanks:*

A. _____

B. _____

C. _____

D. _____

E. _____

F. _____

G. _____

H. _____

I. _____

J. _____

K. _____

L. _____

EXERCISE 4.8:   Nervous System—Cranial Nerves, CN I Olfactory

• *Insert* Anatomy & Physiology | Revealed® **Nervous System** CD, *or, if you are already in the* **Dissection** *section, click the* **CHANGE VIEW** *button at the top of the screen, and skip the next step.*

• *In the* **Home screen**, *select the* **Dissection** *button in the left portion of the screen. You may click either on the* **Dissection** *button or on the word itself.*

• *In the* **SELECT A VIEW** *window that appears, click on the* **Select topic** *button.*

• *Choose* **Cranial nerves** *from the menu, if it is not already selected.*

• *Click the* **Select view** *menu, and choose* **CN I Olfactory**.

• *The* **GO** *button will flash green. Click on it.*

• *Click on* **TAG 3**, *and the following image will appear:*

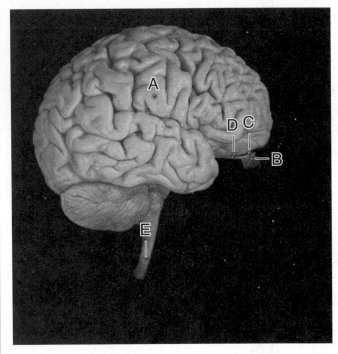

• *Mouse-over the blue pins on the screen to find the information necessary to fill in the following blanks:*

A. _____

B. _____

C. _____

D. _____

E. _____

- *Click on* **TAG 4**, *and the following image will appear:*

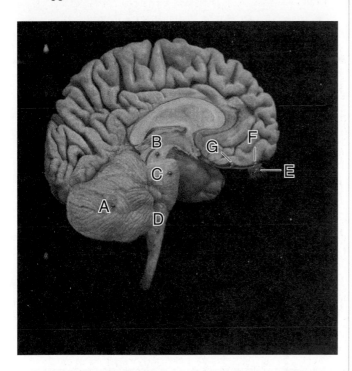

- *Mouse-over the blue pins on the screen to find the information necessary to fill in the following blanks:*

A. _____

B. _____

C. _____

D. _____

E. _____

F. _____

G. _____

CHECK POINT:

Cranial Nerves, CN I Olfactory

1. Which nerves have their axons projecting through the cribriform plate?
2. Where do these nerves synapse?
3. The olfactory tracts travel caudally to which portion of which lobe of the brain?

EXERCISE 4.9:   Nervous System—Cranial Nerves, CN II Optic

- *Insert* Anatomy & Physiology | Revealed® **Nervous System** *CD, or, if you are already in*

the **Dissection** *section, click the* **CHANGE VIEW** *button at the top of the screen, and skip the next step.*

- *In the* **Home screen**, *select the* **Dissection** *button in the left portion of the screen. You may click either on the* **Dissection** *button or on the word itself.*

- *In the* **SELECT A VIEW** *window that appears, click on the* **Select topic** *button.*

- *Choose* **Cranial nerves** *from the menu, if it is not already selected.*

- *Click the* **Select view** *menu, and choose* **CN II Optic**.

- *The* **GO** *button will flash green. Click on it.*

- *Click on* **TAG 4**, *and the following image will appear:*

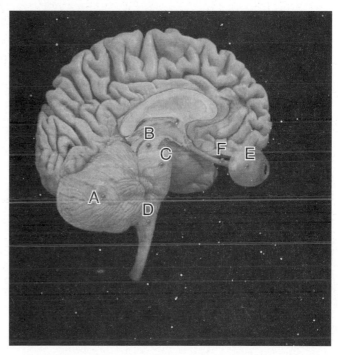

- *Mouse-over the blue pins on the screen to find the information necessary to fill in the following blanks:*

A. _____

B. _____

C. _____

D. _____

E. _____

F. _____

Animation: Vision

Before beginning the following exercises, view the *Anatomy & Physiology | Revealed®* animation covering the divisions of the brain.

- *Insert the* Anatomy & Physiology | Revealed® **Nervous System** *CD or, if you are in the* **Dissection** *section, click on the* **ANIMATIONS** *button at the bottom of the screen.*

- *In the* **Select topic** *menu, select* **Physiology**.

- *From the* **Select animation** *menu, select* **Vision**.

- *Click the* **Play** *button, and the animation will run in the* **IMAGE AREA**.

- *After viewing the animation, answer the following questions:*

1. What are the three structures involved in vision?

2. The optic nerve runs between what two structures?

3. Describe the two parts of the retina.

4. What neurotransmitter do the photoreceptors release in the dark?

5. Describe the light pathway from the eye to the brain.

6. Name the structure where the optic nerves converge.

7. Images are perceived in which lobe of the brain?

- *Click on the* **DISSECTION** *button at the bottom of the screen to return to the dissection view.*

EXERCISE 4.10: Nervous System—Cranial Nerves, CN III Oculomotor

- *Insert* Anatomy & Physiology | Revealed® **Nervous System** *CD, or, if you are already in the* **Dissection** *section, click the* **CHANGE VIEW** *button at the top of the screen, and skip the next step.*

- *In the* **Home screen**, *select the* **Dissection** *button in the left portion of the screen. You may click either on the* **Dissection** *button or on the word itself.*

- *In the* **SELECT A VIEW** *window that appears, click on the* **Select topic** *button.*

- *Choose* **Cranial nerves** *from the menu, if it is not already selected.*

- *Click the* **Select view** *menu, and choose* **CN III Oculomotor**.

- *The* **GO** *button will flash green. Click on it.*

- *Click on* **TAG 4**, *and the following image will appear:*

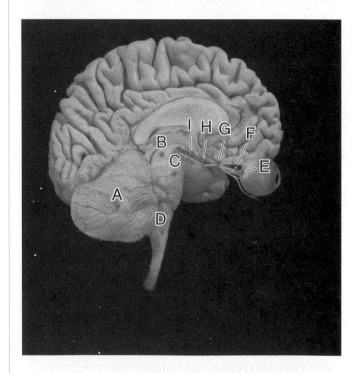

- *Mouse-over the blue pins on the screen to find the information necessary to fill in the following blanks:*

A. _____

B. _____

C. _____

D._____

E._____

F._____

G._____

H._____

I._____

- *Mouse-over the blue pins on the screen to find the information necessary to fill in the following blanks:*

A._____

B._____

C._____

D._____

E._____

EXERCISE 4.11: Nervous System—Cranial Nerves, CN IV Trochlear

- *Insert* Anatomy & Physiology | Revealed® **Nervous System** *CD, or, if you are already in the* **Dissection** *section, click the* **CHANGE VIEW** *button at the top of the screen, and skip the next step.*

- *In the* **Home screen**, *select the* **Dissection** *button in the left portion of the screen. You may click either on the* **Dissection** *button or on the word itself.*

- *In the* **SELECT A VIEW** *window that appears, click on the* **Select topic** *button.*

- *Choose* **Cranial nerves** *from the menu, if it is not already selected.*

- *Click the* **Select view** *menu, and choose* **CN IV Trochlear**.

- *The* **GO** *button will flash green. Click on it.*

- *Click on* **TAG 4**, *and the following image will appear:*

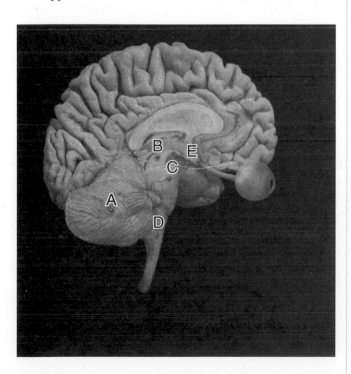

EXERCISE 4.12: Nervous System—Cranial Nerves, CN V Trigeminal

- *Insert* Anatomy & Physiology | Revealed® **Nervous System** *CD, or, if you are already in the* **Dissection** *section, click the* **CHANGE VIEW** *button at the top of the screen, and skip the next step.*

- *In the* **Home screen**, *select the* **Dissection** *button in the left portion of the screen. You may click either on the* **Dissection** *button or on the word itself.*

- *In the* **SELECT A VIEW** *window that appears, click on the* **Select topic** *button.*

- *Choose* **Cranial nerves** *from the menu, if it is not already selected.*

- *Click the* **Select view** *menu, and choose* **CN V Trigeminal**.

- *The* **GO** *button will flash green. Click on it.*

- *Click on* **TAG 4**, *and the following image will appear:*

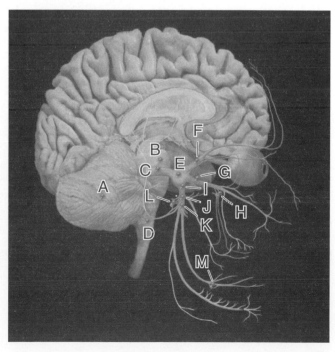

• *Mouse-over the blue pins on the screen to find the information necessary to fill in the following blanks:*

A. _____

B. _____

C. _____

D. _____

E. _____

F. _____

G. _____

H. _____

I. _____

J. _____

K. _____

L. _____

M. _____

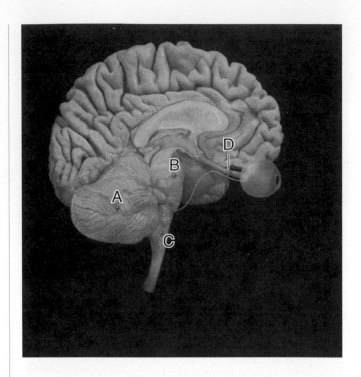

EXERCISE 4.13:   Nervous System—Cranial Nerves, CN VI Abducens

• *Insert* Anatomy & Physiology | Revealed® **Nervous System** *CD, or, if you are already in the* **Dissection** *section, click the* **CHANGE VIEW** *button at the top of the screen, and skip the next step.*

• *In the* **Home screen**, *select the* **Dissection** *button in the left portion of the screen. You may click either on the* **Dissection** *button or on the word itself.*

• *In the* **SELECT A VIEW** *window that appears, click on the* **Select topic** *button.*

• *Choose* **Cranial nerves** *from the menu, if it is not already selected.*

• *Click the* **Select view** *menu, and choose* **CN VI Abducens**.

• *The* **GO** *button will flash green. Click on it.*

• *Click on* **TAG 4**, *and the image at top right will appear.*

• *Mouse-over the blue pins on the screen to find the information necessary to fill in the following blanks:*

A. _____

B. _____

C. _____

D. _____

EXERCISE 4.14:   Nervous System—Cranial Nerves, CN VII Facial

• *Insert* Anatomy & Physiology | Revealed® **Nervous System** *CD, or, if you are already in the* **Dissection** *section, click the* **CHANGE VIEW** *button at the top of the screen, and skip the next step.*

• *In the* **Home screen**, *select the* **Dissection** *button in the left portion of the screen. You may click either on the* **Dissection** *button or on the word itself.*

• *In the* **SELECT A VIEW** *window that appears, click on the* **Select topic** *button.*

• *Choose* **Cranial nerves** *from the menu, if it is not already selected.*

• *Click the* **Select view** *menu and choose* **CN VII Facial**.

• *The* **GO** *button will flash green. Click on it.*

• *Click on* **TAG 4**, *and the following image will appear:*

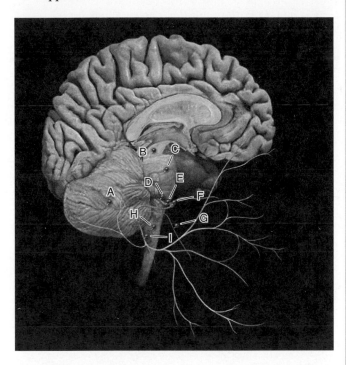

• *Mouse-over the blue pins on the screen to find the information necessary to fill in the following blanks:*

A. _____

B. _____

C. _____

D. _____

E. _____

F. _____

G. _____

H. _____

I. _____

EXERCISE 4.15:   Nervous System—Cranial Nerves, CN VIII Vestibulocochlear

• *Insert* Anatomy & Physiology | Revealed® **Nervous System** *CD, or, if you are already in the* **Dissection** *section, click the* **CHANGE VIEW** *button at the top of the screen, and skip the next step.*

• *In the* **Home screen**, *select the* **Dissection** *button in the left portion of the screen. You may click either on the* **Dissection** *button or on the word itself.*

• *In the* **SELECT A VIEW** *window that appears, click on the* **Select topic** *button.*

• *Choose* **Cranial nerves** *from the menu, if it is not already selected.*

• *Click the* **Select view** *menu, and choose* **CN VIII Vestibulocochlear**.

• *The* **GO** *button will flash green. Click on it.*

• *Click on* **TAG 4**, *and the following image will appear:*

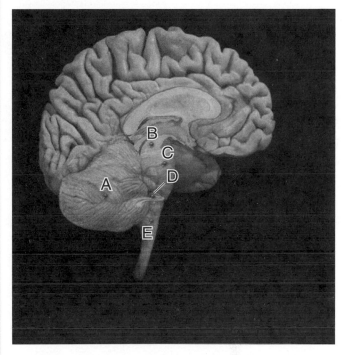

• *Mouse-over the blue pins on the screen to find the information necessary to fill in the following blanks:*

A. _____

B. _____

C. _____

D. _____

E. _____

Animation: Hearing

Before beginning the following exercises, view the *Anatomy & Physiology | Revealed®* animation covering hearing.

• *Insert the* Anatomy & Physiology | Revealed® **Nervous System** *CD or, if you are in the* **Dissection** *section, click on the* **ANIMATIONS** *button at the bottom of the screen.*

- *In the **Select topic** menu, select **Physiology**.*
- *From the **Select animation** menu, select **Hearing**.*
- *Click the **Play** button and the animation will run in the **IMAGE AREA**.*
- *After viewing the animation, answer the following questions:*

1. What structure do sound waves strike and cause to vibrate?

2. The vibrations are transferred to the three bones of the middle ear. From lateral to medial, what are those bones?

3. What structure transfers this vibration to the oval window?

4. The vibrations are then transferred to a fluid-filled chamber of the inner ear. What is that fluid, and what is the name of that chamber?

5. What is the difference in location for the detection of high-pitched and low-pitched sounds?

- *Click on the **DISSECTION** button at the bottom of the screen to return to the dissection view.*

EXERCISE 4.16: Nervous System—Cranial nerves, CN IX Glossopharyngeal

- *Insert Anatomy & Physiology | Revealed® **Nervous System** CD, or, if you are already in the **Dissection** section, click the **CHANGE VIEW** button at the top of the screen, and skip the next step.*
- *In the **Home screen**, select the **Dissection** button in the left portion of the screen. You may click either on the **Dissection** button or on the word itself.*
- *In the **SELECT A VIEW** window that appears, click on the **Select topic** button.*
- *Choose **Cranial nerves** from the menu, if it is not already selected.*

- *Click the **Select view** menu, and choose **CN IX Glossopharyngeal** .*
- *The **GO** button will flash green. Click on it.*
- *Click on **TAG 4**, and the following image will appear:*

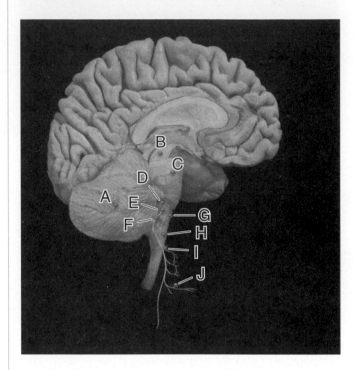

- *Mouse-over the blue pins on the screen to find the information necessary to fill in the following blanks:*

A. _____

B. _____

C. _____

D. _____

E. _____

F. _____

G. _____

H. _____

I. _____

J. _____

EXERCISE 4.17: Nervous System—Cranial Nerves, CN X Vagus

- *Insert Anatomy & Physiology | Revealed® **Nervous System** CD, or, if you are already in the **Dissection** section, click the **CHANGE VIEW** button at the top of the screen, and skip the next step.*

- *In the **Home screen**, select the **Dissection** button in the left portion of the screen. You may click either on the **Dissection** button or on the word itself.*

- *In the **SELECT A VIEW** window that appears, click on the **Select topic** button.*

- *Choose **Cranial nerves** from the menu, if it is not already selected.*

- *Click the **Select view** menu, and choose **CN X Vagus** .*

- *The **GO** button will flash green. Click on it.*

- *Click on **TAG 4**, and the following image will appear:*

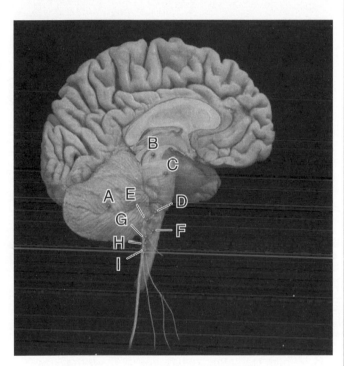

- *Mouse-over the blue pins on the screen to find the information necessary to fill in the following blanks:*

A. _____

B. _____

C. _____

D. _____

E. _____

F. _____

G. _____

H. _____

I. _____

Cranial Nerves, CN XI, Accessory

EXERCISE 4.18:   Nervous System—Cranial Nerves, CN XI Accessory

- *Insert Anatomy & Physiology | Revealed® **Nervous System** CD, or, if you are already in the **Dissection** section, click the **CHANGE VIEW** button at the top of the screen, and skip the next step.*

- *In the **Home screen**, select the **Dissection** button in the left portion of the screen. You may click either on the **Dissection** button or on the word itself.*

- *In the **SELECT A VIEW** window that appears, click on the **Select topic** button.*

- *Choose **Cranial nerves** from the menu, if it is not already selected.*

- *Click the **Select view** menu, and choose **CN XI Accessory**.*

- *The **GO** button will flash green. Click on it.*

- *Click on **TAG 4**, and the following image will appear:*

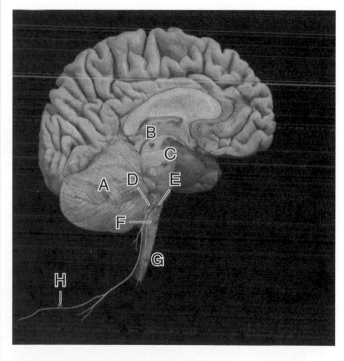

- *Mouse-over the blue pins on the screen to find the information necessary to fill in the following blanks:*

A. _____

B. _____

C. _____

D. _____

E. _____

F. _____

G. _____

H. _____

EXERCISE 4.19:   Nervous System—Cranial Nerves, CN XII Hypoglossal

- *Insert* Anatomy & Physiology | Revealed® **Nervous System** *CD, or, if you are already in the* **Dissection** *section, click the* **CHANGE VIEW** *button at the top of the screen, and skip the next step.*

- *In the* **Home screen***, select the* **Dissection** *button in the left portion of the screen. You may click either on the* **Dissection** *button or on the word itself.*

- *In the* **SELECT A VIEW** *window that appears, click on the* **Select topic** *button.*

- *Choose* **Cranial nerves** *from the menu, if it is not already selected.*

- *Click the* **Select view** *menu, and choose* **CN XII Hypoglossal***.*

- *The* **GO** *button will flash green. Click on it.*

- *Click on* **TAG 4***, and the image at top right will appear.*

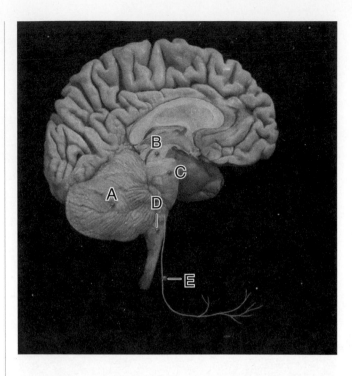

- *Mouse-over the blue pins on the screen to find the information necessary to fill in the following blanks:*

A. _____

B. _____

C. _____

D. _____

E. _____

IN REVIEW

What Have I Learned?

The following questions cover the material that you have just learned: the cranial nerves. Use the **STRUCTURE INFORMATION** to answer these questions:

1. Which cranial nerve is composed of the ophthalmic, the maxillary, and the mandibular nerves?

2. Name the cranial nerve responsible for the intrinsic and extrinsic muscles of the tongue, as well as the control of movements and shape of the tongue.

3. Which cranial nerve has sensory fibers that monitor blood pressure at the carotid sinus?

4. Which cranial nerve is responsible for the constriction of the pupillary sphincter and accommodation of the lens for near vision?

5. Which cranial nerve is responsible for taste from the anterior two third of the tongue and the muscles of facial expression?

6. Which cranial nerve has olfactory hairs on the surface of the nasal mucosa?

7. Which cranial nerve controls the lateral rectus muscle of the eye?

8. Which cranial nerve is involved with hearing and balance?

9. Which cranial nerve is responsible for vision?

10. Which cranial nerve controls the extraocular muscle of the superior oblique?

11. Which cranial nerve controls the muscles of the palate, pharynx, and larynx?

12. Which cranial nerve is the only one that extends beyond the head and neck?

Animation: Action potentials and synapses

Be sure to view the following animations to gain a deeper understanding of the physiology of action potentials and synapses.

- *Insert the* Anatomy & Physiology | Revealed® **Nervous System** *CD or, if you are in the* **Dissection** *section, click on the* **Animations** *button at the bottom of the screen.*

- *In the* **Select topic** *menu, select* **Physiology**.

- *From the* **Select animation** *menu, select* **Action potential generation, Action potential propagation**, *or* **Chemical synapse**.

- *Click the* **Play** *button and the animations will run in the* **IMAGE AREA**.

Spinal Cord

Exercise 4.20: Nervous System—Spinal Cord, Overview

- *Insert* Anatomy & Physiology | Revealed® **Nervous System** *CD, or, if you are already in the* **Dissection** *section, click the* **CHANGE VIEW** *button at the top of the screen, and skip the next step.*

- *In the* **Home screen**, *select the* **Dissection** *button in the left portion of the screen. You may click either on the* **Dissection** *button or on the word itself.*

- *In the* **SELECT A VIEW** *window that appears, click on the* **Select topic** *button.*

- *Choose* **Spinal cord** *from the menu.*

- *Click the* **Select view** *menu, and choose* **Overview**.

- *The* **GO** *button will flash green. Click on it.*

- *Click on* **TAG 1**, *and the following image will appear:*

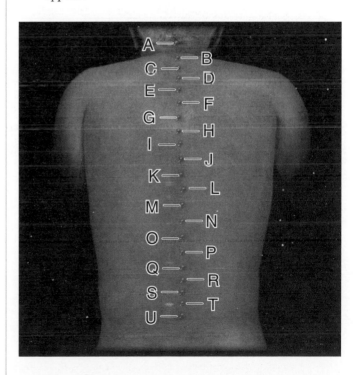

- *Mouse-over the blue pins on the screen to find the information necessary to fill in the following blanks:*

A. _____

B. _____

C. _____

D. _____

E. _____

F. _____

G. _____

H. _____

I. _____

J. _____

K. _____

L. _____

M. _____

N. _____

O. _____

P. _____

Q. _____

R. _____

S. _____

T. _____

U. _____

H E A D S   U P !

Mouse-over and then <u>click on the blue pins</u> on the screen to find the information necessary to answer the following questions.

C H E C K   P O I N T :

Spinal Cord, Overview

1. What is a dermatome?
2. Which dermatome includes the skin of the foot, including the middle three toes?
3. Which dermatome includes the skin over the knee and on the medial foot, including the great toe?

• *Click on* **TAG 3**, *and the following image will appear:*

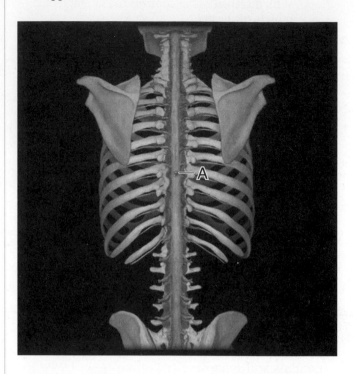

• *Mouse-over the blue pin on the screen to find the information necessary to fill in the following blank:*

A. _____

C H E C K   P O I N T :

Spinal Cord, Overview, cont'd

4. Name the outermost tough connective tissue that surrounds the brain and spinal cord.

• *Click on* **TAG 4**, *and the following image will appear:*

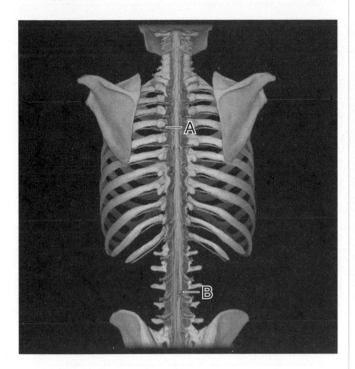

• *Mouse-over the blue pins on the screen to find the information necessary to fill in the following blanks:*

A. _____

B. _____

CHECK POINT:

Spinal Cord, Overview, cont'd

5. Name the large bundle of dorsal and ventral roots for spinal nerves below L2.
6. Name the structure that contains the sensory ganglion for each dorsal root.
7. Name the structure whose Latin name means "horse tail."

• *Click on* **TAG 5**, *and the following image will appear:*

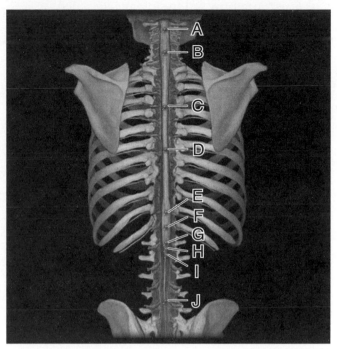

• *Mouse-over the blue pins on the screen to find the information necessary to fill in the following blanks:*

A. _____

B. _____

C. _____

D. _____

E. _____

F. _____

G. _____

H. _____

I. _____

J. _____

CHECK POINT:

Spinal Cord, Overview, cont'd

8. What is the cervical enlargement?
9. Name the structure that contains neurons for lower - limb innervation.
10. Name the tapered inferior end of the spinal cord.

EXERCISE 4.21: Nervous System—Spinal Cord, Typical Spinal Nerve

- *Insert Anatomy & Physiology | Revealed®* **Nervous System** *CD, or, if you are already in the* **Dissection** *section, click the* **CHANGE VIEW** *button at the top of the screen, and skip the next step.*

- *In the* **Home screen**, *select the* **Dissection** *button in the left portion of the screen. You may click either on the* **Dissection** *button or on the word itself.*

- *In the* **SELECT A VIEW** *window that appears, click on the* **Select topic** *button.*

- *Choose* **Spinal cord** *from the menu, if it is not already selected.*

- *Click the* **Select view** *menu, and choose* **Typical spinal nerve**.

- *The* **GO** *button will flash green. Click on it.*

- *Click on* **TAG 1**, *and the following image will appear:*

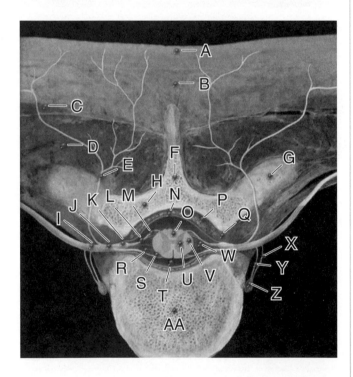

- *Mouse-over the blue pins on the screen to find the information necessary to fill in the following blanks:*

A. _____

B. _____

C. _____

D. _____

E. _____

F. _____

G. _____

H. _____

I. _____

J. _____

K. _____

L. _____

M. _____

N. _____

O. _____

P. _____

Q. _____

R. _____

S. _____

T. _____

U. _____

V. _____

W. _____

X. _____

Y. _____

Z. _____

AA. _____

CHECK POINT:

Typical Spinal Nerve

1. In what respect is the spinal cord structurally opposite to the brain?
2. What structure makes up the core of the spinal cord?
3. Name the structure that consists of ascending and descending bundles of myelinated axons.

Animation: Typical Spinal Nerve

Before continuing with the following exercises, view the *Anatomy & Physiology | Revealed®* animation covering a Typical Spinal Nerve.

- *Click on the* **ANIMATIONS** *button at the bottom of the screen.*
- *In the* **Select topic** *menu, select* **Anatomy**.
- *From the* **Select animation** *menu, select* **Typical spinal nerve**.
- *Click the* **Play** *button, and the animation will run in the* **IMAGE AREA**.
- *After viewing the animation, answer the following questions:*

1. How many spinal nerves exist for each vertebral level?

2. What structures connect each spinal nerve to the spinal cord?

3. What structure is made up of bundles of nerve fibers carrying sensory information from the skin to the spinal cord?

4. Where are the cell bodies (somas) of these sensory nerve fibers located?

5. Name the structure consisting of bundles of motor (efferent) fibers carrying impulses away from the spinal cord to the skeletal muscles.

6. Where are the cell bodies (somas) associated with these nerve fibers located?

7. What structures unite to form the spinal nerve?

8. The spinal nerves exit the vertebral column through what structure?

9. What is a mixed nerve?

10. Name the two branches that form from each spinal nerve.

11. What structures innervate the muscles and skin of the back?

12. What structures innervate the muscles and skin of the lateral and ventral trunk and the limbs?

- *Click on the* **Dissection** *button at the bottom of the screen to return to the dissection view.*

EXERCISE 4.22:   Nervous System—Spinal Cord, Cervical Region

- *Insert* Anatomy & Physiology | Revealed® **Nervous System** *CD, or, if you are already in the* **Dissection** *section, click the* **CHANGE VIEW** *button at the top of the screen, and skip the next step.*
- *In the* **Home screen**, *select the* **Dissection** *button in the left portion of the screen. You may click either on the* **Dissection** *button or on the word itself.*
- *In the* **SELECT A VIEW** *window that appears, click on the* **Select topic** *button.*
- *Choose* **Spinal cord** *from the menu, if it is not already selected.*
- *Click the* **Select view** *menu, and choose* **Cervical region**.
- *The* **GO** *button will flash green. Click on it.*

- *Click on* **TAG 3**, *and the following image will appear:*

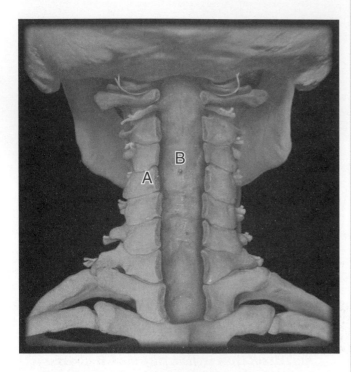

- *Mouse-over the blue pins on the screen to find the information necessary to fill in the following blanks:*

A. _____

B. _____

- *Click on* **TAG 4**, *and the following image will appear:*

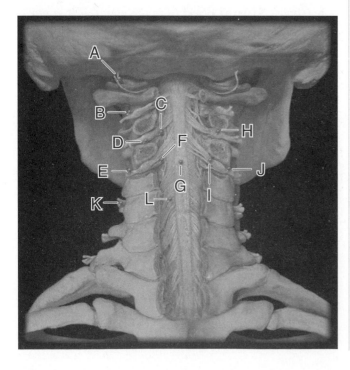

- *Mouse-over the blue pins on the screen to find the information necessary to fill in the following blanks:*

A. _____

B. _____

C. _____

D. _____

E. _____

F._____

G. _____

H._____

I._____

J._____

K. _____

L. _____

CHECK POINT:

Spinal Cord, Cervical Region

1. Name the series of small nerves branching from the dorsal length of the spinal cord.
2. Name the afferent (sensory) branch of each spinal nerve.
3. Name the two terminal branches of each spinal nerve.

EXERCISE 4.23:   Nervous System—Spinal Cord, Thoracic Region

- *Insert Anatomy & Physiology | Revealed®* **Nervous System** *CD, or, if you are already in the* **Dissection** *section, click the* **CHANGE VIEW** *button at the top of the screen, and skip the next step.*

- *In the* **Home screen**, *select the* **Dissection** *button in the left portion of the screen. You may click either on the* **Dissection** *button or on the word itself.*

- *In the* **SELECT A VIEW** *window that appears, click on the* **Select topic** *button.*

- *Choose* **Spinal cord** *from the menu, if it is not already selected.*

- *Click the* **Select view** *menu, and choose* **Thoracic region**.

- *The* **GO** *button will flash green. Click on it.*

- *Click on* **TAG 3**, *and the following image will appear:*

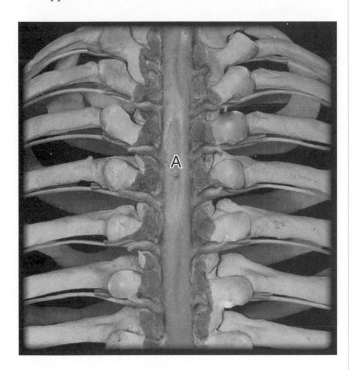

- *Mouse-over the blue pin on the screen to find the information necessary to fill in the following blank:*

   A. _____

- *Click on* **TAG 4**, *and the following image will appear:*

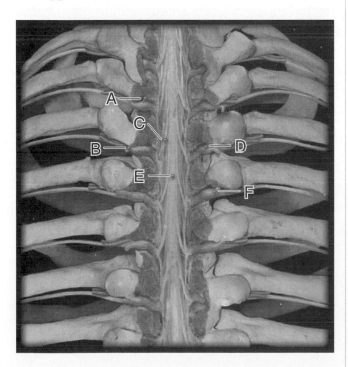

- *Mouse-over the blue pins on the screen to find the information necessary to fill in the following blanks:*

A. _____

B. _____

C. _____

D. _____

E. _____

F._____

EXERCISE 4.24:   Nervous System—Spinal Cord, Lumbar region

- *Insert Anatomy & Physiology | Revealed® **Nervous System** CD, or, if you are already in the **Dissection** section, click the **CHANGE VIEW** button at the top of the screen, and skip the next step.*

- *In the **Home screen**, select the **Dissection** button in the left portion of the screen. You may click either on the **Dissection** button or on the word itself.*

- *In the **SELECT A VIEW** window that appears, click on the **Select topic** button.*

- *Choose **Spinal cord** from the menu, if it is not already selected.*

- *Click the **Select view** menu, and choose **Lumbar region**.*

- *The **GO** button will flash green. Click on it.*

- *Click on* **TAG 3**, *and the following image will appear:*

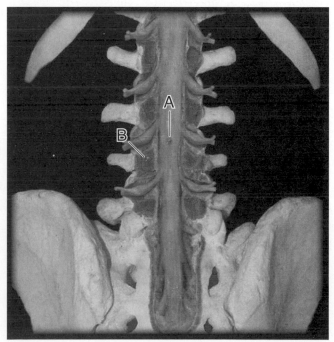

- *Mouse-over the blue pins on the screen to find the information necessary to fill in the following blanks:*

A. _____

B. _____

- *Click on **TAG 4**, and the following image will appear:*

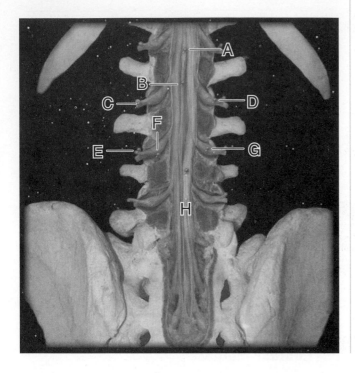

- *Mouse-over the blue pins on the screen to find the information necessary to fill in the following blanks:*

A. _____

B. _____

C. _____

D. _____

E. _____

F. _____

G. _____

H. _____

CHECK POINT:

Spinal Cord, Lumbar Region

1. What is the structure of the pia mater caudal to the spinal cord?

IN REVIEW

What Have I Learned?

The following questions cover the material that you have just learned: the spinal cord. Use the **STRUCTURE INFORMATION** to answer these questions:

1. Name the structure that consists of a filament of pia mater that forms at the conus medullaris, passes through the sacral hiatus, and attaches to the coccyx.

2. In adults, where does the spinal cord end?

3. Name the structures that contribute to the lumbosacral enlargement.

4. Name the bony structure through which each spinal nerve passes.

5. Name the structure that is the sensory ganglion of each dorsal root.

6. Which terminal branch of the spinal nerves innervates the skin and deep muscles of the back, neck, and head?

7. What do the dorsal and ventral roots at the same spinal cord level unite to form?

EXERCISE 4.25:   Nervous System—Peripheral
Nerves, Cervical Plexus

- *Insert Anatomy & Physiology | Revealed®* **Nervous System** *CD, or, if you are already in the* **Dissection** *section, click the* **CHANGE VIEW** *button at the top of the screen, and skip the next step.*

- *In the* **Home screen**, *select the* **Dissection** *button in the left portion of the screen. You may click either on the* **Dissection** *button or on the word itself.*

- *In the* **SELECT A VIEW** *window that appears, click on the* **Select topic** *button.*

- *Choose* **Peripheral nerves** *from the menu.*

- *Click the* **Select view** *menu, and choose* **Cervical plexus**.

- *The* **GO** *button will flash green. Click on it.*

- *Click on* **TAG 1**, *and the following image will appear:*

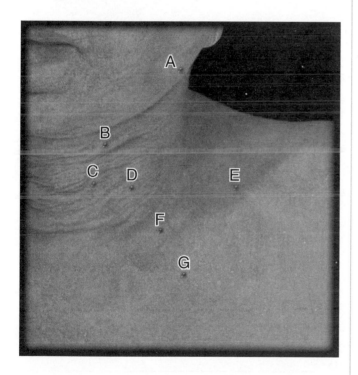

- *Mouse-over the blue pins on the screen to find the information necessary to fill in the following blanks:*

A. _____

B. _____

C. _____

D. _____

E. _____

F. _____

G. _____

- *Click on* **TAG 2**, *and the following image will appear:*

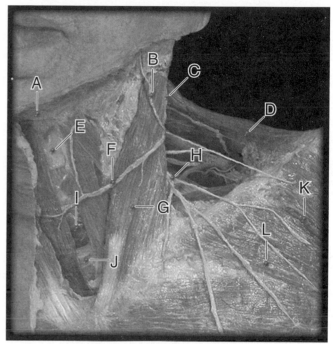

- *Mouse-over the blue pins on the screen to find the information necessary to fill in the following blanks:*

A. _____

B. _____

C. _____

D. _____

E. _____

F. _____

G. _____

H. _____

I. _____

J. _____

K. _____

L. _____

CHECK POINT:

Peripheral Nerves, Cervical Plexus

1. Name the four sensory branches of the cervical plexus.
2. Which of these four innervates the skin over the anterior and lateral neck?
3. Which of these branches innervates the lateral scalp and posterior auricle of the ear?

- *Click on* **TAG 3**, *and the following image will appear:*

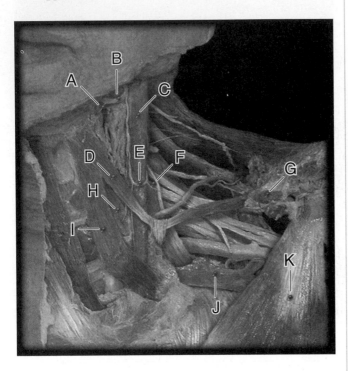

- *Mouse-over the blue pins on the screen to find the information necessary to fill in the following blanks:*

A. _____

B. _____

C. _____

D. _____

E. _____

F. _____

G. _____

H. _____

I. _____

J. _____

K. _____

CHECK POINT:

Peripheral Nerves, Cervical Plexus, cont'd

4. Which nerve innervates the extrinsic and intrinsic muscles of the tongue?
5. Which nerve fibers "hitchhike" on this nerve?
6. Which nerve's Latin name means "cervical loop"?

- *Click on* **TAG 4**, *and the following image will appear:*

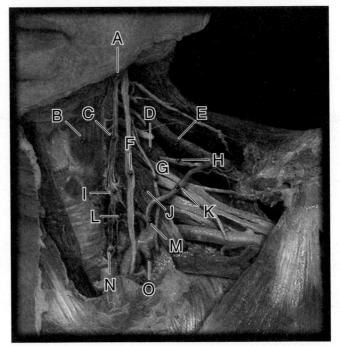

- *Mouse-over the blue pins on the screen to find the information necessary to fill in the following blanks:*

A. _____

B. _____

C. _____

D. _____

E. _____

F. _____

G. _____

H. _____

I. _____

J. _____

K. _____

L. _____

M. _____

N. _____

O. _____

C. _____

D. _____

E. _____

F. _____

G. _____

H. _____

CHECK POINT:

Peripheral Nerves, Cervical Plexus, cont'd

7. Name the three sympathetic ganglia in the neck area.
8. Which of these distributes all postganglionic sympathetic nerve fibers to the head?
9. What is the stellate ganglion?

CHECK POINT:

Peripheral Nerves, Cervical Plexus, cont'd

10. Name the structure that consists of nerves distributed to the upper limb?
11. What are the ventral rami of spinal nerves C1–4 referred to as?

- *Click on* **TAG 5**, *and the following image will appear:*

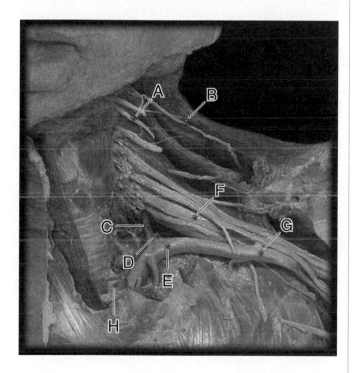

- *Mouse-over the blue pins on the screen to find the information necessary to fill in the following blanks:*

A. _____

B. _____

EXERCISE 4.26: Nervous System—Peripheral Nerves, Brachial Plexus

- *Insert* Anatomy & Physiology | Revealed® **Nervous System** *CD, or, if you are already in the* **Dissection** *section, click the* **CHANGE VIEW** *button at the top of the screen, and skip the next step.*

- *In the* **Home screen**, *select the* **Dissection** *button in the left portion of the screen. You may click either on the* **Dissection** *button or on the word itself.*

- *In the* **SELECT A VIEW** *window that appears, click on the* **Select topic** *button.*

- *Choose* **Peripheral nerves** *from the menu, if it is not already selected.*

- *Click the* **Select view** *menu, and choose* **Brachial plexus**.

- *The* **GO** *button will flash green. Click on it.*

• *Click on* **TAG 1**, *and the following image will appear:*

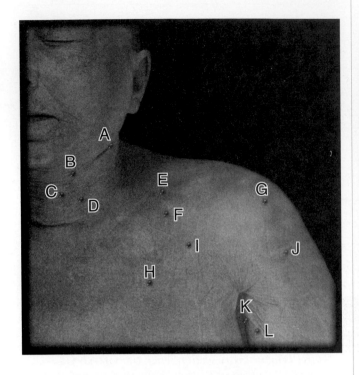

• *Mouse-over the blue pins on the screen to find the information necessary to fill in the following blanks:*

A. _____

B. _____

C. _____

D. _____

E. _____

F. _____

G. _____

H. _____

I. _____

J. _____

K. _____

L. _____

CHECK POINT:

Peripheral Nerves, Brachial Plexus

1. Name the nerve that supplies cutaneous innervation to the skin of and around the auricle of the ear.

• *Click on* **TAG 3**, *and the following image will appear:*

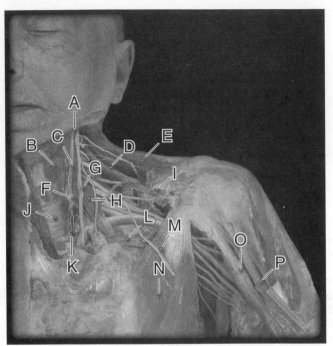

• *Mouse-over the blue pins on the screen to find the information necessary to fill in the following blanks:*

A. _____

B. _____

C. _____

D. _____

E. _____

F. _____

G. _____

H. _____

I. _____

J. _____

K. _____

L. _____

M. _____

N. _____

O. _____

P. _____

- *Click on* **TAG 4**, *and the following image will appear:*

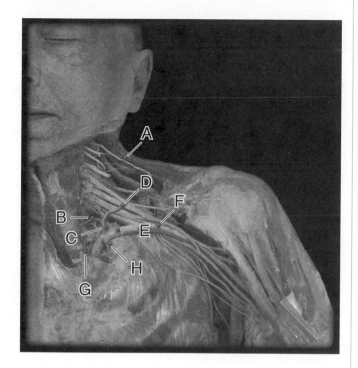

- *Mouse-over the blue pins on the screen to find the information necessary to fill in the following blanks:*

A. _____

B. _____

C. _____

D. _____

E. _____

F. _____

G. _____

H. _____

- *Click on* **TAG 5**, *and the following image will appear:*

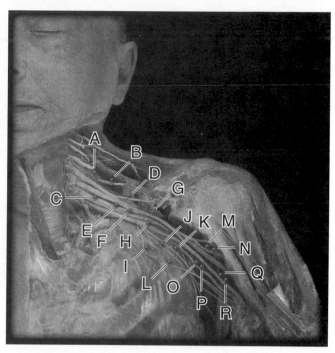

- *Mouse-over the blue pins on the screen to find the information necessary to fill in the following blanks:*

A. _____

B. _____

C. _____

D. _____

E. _____

F. _____

G. _____

H. _____

I. _____

J. _____

K. _____

L. _____

M. _____

N. _____

O. _____

P. _____

Q. _____

R. _____

CHECK POINT:

Pheripheral Nerves, Brachial Plexus, cont'd

2. Which nerve is stimulated when you strike your "funny bone"?
3. Name a nerve that innervates the pectoralis major and minor muscles.
4. Which brachial plexus roots contribute to the long thoracic nerve?

- *Click on **TAG 6**, and the following image will appear:*

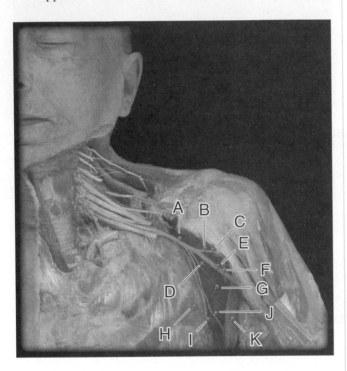

- *Mouse-over the blue pins on the screen to find the information necessary to fill in the following blanks:*

A. _____

B. _____

C. _____

D. _____

E. _____

F. _____

G. _____

H. _____

I. _____

J. _____

K. _____

EXERCISE 4.27: Nervous System—Peripheral Nerves, Upper Limb, Anterior

- *Insert Anatomy & Physiology | Revealed® **Nervous System** CD, or, if you are already in the **Dissection** section, click the **CHANGE VIEW** button at the top of the screen, and skip the next step.*

- *In the **Home screen**, select the **Dissection** button in the left portion of the screen. You may click either on the **Dissection** button or on the word itself.*

- *In the **SELECT A VIEW** window that appears, click on the **Select topic** button.*

- *Choose **Peripheral nerves** from the menu, if it is not already selected.*

- *Click the **Select view** menu, and choose **Upper limb-anterior**.*

- *The **GO** button will flash green. Click on it.*

- *Click on **TAG 1**, and the following image will appear:*

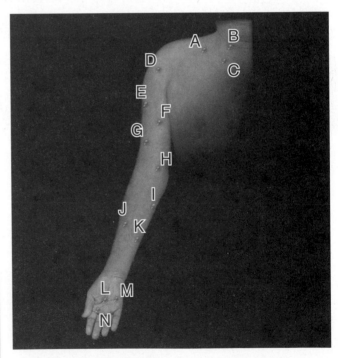

- *Mouse-over the blue pins on the screen to find the information necessary to fill in the following blanks:*

A. _____

B. _____

C. _____

D._____

E._____

F._____

G._____

H._____

I._____

J._____

K._____

L._____

M._____

N._____

CHECK POINT:

Pheripheral Nerves, Upper Limb, Anterior

1. Which nerve supplies the cutaneous innervation to the skin over the medial arm and forearm?
2. Which nerve supplies the cutaneous innervation to the skin over finger V?
3. Name two nerves responsible for the cutaneous innervation of the shoulder.

• *Click on* **TAG 3**, *and the following image will appear:*

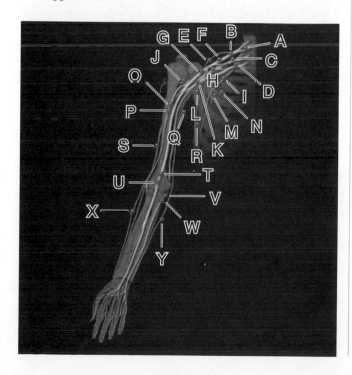

• *Mouse-over the blue pins on the screen to find the information necessary to fill in the following blanks:*

A._____

B._____

C._____

D._____

E._____

F._____

G._____

H._____

I._____

J._____

K._____

L._____

M._____

N._____

O._____

P._____

Q._____

R._____

S._____

T._____

U._____

V._____

W._____

X._____

Y._____

CHECK POINT:

Pheripheral Nerves, Upper Limb, Anterior, cont'd

4. Name the nerve that innervates the joints of the hand.
5. Which division of the brachial plexus gives rise to nerves that distribute to the anterior aspect of the upper limb?
6. Name the nerve responsible for the innervation of the glenohumeral joint.

EXERCISE 4.28:   Nervous System—Peripheral
Nerves, Upper Limb,
Posterior View

- *Insert Anatomy & Physiology | Revealed*® **Nervous System** *CD, or, if you are already in the* **Dissection** *section, click the* **CHANGE VIEW** *button at the top of the screen, and skip the next step.*

- *In the* **Home screen**, *select the* **Dissection** *button in the left portion of the screen. You may click either on the* **Dissection** *button or on the word itself.*

- *In the* **SELECT A VIEW** *window that appears, click on the* **Select topic** *button.*

- *Choose* **Peripheral nerves** *from the menu, if it is not already selected.*

- *Click the* **Select view** *menu, and choose* **Upper limb-posterior**.

- *The* **GO** *button will flash green. Click on it.*

- *Click on* **TAG 1**, *and the following image will appear:*

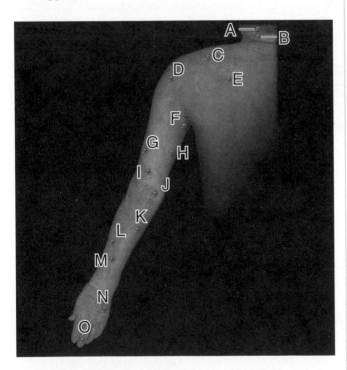

- *Mouse-over the blue pins on the screen to find the information necessary to fill in the following blanks:*

A. _____

B. _____

C. _____

D. _____

E. _____

F. _____

G. _____

H. _____

I. _____

J. _____

K. _____

L. _____

M. _____

N. _____

O. _____

- *Click on* **TAG 3**, *and the following image will appear:*

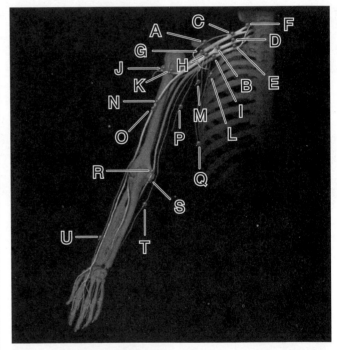

- *Mouse-over the blue pins on the screen to find the information necessary to fill in the following blanks:*

A. _____

B. _____

C. _____

D. _____

E. _____

F. _____

G._____

H._____

I._____

J._____

K._____

L._____

M._____

N._____

O._____

P._____

Q._____

R._____

S._____

T._____

U._____

CHECK POINT:

Peripheral Nerves, Upper Limb, Posterior view

1. Name a nerve that innervates the serratus anterior muscle.
2. Name a nerve that innervates the supraspinatus and infraspinatus muscles.
3. Name a nerve that innervates the subscapularis and teres major muscles.

EXERCISE 4.29: Nervous System—Peripheral Nerves, Trunk

• *Insert* Anatomy & Physiology | Revealed® **Nervous System** *CD, or, if you are already in the* **Dissection** *section, click the* **CHANGE VIEW** *button at the top of the screen, and skip the next step.*

• *In the* **Home screen,** *select the* **Dissection** *button in the left portion of the screen. You may click either on the* **Dissection** *button or on the word itself.*

• *In the* **SELECT A VIEW** *window that appears, click on the* **Select topic** *button.*

• *Choose* **Peripheral Nerves** *from the menu, if it is not already selected.*

• *Click the* **Select view** *menu, and choose* **Trunk**.

• *The* **GO** *button will flash green. Click on it.*

• *Click on* **TAG 1**, *and the following screen will appear:*

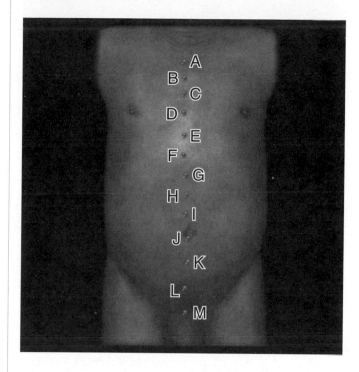

• *Mouse-over the blue pins on the screen to find the information necessary to fill in the following blanks:*

A._____

B._____

C._____

D._____

E._____

F._____

G._____

H._____

I._____

J._____

K._____

L._____

M._____

• *Click on* **TAG 3**, *and the following image will appear:*

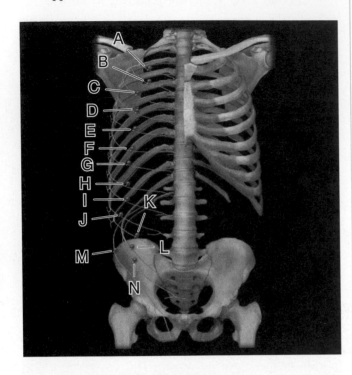

• *Mouse-over the blue pins on the screen to find the information necessary to fill in the following blanks:*

A. _____

B. _____

C. _____

D. _____

E. _____

F. _____

G. _____

H. _____

I. _____

J. _____

K. _____

L. _____

M. _____

N. _____

CHECK POINT:

Peripheral Nerves, Trunk

1. Name three nerves that innervate the skin of the abdomen.
2. Which group of nerves are located within the intercostal space between the innermost and the internal intercostal muscles?
3. Name three of these nerves.

EXERCISE 4.30:   Nervous System—Peripheral Nerves, Lumbosacral Plexus

• *Insert* Anatomy & Physiology | Revealed® **Nervous System** *CD, or, if you are already in the* **Dissection** *section, click the* **CHANGE VIEW** *button at the top of the screen, and skip the next step.*

• *In the* **Home screen**, *select the* **Dissection** *button in the left portion of the screen. You may click either on the* **Dissection** *button or on the word itself.*

• *In the* **SELECT A VIEW** *window that appears, click on the* **Select topic** *button.*

• *Choose* **Peripheral nerves** *from the menu, if it is not already selected.*

• *Click the* **Select view** *menu, and choose* **Lumbosacral plexus**.

• *The* **GO** *button will flash green. Click on it.*

• *Click on* **TAG 1**, *and the following image will appear:*

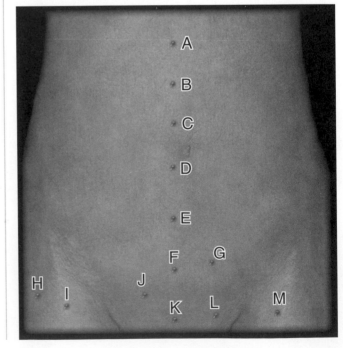

- *Mouse-over the blue pins on the screen to find the information necessary to fill in the following blanks:*

A. _____

B. _____

C. _____

D. _____

E. _____

F. _____

G. _____

H. _____

I. _____

J. _____

K. _____

L. _____

M. _____

- *Click on **TAG 2**, and the following image will appear:*

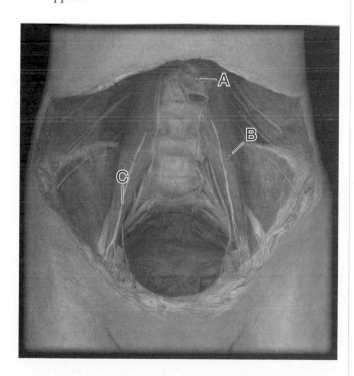

- *Mouse-over the blue pins on the screen to find the information necessary to fill in the following blanks:*

A. _____

B. _____

C. _____

- *Click on **TAG 3**, and the following image will appear:*

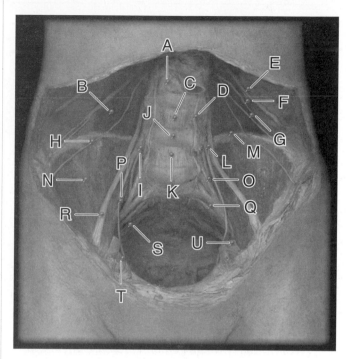

- *Mouse-over the blue pins on the screen to find the information necessary to fill in the following blanks:*

A. _____

B. _____

C. _____

D. _____

E. _____

F. _____

G. _____

H. _____

I. _____

J. _____

K. _____

L. _____

M. _____

N. _____

O. _____

P. _____

Q. _____

R. _____

S. _____

T. _____

U. _____

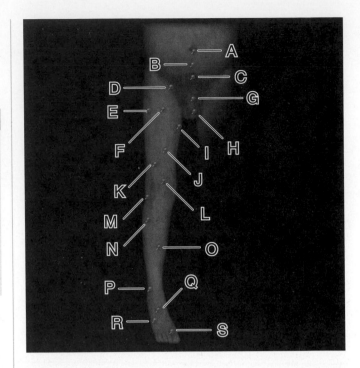

CHECK POINT:

Peripheral Nerves, Lumbosacral Plexus

1. Nerves derived from what structure distribute to the lower anterior abdominal wall, spermatic cord, thigh, medial leg and foot, and the sacral plexus?
2. Name a nerve that supplies sensory innervation to the lateral thigh.
3. Name the seven terminal branches of the lumbar plexus.

EXERCISE 4.31:   Nervous System—Peripheral Nerves, Lower Limb, Anterior View

- *Insert* Anatomy & Physiology | Revealed® **Nervous System** *CD, or, if you are already in the* **Dissection** *section, click the* **CHANGE VIEW** *button at the top of the screen, and skip the next step.*

- *In the* **Home screen***, select the* **Dissection** *button in the left portion of the screen. You may click either on the* **Dissection** *button or on the word itself.*

- *In the* **SELECT A VIEW** *window that appears, click on the* **Select topic** *button.*

- *Choose* **Peripheral nerves** *from the menu, if it is not already selected.*

- *Click the* **Select view** *menu, and choose* **Lower limb-anterior**.

- *The* **GO** *button will now flash green. Click on it.*

- *Click on* **TAG 1***, and the image at top right will appear.*

- *Mouse-over the blue pins on the screen to find the information necessary to fill in the following blanks:*

A. _____

B. _____

C. _____

D. _____

E. _____

F._____

G._____

H._____

I. _____

J._____

K. _____

L. _____

M. _____

N. _____

O._____

P. _____

Q._____

R. _____

S. _____

CHECK POINT:

Peripheral Nerves, Lower Limb, Anterior View

1. Name the nerve providing cutaneous innervation to the lateral thigh.
2. Name the nerve providing cutaneous innervation to the anterior thigh, medial leg, and the medial margin of the foot.
3. Name the nerve providing cutaneous innervation to the medial leg and medial margin of the foot.

- *Click on* **TAG 2**, *and the following image will appear:*

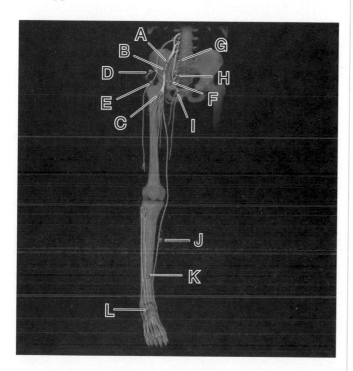

- *Mouse-over the blue pins on the screen to find the information necessary to fill in the following blanks:*

A. _____

B. _____

C. _____

D. _____

E. _____

F. _____

G. _____

H. _____

I. _____

J. _____

K. _____

L. _____

CHECK POINT:

Peripheral Nerves, Lower Limb, Anterior View, cont'd

4. Name the branch of the femoral nerve that provides sensory innervation to the medial leg and the medial margin of the foot.
5. Name the nerve providing motor innervation to the gluteus medius, gluteus minimus, and the tensor fascia lata muscles.
6. Name the nerve providing motor and sensory innervation to the muscles of the lateral leg.

EXERCISE 4.32: Nervous System—Peripheral Nerves, Lower Limb, Posterior View

- *Insert* Anatomy & Physiology | Revealed® **Nervous System** *CD, or, if you are already in the* **Dissection** *section, click the* **CHANGE VIEW** *button at the top of the screen, and skip the next step.*

- *In the* **Home Screen**, *select the* **Dissection** *button in the left portion of the screen. You may click either on the* **Dissection** *button or on the word itself.*

- *In the* **SELECT A VIEW** *window that appears, click on the* **Select topic** *button.*

- *Choose* **Peripheral nerves** *from the menu, if it is not already selected.*

- *Click the* **Select view** *menu, and choose* **Lower limb-posterior**.

- *The* **GO** *button will flash green. Click on it.*

• *Click on* **TAG 1**, *and the following image will appear:*

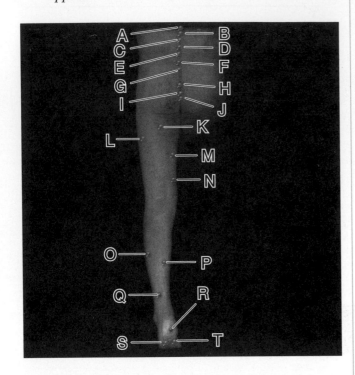

• *Mouse-over the blue pins on the screen to find the information necessary to fill in the following blanks:*

A. _____

B. _____

C. _____

D. _____

E. _____

F. _____

G. _____

H. _____

I. _____

J. _____

K. _____

L. _____

M. _____

N. _____

O. _____

P. _____

Q. _____

R. _____

S. _____

T. _____

CHECK POINT:

Peripheral Nerves, Lower Limb, Posterior View

1. Name the nerve providing cutaneous inner- vation to the posterior distal and lateral proximal leg and the lateral margin of the foot.
2. Name the two nerves providing cutaneous innervation to the sole of the foot.
3. Name the nerve providing innervation to the skin of the medial thigh.

• *Click on* **TAG 2**, *and the following image will appear:*

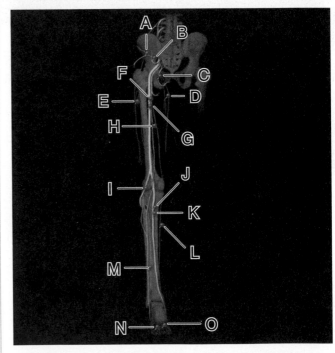

• *Mouse-over the blue pins on the screen to find the information necessary to fill in the following blanks:*

A. _____

B. _____

C. _____

D. _____

E. _____

F. _____

G. _____

H. _____

I. _____

J. _____

K. _____

L. _____

M. _____

N. _____

O. _____

CHECK POINT:

Peripheral Nerves, Lower Limb, Posterior View, cont'd

4. Name the nerve supplying motor innervation to the anterior and lateral leg muscles and the muscles of the dorsum of the foot.
5. Name the nerve supplying motor innervation to the adductor muscles of the medial thigh.
6. Name the nerve supplying sensory innervation to the medial leg and the medial margin of the foot.

Autonomic Nervous System

EXERCISE 4.33: Nervous System— Sympathetic (ANS), Overview

- *Insert* Anatomy & Physiology | Revealed® **Nervous System** *CD, or, if you are already in the* **Dissection** *section, click the* **CHANGE VIEW** *button at the top of the screen, and skip the next step.*

- *In the* **Home screen**, *select the* **Dissection** *button in the left portion of the screen. You may click either on the* **Dissection** *button or on the word itself.*

- *In the* **SELECT A VIEW** *window that appears, click on the* **Select topic** *button.*

- *Choose* **Sympathetic (ANS)** *from the menu.*

- *Click the* **Select view** *menu, and choose* **Overview**.

- *The* **GO** *button will flash green. Click on it.*

- *Click on* **TAG 1**, *and the following image will appear:*

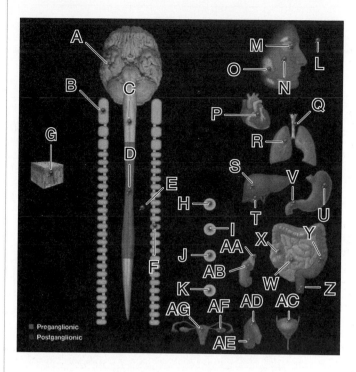

- *Mouse-over the blue pins on the screen to find the information necessary to fill in the following blanks:*

A. _____

B. _____

C. _____

D. _____

E. _____

F. _____

G. _____

H. _____

I. _____

J. _____

K. _____

L. _____

M. _____

N. _____

O. _____

P. _____

Q. _____

R. _____

S. _____

T. _____

U. _____

V. _____

W. _____

X. _____

Y. _____

Z. _____

AA. _____

AB. _____

AC. _____

AD. _____

AE. _____

AF. _____

AG. _____

CHECK POINT:

Sympathetic (ANS), Overiview

1. Name the structure also known as the sympathetic trunk.
2. Name the sympathetic ganglion that distributes postganglionic neuronal processes to the stomach, duodenum, and spleen.
3. Describe the sympathetic postganglionic neuron and pathway of the adrenal medulla.

EXERCISE 4.34: Nervous System—Sympathetic (ANS), Thoracic Region

- *Insert* Anatomy & Physiology | Revealed® **Nervous System** *CD, or, if you are already in the* **Dissection** *section, click the* **CHANGE VIEW** *button at the top of the screen, and skip the next step.*

- *In the* **Home screen**, *select the* **Dissection** *button in the left portion of the screen. You may click either on the* **Dissection** *button or on the word itself.*

- *In the* **SELECT A VIEW** *window that appears, click on the* **Select topic** *button.*

- *Choose* **Sympathetic (ANS)** *from the menu, if it is not already selected.*

- *Click the* **Select view** *menu, and choose* **Thoracic region**.

- *The* **GO** *button will flash green. Click on it.*
- *Click on* **TAG 4**, *and the following image will appear:*

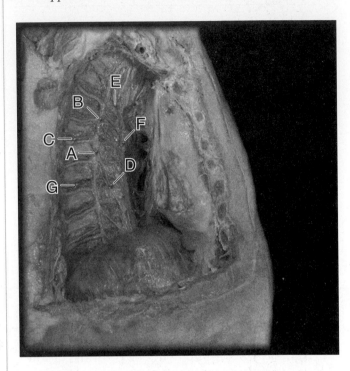

- *Mouse-over the blue pins on the screen to find the information necessary to fill in the following blanks:*

A. _____

B. _____

C. _____

D. _____

E. _____

F. _____

G. _____

CHECK POINT:

Sympathetic (ANS), Thoracic Region

1. Name the location of the postganglionic sympathetic neuronal cell bodies (somas).
2. Which nerves are responsible for motor innervation of all thoracic muscles?
3. Name the three nerves that contain preganglionic sympathetic nerve fibers that enter the abdomen to synapse in prevertebral ganglia.

EXERCISE 4.35: Nervous System—
Parasympathetic (ANS),
Overview

- *Insert* Anatomy & Physiology | Revealed® **Nervous System** *CD, or, if you are already in the* **Dissection** *section, click the* **CHANGE VIEW** *button at the top of the screen, and skip the next step.*

- *In the* **Home screen**, *select the* **Dissection** *button in the left portion of the screen. You may click either on the* **Dissection** *button or on the word itself.*

- *In the* **SELECT A VIEW** *window that appears, click on the* **Select topic** *button.*

- *Choose* **Parasympathetic (ANS)** *from the menu.*

- *Click the* **Select view** *menu, and choose* **Overview**.

- *The* **GO** *button will flash green. Click on it.*

- *Click on* **TAG 1**, *and the following image will appear:*

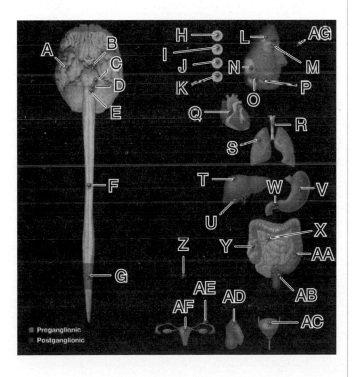

- *Mouse-over the blue pins on the screen to find the information necessary to fill in the following blanks:*

A. _____

B. _____

C. _____

D. _____

E. _____

F. _____

G. _____

H. _____

I. _____

J. _____

K. _____

L. _____

M. _____

N. _____

O. _____

P. _____

Q. _____

R. _____

S. _____

T. _____

U. _____

V. _____

W. _____

X. _____

Y. _____

Z. _____

AA. _____

AB. _____

AC. _____

AD. _____

AE. _____

AF. _____

AG. _____

Parasympathetic (ANS), Overview

1. Where do the parasympathetic fibers terminate in the heart?
2. Name the nerves that distribute to the pelvic viscera via blood vessels.
3. The postganglionic fibers from which ganglion are distributed along branches of the auriculotemporal nerve to the parotid gland?

EXERCISE 4.36: Nervous System—Parasympathetic (ANS), Inferior Brain

- *Insert* Anatomy & Physiology | Revealed® **Nervous System** *CD, or, if you are already in the* **Dissection** *section, click the* **CHANGE VIEW** *button at the top of the screen, and skip the next step.*

- *In the* **Home screen**, *select the* **Dissection** *button in the left portion of the screen. You may click either on the* **Dissection** *button or on the word itself.*

- *In the* **SELECT A VIEW** *window that appears, click on the* **Select topic** *button.*

- *Choose* **Parasympathetic (ANS)** *from the menu, if it is not already selected.*

- *Click the* **Select view** *menu, and choose* **Inferior brain**.

- *The* **GO** *button will flash green. Click on it.*

- *Click on* **TAG 1**, *and the following image will appear:*

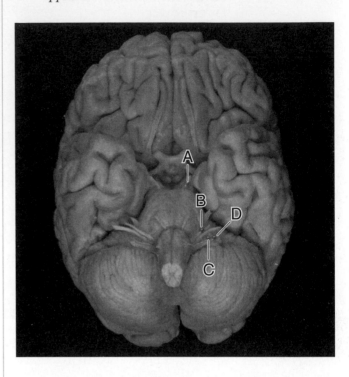

- *Mouse-over the blue pins on the screen to find the information necessary to fill in the following blanks:*

A. _____

B. _____

C. _____

D. _____

Parasympathetic (ANS), Inferior Brain

1. Which cranial nerve has its postganglionic parasympathetic cell bodies (somas) located in the ciliary ganglia?
2. Name the cranial nerve responsible for taste from the epiglottis. (That's right, the epiglottis!)
3. Name the cranial nerve responsible for taste from the anterior two-thirds of the tongue.

IN REVIEW

What Have I Learned?

The following questions cover the material that you have just learned: the peripheral nerves and autonomic nervous system. Use the **STRUCTURE INFORMATION** to answer these questions:

1. What structure is also known as the sympathetic chain ganglia?

2. What nerve carries parasympathetic impulses to the smooth muscle of the thoracic and abdominal viscera?

3. Which nerve supplies all motor innervation to the diaphragm?

4. Name a nerve that provides motor innervation to the cremaster muscle of the male.

5. Name the sympathetic ganglion that distributes postganglionic neuronal processes to the kidneys and gonads.

6. Which division of the ANS has its terminal ganglia near or in the wall of the innervated organ?

7. Which division of the ANS has its terminal ganglia near the spinal cord?

8. Name a nerve that passes through the spermatic cord.

The Senses

EXERCISE 4.37:   Nervous System—Taste, Inferior Brain

- *Insert* Anatomy & Physiology | Revealed® **Nervous System** *CD, or, if you are already in the* **Dissection** *section, click the* **CHANGE VIEW** *button at the top of the screen, and skip the next step.*

- *In the* **Home screen**, *select the* **Dissection** *button in the left portion of the screen. You may click either on the* **Dissection** *button or on the word itself.*

- *In the* **SELECT A VIEW** *window that appears, click on the* **Select topic** *button.*

- *Choose* **Taste** *from the menu.*

- *Click the* **Select view** *menu, and choose* **Inferior brain**.

- *The* **GO** *button will flash green. Click on it.*

- *Click on* **TAG 1**, *and the screen at right will appear.*

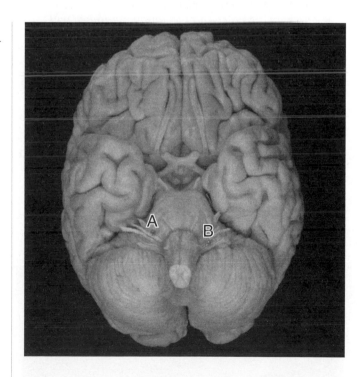

- *Mouse-over the blue pins on the screen to find the information necessary to fill in the following blanks:*

A. _____

B. _____

Nervous System—Taste, Tongue—Superior View

- *Insert* Anatomy & Physiology | Revealed® **Nervous System** *CD, or, if you are already in the* **Dissection** *section, click the* **CHANGE VIEW** *button at the top of the screen, and skip the next step.*

- *In the* **Home screen**, *select the* **Dissection** *button in the left portion of the screen. You may click either on the* **Dissection** *button or on the word itself.*

- *In the* **SELECT A VIEW** *window that appears, click on the* **Select topic** *button.*

- *Choose* **Taste** *from the menu, if it is not already selected.*

- *Click the* **Select view** *menu, and choose* **Tongue—superior**.

- *The* **GO** *button will flash green. Click on it.*

- *Click on* **TAG 2**, *and the following image will appear:*

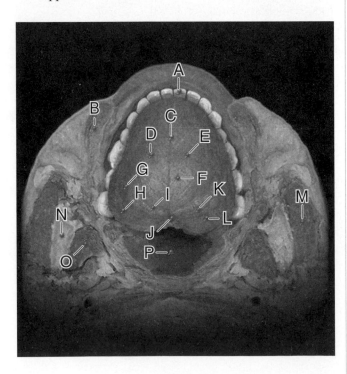

- *Mouse-over the blue pins on the screen to find the information necessary to fill in the following blanks:*

A. _____

B. _____

C. _____

D. _____

E. _____

F. _____

G. _____

H. _____

I. _____

J. _____

K. _____

L. _____

M. _____

N. _____

O. _____

P. _____

CHECK POINT:

Taste

1. Name the shallow median longitudinal groove on the anterior part of the tongue.
2. Name the nerve that innervates the posterior tongue, pharynx, and middle ear.
3. Name the eight-to-twelve large, flat-topped papilla containing numerous taste buds.

Nervous System—Taste, Histology—Vallate Papilla

- *Insert* Anatomy & Physiology | Revealed® **Nervous System** *CD, or, if you are already in the* **Dissection** *section, click the* **CHANGE VIEW** *button at the top of the screen, and skip the next step.*

- *In the* **Home screen**, *select the* **Dissection** *button in the left portion of the screen. You may click either on the* **Dissection** *button or on the word itself.*

- *In the* **SELECT A VIEW** *window that appears, click on the* **Select topic** *button.*

- *Choose* **Taste** *from the menu, if it is not already selected.*

- *Click the* **Select view** *menu, and choose* **Histology—vallate papilla**.

- *The* **GO** *button will flash green. Click on it.*

- *Click on* **TAG 1**, *and the following image will appear:*

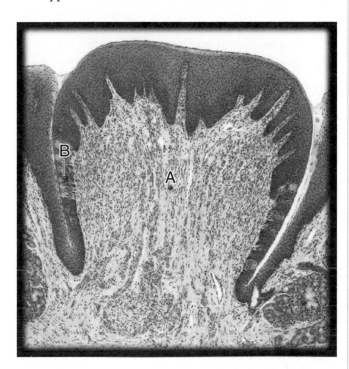

- *Mouse-over the blue pins on the screen to find the information necessary to fill in the following blanks:*

A. _____

B. _____

C. _____

CHECK POINT:

Taste, Histology—Vallate Papilla

1. Name the nerve that receives special sensory information from the taste buds.
2. In what structure are the taste buds located?
3. Name the five primary taste sensations.

EXERCISE 4.40: Nervous System—Taste, Histology—Taste Bud

- *Insert* Anatomy & Physiology | Revealed® **Nervous System** *CD, or, if you are already in the* **Dissection** *section, click the* **CHANGE VIEW** *button at the top of the screen, and skip the next step.*

- *In the* **Home screen**, *select the* **Dissection** *button in the left portion of the screen. You may click either on the* **Dissection** *button or on the word itself.*

- *In the* **SELECT A VIEW** *window that appears, click on the* **Select topic** *button.*

- *Choose* **Taste** *from the menu, if it is not already selected.*

- *Click the* **Select view** *menu, and choose* **Histology—taste bud**.

- *The* **GO** *button will flash green. Click on it.*

- *Click on* **TAG 1**, *and the following image will appear:*

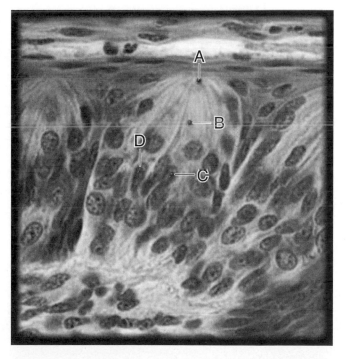

- *Mouse-over the blue pins on the screen to find the information necessary to fill in the following blanks:*

A. _____

B. _____

C. _____

D. _____

CHECK POINT:

Taste, Histology—Taste Bud

1. Name the structure that serves as a receptor surface for taste molecules.
2. What type of cells are taste cells?
3. What is the site of taste reception?

EXERCISE 4.41: Nervous System—Smell, Inferior Brain

- *Insert* Anatomy & Physiology | Revealed® **Nervous System** CD, or, if you are already in the **Dissection** section, click the **CHANGE VIEW** button at the top of the screen, and skip the next step.

- *In the* **Home screen**, *select the* **Dissection** button in the left portion of the screen. You may click either on the **Dissection** button or on the word itself.

- *In the* **SELECT A VIEW** window that appears, click on the **Select topic** button.

- *Choose* **Smell** from the menu.

- *Click the* **Select view** menu, and choose **Inferior brain**.

- *The* **GO** button will flash green. Click on it.

- *Click on* **TAG 1**, and the following image will appear:

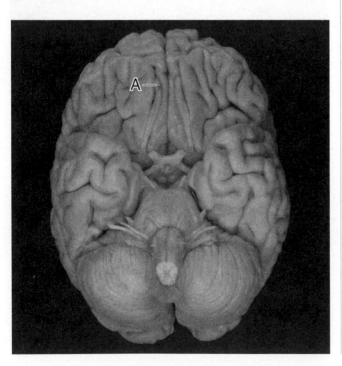

- *Mouse-over the blue pin on the screen to find the information necessary to fill in the following blank:*

  A. _____

CHECK POINT:

Smell, Inferior Brain

1. Name the type of neurons found in the mucous membranes of the nasal cavities.
2. What bony structure do these neurons pass through to synapse with the olfactory nerve?
3. With what specific portion of the olfactory neuron do these neurons synapse?

EXERCISE 4.42: Nervous System—Smell, Nasal Cavity-Lateral View

- *Insert* Anatomy & Physiology | Revealed® **Nervous System** CD, or, if you are already in the **Dissection** section, click the **CHANGE VIEW** button at the top of the screen, and skip the next step.

- *In the* **Home screen**, *select the* **Dissection** button in the left portion of the screen. You may click either on the **Dissection** button or on the word itself.

- *In the* **SELECT A VIEW** window that appears, click on the **Select topic** button.

- *Choose* **Smell** from the menu, if it is not already selected

- *Click the* **Select view** menu, and choose **Nasal cavity—lateral**.

- *The* **GO** button will flash green. Click on it.

- *Click on* **TAG 3**, *and the following image will appear.*

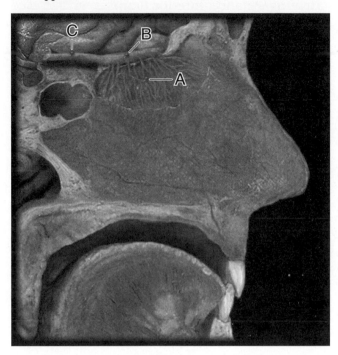

- *Mouse-over the blue pins on the screen to find the information necessary to fill in the following blanks:*

A. _____

B. _____

C. _____

CHECK POINT:

Smell, Nasal Cavity—Lateral View

1. Where in the brain does the olfactory nerve connect?
2. What structure detects odorants in the nasal cavities?
3. Name the expanded anterior end of the olfactory tract.

EXERCISE 4.43:   Nervous System—Smell, Histology—Olfactory Mucosa

- *Insert* Anatomy & Physiology | Revealed® **Nervous System** *CD, or, if you are already in the* **Dissection** *section, click the* **CHANGE VIEW** *button at the top of the screen, and skip the next step.*

- *In the* **Home screen**, *select the* **Dissection** *button in the left portion of the screen. You may click either on the* **Dissection** *button or on the word itself.*

- *In the* **SELECT A VIEW** *window that appears, click on the* **Select topic** *button.*

- *Choose* **Smell** *from the menu, if it is not already selected.*

- *Click the* **Select view** *menu, and choose* **Histology—olfactory mucosa**.

- *The* **GO** *button will flash green. Click on it.*

- *Click on* **TAG 1**, *and the following image will appear:*

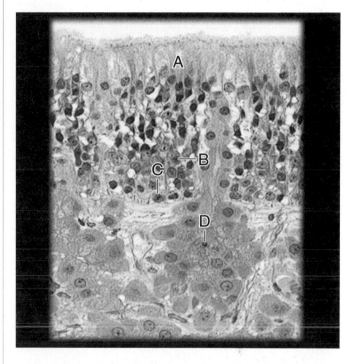

- *Mouse-over the blue pins on the screen to find the information necessary to fill in the following blanks:*

A. _____

B. _____

C. _____

D. _____

CHECK POINT:

Smell, Histology—Olfactor, Mucosa

1. Name the structures that produce mucus in the nasal cavities.
2. What is the function of the mucus?
3. Name the cells that produce new olfactory receptors.

EXERCISE 4.44: Nervous System—Hearing/ Balance, Inferior Brain

- *Insert* Anatomy & Physiology | Revealed® **Nervous System** *CD, or, if you are already in the* **Dissection** *section, click the* **CHANGE VIEW** *button at the top of the screen, and skip the next step.*

- *In the* **Home screen**, *select the* **Dissection** *button in the left portion of the screen. You may click either on the* **Dissection** *button or on the word itself.*

- *In the* **SELECT A VIEW** *window that appears, click on the* **Select topic** *button.*

- *Choose* **Hearing/balance** *from the menu.*

- *Click the* **Select view** *menu, and choose* **Inferior brain**.

- *The* **GO** *button will flash green. Click on it.*

- *Click on* **TAG 1**, *and the following image will appear:*

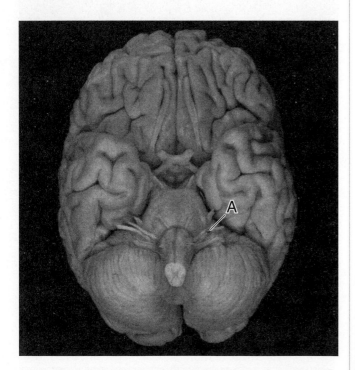

- *Mouse-over the blue pin on the screen to find the information necessary to fill in the following blank:*

  A. _____

CHECK POINT:

Hearing/Balance, Inferior Brain

1. What two special senses are provided by the vestibulocochlear nerve?
2. What bony structure allows internal passage of the vestibulocochlear nerve?
3. What is the CNS connection for the vestibulocochlear nerve?

EXERCISE 4.45: Nervous System—Hearing/ Balance, Ear—Anterior View

- *Insert* Anatomy & Physiology | Revealed® **Nervous System** *CD, or, if you are already in the* **Dissection** *section, click the* **CHANGE VIEW** *button at the top of the screen, and skip the next step.*

- *In the* **Home screen**, *select the* **Dissection** *button in the left portion of the screen. You may click either on the* **Dissection** *button or on the word itself.*

- *In the* **SELECT A VIEW** *window that appears, click on the* **Select topic** *button.*

- *Choose* **Hearing/balance** *from the menu, if it is not already selected.*

- *Click the* **Select view** *menu, and choose* **Ear— anterior**.

- *The* **GO** *button will flash green. Click on it.*

- *Click on* **TAG 1**, *and the following image will appear:*

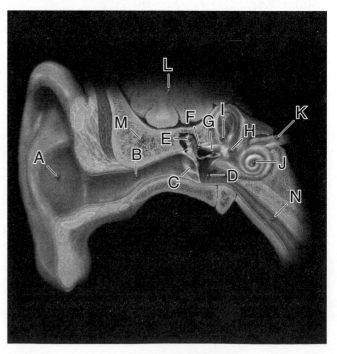

- *Mouse-over the blue pins on the screen to find the information necessary to fill in the following blanks:*

A. _____

B. _____

C. _____

D. _____

E. _____

F. _____

G. _____

H. _____

I. _____

J. _____

K. _____

L. _____

M. _____

N. _____

CHECK POINT:

Hearing/Balance, Ear—Anterior View

1. Name the organ of hearing.
2. Name the structure of the ear that senses linear movement.
3. Name the organ of equilibrium.

EXERCISE 4.46: Nervous System—Hearing/Balance, Histology—Cochlea

- *Insert Anatomy & Physiology | Revealed® **Nervous System** CD, or, if you are already in the **Dissection** section, click the **CHANGE VIEW** button at the top of the screen, and skip the next step.*

- *In the **Home screen**, select the **Dissection** button in the left portion of the screen. You may click either on the **Dissection** button or on the word itself.*

- *In the **SELECT A VIEW** window that appears, click on the **Select topic** button.*

- *Choose **Hearing/balance** from the menu, if it is not already selected.*

- *Click the **Select view** menu, and choose **Histology—cochlea**.*

- *The **GO** button will flash green. Click on it.*

- *Click on **TAG 1**, and the following image will appear:*

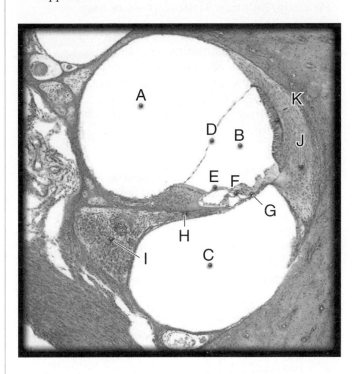

- *Mouse-over the blue pins on the screen to find the information necessary to fill in the following blanks:*

A. _____

B. _____

C. _____

D. _____

E. _____

F. _____

G. _____

H. _____

I. _____

J. _____

K. _____

Hearing/Balance, Histology—Cochlea

1. Name the three fluid-filled chambers of the cochlea, from superior to inferior.
2. What specific structure of the cochlea is the site of conversion of sound vibrations into electrochemical signals?
3. Name the structure that supports this structure.

EXERCISE 4.47:   Nervous System—Hearing/ Balance, Histology— Organ of Corti

- *Insert* Anatomy & Physiology | Revealed® **Nervous System** *CD, or, if you are already in the* **Dissection** *section, click the* **CHANGE VIEW** *button at the top of the screen, and skip the next step.*

- *In the* **Home screen***, select the* **Dissection** *button in the left portion of the screen. You may click either on the* **Dissection** *button or on the word itself.*

- *In the* **SELECT A VIEW** *window that appears, click on the* **Select topic** *button.*

- *Choose* **Hearing/balance** *from the menu, if it is not already selected.*

- *Click the* **Select view** *menu, and choose* **Histology—Organ of Corti***.*

- *The* **GO** *button will flash green. Click on it.*

- *Click on* **TAG 1***, and the following image will appear:*

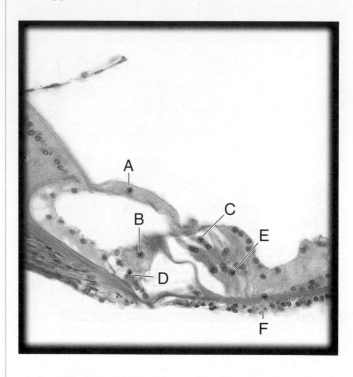

- *Mouse-over the blue pins on the screen to find the information necessary to fill in the following blanks:*

A. _____

B. _____

C. _____

D. _____

E. _____

F. _____

HEADS UP!

Now is a good time to review the Animation: Hearing.

Hearing/Balance, Histology—Organ of Corti

1. Name the structure with the function of transmitting vibrations to embedded stereocilia of sensory cells.
2. What specific structures function as the receptors for hearing?
3. Describe the function of stereocilia.

EXERCISE 4.48: Nervous System—Vision, Inferior Brain

- *Insert* Anatomy & Physiology | Revealed® **Nervous System** *CD, or, if you are already in the* **Dissection** *section, click the* **CHANGE VIEW** *button at the top of the screen, and skip the next step.*

- *In the* **Home screen***, select the* **Dissection** *button in the left portion of the screen. You may click either on the* **Dissection** *button or on the word itself.*

- *In the* **SELECT A VIEW** *window that appears, click on the* **Select topic** *button.*

- *Choose* **Vision** *from the menu.*

- *Click the* **Select view** *menu, and choose* **Inferior brain***.*

- *The* **GO** *button will flash green. Click on it.*

- *Click on* **TAG 1***, and the following image will appear:*

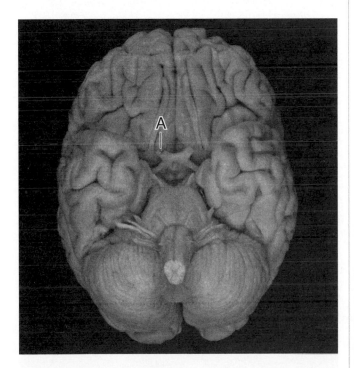

- *Mouse-over the blue pin on the screen to find the information necessary to fill in the following blank:*

    A. _____

EXERCISE 4.49: Nervous System—Vision, Orbit—Lateral View

- *Insert* Anatomy & Physiology | Revealed® **Nervous System** *CD, or, if you are already in the* **Dissection** *section, click the* **CHANGE VIEW** *button at the top of the screen, and skip the next step.*

- *In the* **Home screen***, select the* **Dissection** *button in the left portion of the screen. You may click either on the* **Dissection** *button or on the word itself.*

- *In the* **SELECT A VIEW** *window that appears, click on the* **Select topic** *button.*

- *Choose* **Vision** *from the menu, if it is not already selected.*

- *Click the* **Select view** *menu, and choose* **Orbit—lateral***.*

- *The* **GO** *button will flash green. Click on it.*

- *Click on* **TAG 3***, and the following image will appear:*

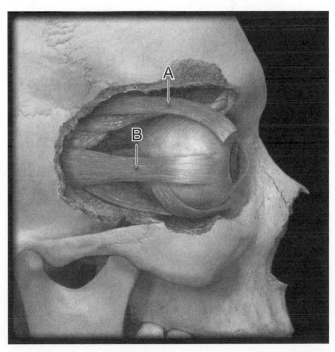

• *Mouse-over the blue pins on the screen to find the information necessary to fill in the following blanks:*

A. _____

B. _____

CHECK POINT:

Vision, Orbit—Lateral View

1. Name the muscle responsible for elevation of the upper eyelid.
2. What cranial nerve supplies the innervation for this muscle?
3. Name the muscle responsible for abduction of the eyeball.

• *Click on* **TAG 4**, *and the following image will appear:*

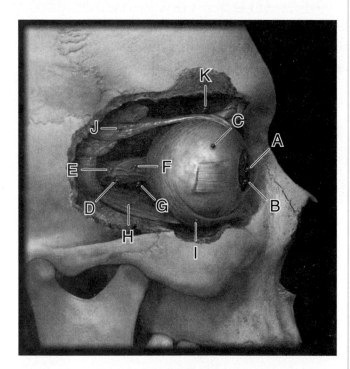

• *Mouse-over the blue pins on the screen to find the information necessary to fill in the following blanks:*

A. _____

B. _____

C. _____

D. _____

E. _____

F. _____

G. _____

H. _____

I. _____

J. _____

K. _____

CHECK POINT:

Vision, Orbit—Lateral View, cont'd

4. Name the muscle responsible for depression and medial rotation of the eyeball.
5. Name the fiber types found in the short ciliary nerves.
6. Name the structure that contains postganglionic fibers distributed to pupillary sphincter and ciliary muscles.

• *Click on* **TAG 5**, *and the following image will appear:*

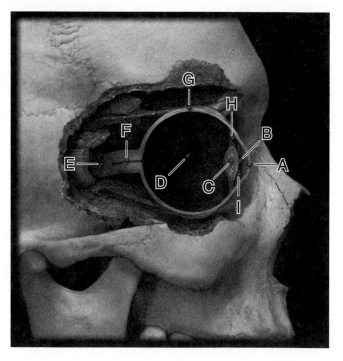

• *Mouse-over the blue pins on the screen to find the information necessary to fill in the following blanks:*

A. _____

B. _____

C. _____

D. _____

E. _____

F. _____

G. _____

H. _____

I. _____

CHECK POINT:

Vision, Orbit—Lateral View, cont'd

7. Name the structure referred to as the "white" of the eye.
8. Name the artery that, when blocked, may cause blindness.
9. Name the internal structure of the eye that focuses incoming light.

EXERCISE 4.50: Nervous System—Vision, Histology—Retina

- *Insert* Anatomy & Physiology | Revealed® **Nervous System** *CD, or, if you are already in the* **Dissection** *section, click the* **CHANGE VIEW** *button at the top of the screen, and skip the next step.*

- *In the* **Home screen**, *select the* **Dissection button** *in the left portion of the screen. You may click either on the* **Dissection** *button or on the word itself.*

- *In the* **SELECT A VIEW** *window that appears, click on the* **Select topic** *button.*

- *Choose* **Vision** *from the menu, if it is not already selected.*

- *Click the* **Select view** *menu, and choose* **Histology—retina**.

- *The* **GO** *button will flash green. Click on it.*

- *Click on* **TAG 1**, *and the following image will appear:*

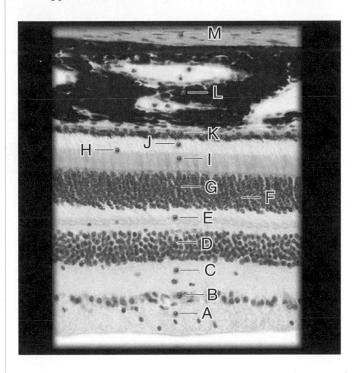

- *Mouse-over the blue pins on the screen to find the information necessary to fill in the following blanks:*

A. _____

B. _____

C. _____

D. _____

E. _____

F. _____

G. _____

H. _____

I. _____

J. _____

K. _____

L. _____

M. _____

HEADS UP!

Now is a good time to review the Animation, Vision.

CHECK POINT:

Vision, Histology—Retina

1. Which retinal cells detect color vision? What light-sensitive pigments do they contain?
2. Name the highly vascular layer of the retina containing melanin.
3. What is the function of the ganglionic layer of the retina?

IN REVIEW

What Have I Learned?

The following questions cover the material that you have just learned: the special senses. Use the **STRUCTURE INFORMATION** to answer these questions:

1. Name the cranial nerves responsible for taste. Name the structures innervated by each of these nerves.

2. Name the ovoid collections of lymphoid tissue covered with mucous membrane on the posterior aspect of the tongue.

3. Name the three subdivisions of the pharynx.

4. Name the structure that is the remnant of the embryonic thyroglossal duct.

5. Name the three types of cells found in taste buds.

6. What is another name for olfactory neurons?

7. Name the cranial nerve responsible for hearing and balance.

8. Name the organ of hearing.

9. Name the two organs of balance.

10. Name the bony canal of the external ear.

11. What is the anatomical name for the eardrum?

12. Name the three smallest bones in the body from lateral to medial.

13. What is the function of these three bones?

14. What is the name for the passage between the tympanic cavity and nasopharynx?

15. What is the function of this passage?

16. The auditory ossicles articulate with what structure of the inner ear?

17. What are the functions of the three fluid-filled chambers of the cochlea?

18. Where does the lens of the eye focus the incoming light?

19. What term refers to changes of the lens' shape when focusing?

20. Name the structure that prevents light scatter in the eye.

CHAPTER **5**

The Cardiovascular System

Your heart and circulatory system accomplish amazing feats! Think about these accomplishments:

- Your heart, roughly the size of your two hands nested together, beats over 100,000 times per day. This equals approximately 36 million times per year.
- Your body contains an average volume of 5 liters of blood, which circulates through your heart, and thus your body, once every minute.
- It has been estimated that your heart pumps 1,900 gallons (7,200 liters) of blood through approximately 60,000–100,000 miles (96,560–63,730 kilometers) of blood vessels every day. That's 2½ to 3 times around the earth at the equator—every day.
- If you live 70 years, your heart will beat some 2.5 billion times. In this 70-year lifetime, it will pump roughly 1 million barrels of blood, enough to fill more than three super tankers.

Pretty impressive accomplishments, wouldn't you say? Let's begin our exploration of the wonders of this amazing cardiovascular system.

We'll begin with an overview of the entire system, and then we'll look at the structures in detail region by region, beginning with the heart. Let's get started.

Animation: Cardiovascular System

Before beginning your study of the cardiovascular system, view the following correlated *Anatomy & Physiology | Revealed®* animations:

- *Insert the* Anatomy & Physiology | Revealed® **Cardiovascular, Lymphatic, and Respiratory Systems** *CD.*
- *In the* **Select system** *menu at the top left of the screen, select* **Cardiovascular**.
- *In the* **Home screen**, *click on the* **Animation** *button in the left portion of the screen.*
- *In the* **Select topic** *menu, select* **Anatomy**.
- *From the* **Select animation** *menu, select* **Cardiovascular system**.
- *Click the* **Play** *button and the animation will run in the* **IMAGE AREA**.
- *After viewing the animation, answer the following questions:*

1. The cardiovascular consists of what three structures?

2. The heart distributes and receives blood through which structures?

3. Define an artery.

4. Gases and nutrients are exchanged with the tissues through which blood vessels?

5. Define a vein.

Animation: Blood Flow Through Heart

- *Return to the **ANIMATION LIST** and click on the second menu down, the **Select animation** menu.*

- *Select **Blood flow through heart**.*

- *Click the **Play** button and the animation will run in the **IMAGE AREA**.*

- *After viewing the animation, answer the following questions:*

1. Name the four chambers of the heart. What are the differences between them?

2. Name the two major vessels that deliver oxygen-poor blood to the heart.

3. What areas of the body do these two vessels drain?

4. What is the route of venous blood from the heart?

5. Trace the route of blood as it flows through the heart. Be sure to list *all* structures involved.

6. Name the only arteries to carry oxygen-poor blood.

7. Name the only veins to carry oxygen-rich blood.

HEADS UP!

The hearts in the following exercise have been removed from the cadaver, and thus will not have all of the surrounding structures visible for reference. The subsequent section of the blood vessels of the thorax will cover the anatomy surrounding the heart in detail.

The Heart

EXERCISE 5.1:   Cardiovascular System, Heart, Internal Features, Anterior View

- *Insert the* Anatomy & Physiology | Revealed® **Cardiovascular, Lymphatic, and Respiratory Systems** *CD.*

- *In the **Select system** menu at the top left of the screen select **Cardiovascular**.*

- *In the **Home screen**, select the **Dissection** button in the left portion of the screen. You may click either on the **Dissection button** or on the word.*

- *In the **SELECT A VIEW** window that appears, click on the **Select topic** button.*

- *Choose **Heart** from the menu.*

- *Click the **Select view** menu, and choose **Internal features - anterior**.*

- *The **GO** button will now flash green. Click on it.*

- *Click on **TAG 1**, and the following image will appear:*

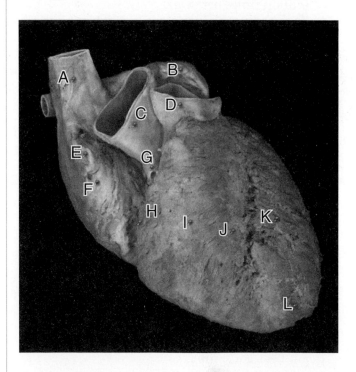

- *Mouse-over the blue pins on the screen to find the information necessary to fill in the following blanks:*

A. _____

B. _____

C. _____

D. _____

E. _____

F. _____

G. _____

H. _____

I. _____

J. _____

K. _____

L. _____

CHECK POINT:

Heart, Internal Features, Anterior View

1. Name the major blood vessel between the heart and the aortic arch.
2. What are its two branches?
3. Name the branches of the right coronary artery.

• *Click on* **TAG 2**, and the following image will appear:

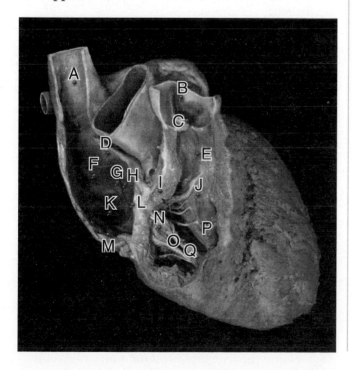

• *Mouse-over the blue pins on the screen to find the information necessary to fill in the following blanks:*

A. _____

B. _____

C. _____

D. _____

E. _____

F. _____

G. _____

H. _____

I. _____

J. _____

K. _____

L. _____

M. _____

N. _____

O. _____

P. _____

Q. _____

CHECK POINT:

Heart, Internal Features, Anterior View

4. What are the two terms for the valve between the right atrium and right ventricle?
5. Name the structure that prevents reflux of blood into the right ventricle.
6. Name the two branches of the pulmonary trunk.

- *Click on* **TAG 3**, *and the following image will appear:*

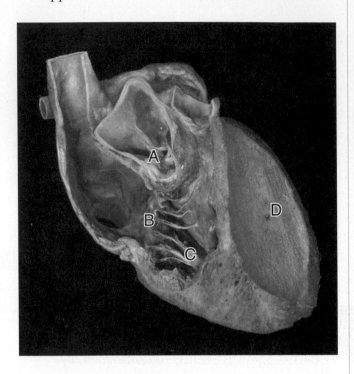

- *Mouse-over the blue pins on the screen to find the information necessary to fill in the following blanks:*

A. _____

B. _____

C. _____

D. _____

CHECK POINT:

Heart, Internal Features, Anterior View

7. Name the structure that prevents reflux of blood into the left ventricle.
8. Name the muscle layer of the heart wall.
9. Where is that muscle layer thinner and thicker?

- *Click on* **TAG 4**, *and the following image will appear:*

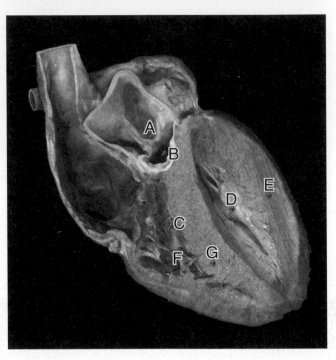

- *Mouse-over the blue pins on the screen to find the information necessary to fill in the following blanks:*

A. _____

B. _____

C. _____

D. _____

E. _____

F._____

G. _____

CHECK POINT:

Heart, Internal Features, Anterior View

10. What heart chamber is responsible for pumping oxygen-rich blood to the body (except the lungs)?
11. What heart chamber is responsible for pumping oxygen-poor blood to the lungs?
12. Name the irregular, muscular elevations on the internal surface of both ventricles.

• *Click on* **TAG 5**, *and the following image will appear:*

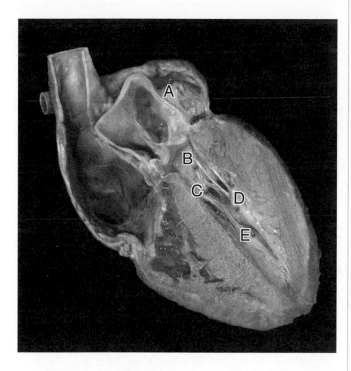

• *Mouse-over the blue pins on the screen to find the information necessary to fill in the following blanks:*

A. _____

B. _____

C. _____

D. _____

E. _____

CHECK POINT:

Heart, Internal Features, Anterior View

13. Name the fibrous strands that attach the free edges of the atrioventricular valve cusps to the papillary muscles.
14. Name the heart chamber that receives oxygen-rich blood from the lungs.
15. Name the blood vessels that bring this blood from the lungs to the heart. How many are there?

• *Click on* **TAG 6**, *and the following image will appear:*

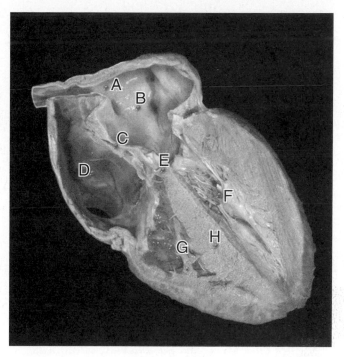

• *Mouse-over the blue pins on the screen to find the information necessary to fill in the following blanks:*

A. _____

B. _____

C. _____

D. _____

E. _____

F. _____

G. _____

H. _____

CHECK POINT:

Heart, Internal Features, Anterior View

16. Name the structure that separates the right and left atria.
17. Name the structure that separates the right and left ventricles.
18. What is the name for the superior membranous part of the structure that separates the right and left ventricles?

EXERCISE 5.2: Cardiovascular System, Heart, Vasculature, Anterior View

- *Insert the* Anatomy & Physiology | Revealed® **Cardiovascular, Lymphatic, and Respiratory Systems** *CD, or, if you are already in the* **Dissection** *section, click the* **CHANGE VIEW** *button at the top of the screen, and skip the next two steps.*

- *In the* **Select system** *menu at the top left of the screen select* **Cardiovascular**.

- *In the* **Home screen**, *select the* **Dissection** *button in the left portion of the screen. You may click either on the* **Dissection button** *or on the word.*

- *In the* **SELECT A VIEW** *window,* **Heart** *appears in the* **Select topic** *menu.*

- *Click the* **Select view** *menu and choose* **Vasculature - anterior**.

- *The* **GO** *button will now flash green. Click on it.*

- *Click on* **TAG 1**, *and the following screen will appear:*

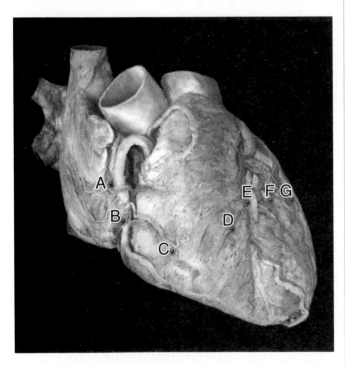

- *Mouse-over the blue pins on the screen to find the information necessary to fill in the following blanks:*

A. _____

B. _____

C. _____

D. _____

E. _____

F. _____

G. _____

Heart, Vasculature, Anterior View

1. Name the vein that ascends the anterior interventricular sulcus.
2. Name the vein that ascends across the anterior surface of the left ventricle.
3. Name the numerous veins that course across the anterior surface of the right ventricle.

- *Click on* **TAG 2**, *and the following image will appear:*

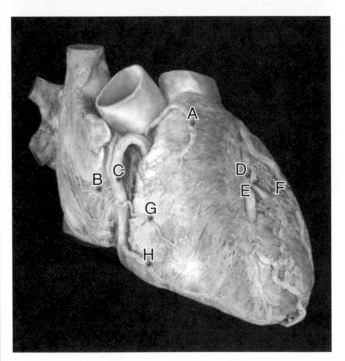

- *Mouse-over the blue pins on the screen to find the information necessary to fill in the following blanks:*

A. _____

B. _____

C. _____

D. _____

E. _____

F._____

G._____

H._____

E. _____

F._____

G._____

H._____

I._____

J._____

K. _____

L. _____

M. _____

N. _____

CHECK POINT:

Heart, Vasculature, Anterior View

4. Name the artery that lies in the right coronary sulcus.
5. What are the branches of this artery?
6. Name the artery that descends along the margin of the left ventricle.

- *Click on* **TAG 3**, *and the following image will appear:*

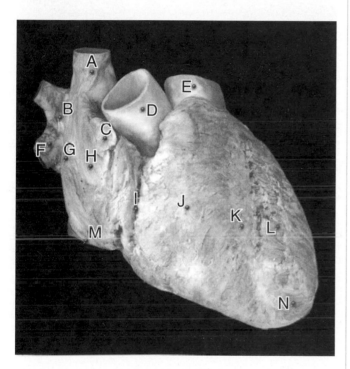

- *Mouse-over the blue pins on the screen to find the information necessary to fill in the following blanks:*

A. _____

B. _____

C. _____

D. _____

CHECK POINT:

Heart, Vasculature, Anterior View

7. Name the major vein that drains everything inferior to the diaphragm.
8. Name the tributary of the superior vena cava.
9. Name the inferolateral point of the heart.

EXERCISE 5.3: Cardiovascular System, Heart, Vasculature, Posterior View

- *Insert the* Anatomy & Physiology | Revealed® **Cardiovascular, Lymphatic, and Respiratory Systems** *CD, or, if you are already in the* **Dissection** *section, click the* **CHANGE VIEW** *button at the top of the screen, and skip the next two steps.*

- *In the* **Select system** *menu at the top left of the screen, select* **Cardiovascular**.

- *In the* **Home screen**, *select the* **Dissection** *button in the left portion of the screen. You may click either on the* **Dissection button** *or on the word.*

- *In the* **SELECT A VIEW** *window,* **Heart** *appears in the* **Select topic** *menu.*

- *Click the* **Select view** *menu and choose* **Vasculature - posterior**.

- *The* **GO** *button will flash green. Click on it.*

• *Click on* **TAG 1**, *and the following screen will appear:*

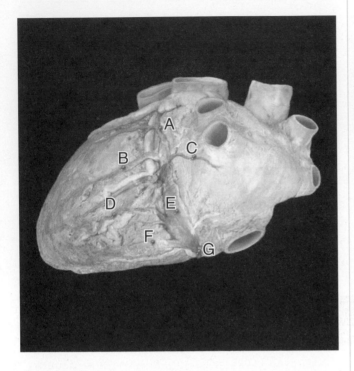

• *Mouse-over the blue pins on the screen to find the information necessary to fill in the following blanks:*

A. _____

B. _____

C. _____

D. _____

E. _____

F. _____

G. _____

CHECK POINT:

Heart, Vasculature, Posterior View

1. Name the structure that receives venous blood from the heart.
2. What are the three major tributaries of this structure?
3. List the structures drained by the great cardiac vein.

• *Click on* **TAG 2**, *and the following image will appear:*

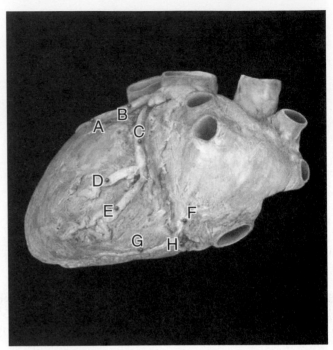

• *Mouse-over the blue pins on the screen to find the information necessary to fill in the following blanks:*

A. _____

B. _____

C. _____

D. _____

E. _____

F. _____

G. _____

H. _____

CHECK POINT:

Heart, Vasculature, Posterior View

4. What is the term that refers to the end-to-end union of blood vessels?
5. Name the arteries located in the posterior interventricular sulcus.
6. Name the artery that descends in the anterior interventricular sulcus.

• *Click on* **TAG 3**, *and the following image will appear:*

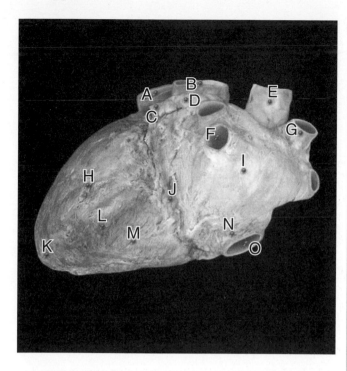

• *Mouse-over the blue pins on the screen to find the information necessary to fill in the following blanks:*

A. _____

B. _____

C. _____

D. _____

E. _____

F. _____

G. _____

H. _____

I. _____

J. _____

K. _____

L. _____

M. _____

N. _____

O. _____

Animation: Conducting System of Heart

Before concluding your study of the heart, view the correlated *Anatomy & Physiology | Revealed*® animations described here:

• *Insert the* Anatomy & Physiology | Revealed® **Cardiovascular, Lymphatic, and Respiratory Systems** *CD, or, if you are in the* **Dissection section**, *click on the* **ANIMATIONS** *button at the bottom of the screen and skip the next two steps.*

• *In the* **Select system** *menu at the top left of the screen select* **Cardiovascular**.

• *In the* **Home screen**, *click on the* **Animation** *button in the left portion of the screen.*

• *In the* **Select topic** *menu, select* **Physiology**.

• *From the* **Select animation** *menu, select* **Conducting system of heart**.

• *Click the* **Play** *button, and the animation will run in the* **IMAGE AREA**.

• *After viewing the animation, answer the following questions:*

1. Where do action potentials associated with heartbeat regulation originate?

2. From their origin, the action potentials travel across the wall of the atrium and to the

   _____.

3. Name the structure in the interventricular septum where the action potentials pass upon leaving the atria.

4. The structure in question 3 then divides into two _____.

5. After passing the apex of the ventricles, the action potentials pass along what structure to the ventricle walls?

6. What causes the ventricular muscle cells to contract in unison, providing a strong contraction?

Animation: Cardiac Cycle

- *Return to the* **ANIMATION LIST**, *and click on the second menu down, the* **Select animation menu.**
- *Select* **Cardiac cycle.**
- *Click the* **Play** *button, and the animation will run in the* **IMAGE AREA.**
- *After viewing the animation, answer the following questions:*

1. A single cardiac cycle is made up of

   _____

   _____.

2. What is the term that refers to the relaxation of a heart chamber?

3. What physically occurs in a heart chamber during relaxation?

4. The term that refers to the contraction of a heart chamber is _____.

5. Atrial depolarization is represented by which wave on an electrocardiogram?

6. What is initiated by atrial depolarization?

7. This phenomenon is represented by what portion of the ECG?

8. Which section of the ECG represents ventricular depolarization?

9. This same section of the ECG in question 8 masks what portion of the cardiac cycle?

10. What is S1? How is it often described?

11. What is represented by the S-T segment of the ECG?

12. The T-wave on the ECG represents _____.

13. What happens in the ventricles at this time?

14. What is S2? How is it often described?

15. What causes the beginning of the next cardiac cycle?

IN REVIEW

What Have I Learned?

The following questions cover the material that you have just learned—the heart. Use the information in the **STRUCTURE INFORMATION** window for the heart to answer these questions:

1. Name the body areas drained by the superior vena cava.

2. Name the small pouchlike extensions of the atria.

3. Name the vessel that conveys oxygen-poor blood from the right ventricle.

4. Name the three blood vessels that drain into the right atrium.

5. What external structure of the heart marks the position of the junction between the atria and the ventricles?

6. What external structure of the heart marks the position of the interventricular septum?

7. What is the term that refers to the blunt tip of the left ventricle?

8. Name the conical elevations of myocardium in the ventricular walls. What is their function?

10. Name the muscle layer of the heart wall.

9. Name the remnant of the fetal blood shunt from the right atrium to the left atrium.

Imaging

Take the time to view the images of the cardiovascular system available on the *Anatomy & Physiology | Revealed*® CD at this time.

The Thorax

EXERCISE 5.4: Cardiovascular System, Thorax, Arteries, Anterior View

- *Insert* Anatomy & Physiology | Revealed® **Cardiovascular, Lymphatic, and Respiratory Systems** *CD, or, if you are already in the* **Dissection** *section, click the* **CHANGE VIEW** *button at the top of the screen, and skip the next two steps.*

- *From the* **Select system** *menu, select* **Cardiovascular**.

- *In the* **Home screen**, *select the* **Dissection** *button in the left portion of the screen. You may click either on the* **Dissection button** *or on the word.*

- *In the* **SELECT A VIEW** *window that appears, click on the* **Select topic** *button.*

- *Choose* **Thorax** *from the menu.*

- *Click the* **Select view** *menu and choose* **Arteries - anterior**.

- *The* **GO** *button will flash green. Click on it.*

- *Click on* **TAG 1**, *and the image at top right will appear:*

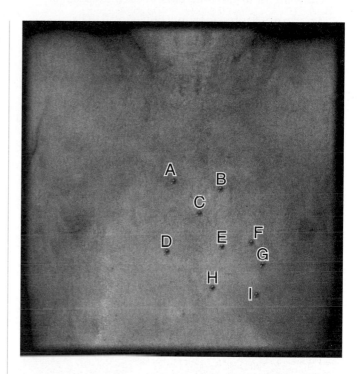

- *Mouse-over the blue pins on the screen to find the information necessary to fill in the following blanks:*

A. _____

B. _____

C. _____

D. _____

E. _____

F. _____

G. _____

H. _____

I. _____

CHECK POINT:

Thorax, Arteries, Anterior View

1. Where do heart valve sounds resonate?
2. What structure forms the right margin of the heart?
3. What two structures form the inferior margin of the heart?

HEADS UP!

Click on all of the blue pins in each exercise for Anatomy & Physiology | Revealed®. *The* **"What Have I Learned?"** *questions will cover the information for all of the blue pins, not just the ones that you identify.*

- *Click on* **TAG 2**, and the following image will appear:

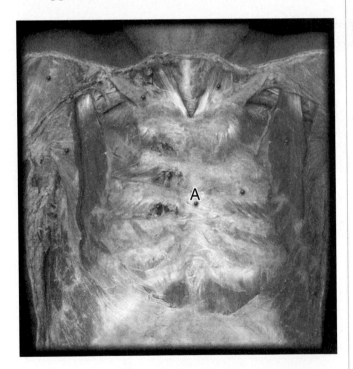

- *Mouse-over the blue pin on the screen to find the information necessary to fill in the following blank:*

A. _____

CHECK POINT:

Thorax, Arteries, Anterior View

4. What artery is also called the internal mammary artery?

- *Click on* **TAG 3**, and the following image will appear:

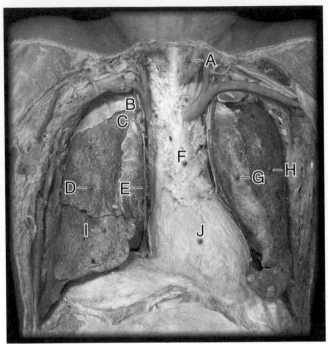

- *Mouse-over the blue pins on the screen to find the information necessary to fill in the following blanks:*

A. _____

B. _____

C. _____

D. _____

E. _____

F. _____

G. _____

H. _____

I. _____

J. _____

Thorax, Arteries, Anterior View

5. Name the artery that descends adjacent to the sternum within the thoracic cavity.

Animation: Pulmonary and Systemic Circulation

- *Click on the* **ANIMATIONS** *button at the bottom of the screen.*

- *In the* **Select topic** *menu, select* **Anatomy**.

- *From the* **Select animation** *menu, select* **Pulmonary and systemic circulation**.

- *Click the* **Play** *button, and the animation will run in the* **IMAGE AREA**.

- *After viewing the animation, answer the following questions:*

1. Name the two divisions of the cardiovascular system.

2. What are the destinations of these two circuits?

3. In the systemic circulation, where does gas exchange occur?

4. In the pulmonary circulation, where does gas exchange occur?

5. Name the blood vessels that carry oxygen-rich blood to the heart. How many are there? Where do they terminate?

- *Click on the* **DISSECTION** *button at the bottom of the screen to return to the dissection view, and click on* **TAG 4** *to view the image at top right.*

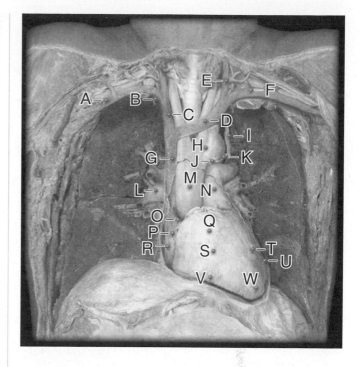

- *Mouse-over the blue pins on the screen to find the information necessary to fill in the following blanks:*

A. _____

B. _____

C. _____

D. _____

E. _____

F. _____

G. _____

H. _____

I. _____

J. _____

K. _____

L. _____

M. _____

N. _____

O. _____

P. _____

Q. _____

R. _____

S. _____

T. _____

U. _____

V. _____

W. _____

G. _____

H. _____

I. _____

J. _____

K. _____

CHECK POINT:

Thorax, Arteries, Anterior View

6. Name the large artery that originates at the aortic valve and ascends 5 mm within the pericardium.
7. Name the arched continuation of the artery in question 6.
8. Name the three branches of the artery in question 7.

- *Click on* **TAG 5**, *and the following image will appear:*

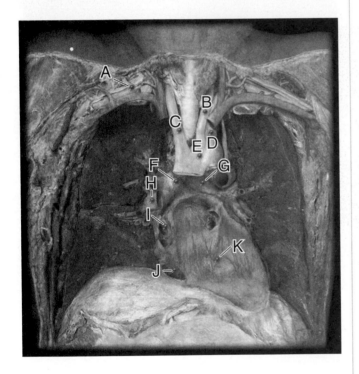

- *Mouse-over the blue pins on the screen to find the information necessary to fill in the following blanks:*

A. _____

B. _____

C. _____

D. _____

E. _____

F. _____

CHECK POINT:

Thorax, Arteries, Anterior View

9. Name the blood vessel also known as the innominate artery. What two vessels are found at its terminus?
10. Name the second artery to branch off of the aortic arch. What are its terminal branches?
11. Name the third artery to branch off of the aortic arch. Name the vessel that is the continuation of this artery beginning at the lateral border of rib 1.

- *Click on* **TAG 6**, *and the following image will appear:*

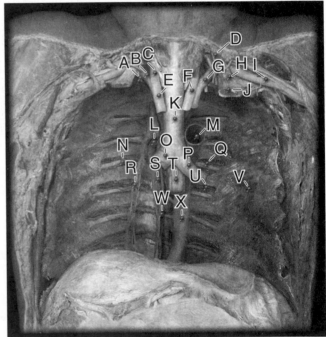

- *Mouse-over the blue pins on the screen to find the information necessary to fill in the following blanks:*

A. _____

B. _____

C. _____

D._____

E._____

F._____

G._____

H._____

I._____

J._____

K._____

L._____

M._____

N._____

O._____

P._____

Q._____

R._____

S._____

T._____

U._____

V._____

W._____

X._____

CHECK POINT:

Thorax, Arteries, Anterior View

12. Name the major artery that passes through the axilla.
13. Name the artery that ascends into the neck from the brachiocephalic trunk.
14. What are the two terminal branches of the artery in question 13?

EXERCISE 5.5: Cardiovascular System, Thorax, Veins, Anterior View

- *Insert* Anatomy & Physiology | Revealed® **Cardiovascular, Lymphatic, and Respiratory Systems** CD, or, if you are already in the **Dissection** section, click the **CHANGE VIEW** button at the top of the screen, and skip the next two steps.

- *From the* **Select system** *menu, select* **Cardiovascular**.

- *In the* **Home screen**, *select the* **Dissection** *button in the left portion of the screen. You may click either on the* **Dissection** *button or on the word.*

- *In the* **SELECT A VIEW** *window,* **Thorax** *appears in the* **Select topic** *button.*

- *Click the* **Select view** *menu, and choose* **Veins - anterior**.

- *The* **GO** *button will now flash green. Click on it.*

- *Click on* **TAG 2**, and the following image will appear:

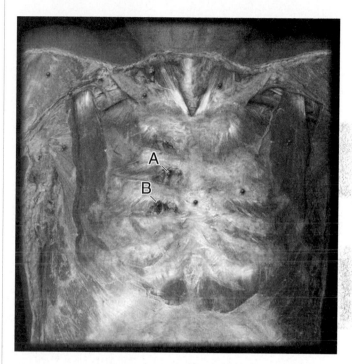

- *Mouse-over the blue pins on the screen to find the information necessary to fill in the following blanks:*

A._____

B._____

CHECK POINT:

Thorax, Veins, Anterior View

1. Name the vein that ascends from the dorsum of the hand to the anterolateral forearm and arm and into the deltopectoral triangle.
2. Name the vein also known as the internal mammary vein.

- *Click on* **TAG 3**, *and the following image will appear:*

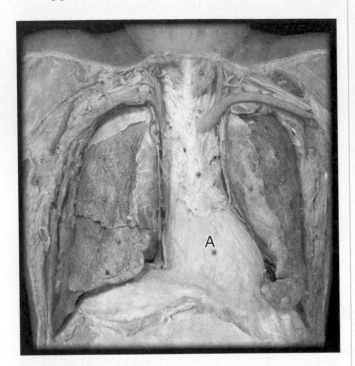

- *Mouse-over the blue pin on the screen to find the information necessary to fill in the following blank:*

A. _____

- *Click on* **TAG 4**, *and the following image will appear:*

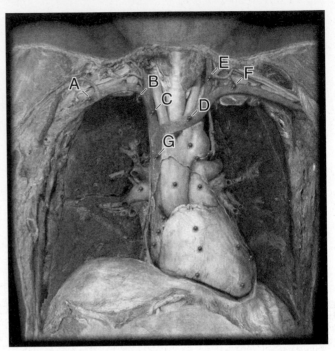

- *Mouse-over the blue pins on the screen to find the information necessary to fill in the following blanks:*

A. _____

B. _____

C. _____

D. _____

E. _____

F. _____

G. _____

CHECK POINT:

Thorax, Veins, Anterior View

4. Name the areas drained by the superior vena cava. Where does it terminate?
5. What two vessels unite to form the superior vena cava?
6. Name the continuation of the sigmoid dural sinus of the cranial cavity that drains the brain, face, and neck.

• *Click on* **TAG 5**, *and the following image will appear:*

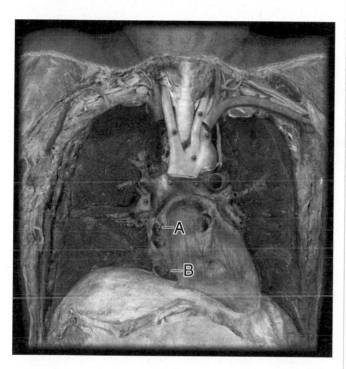

• *Mouse-over the blue pins on the screen to find the information necessary to fill in the following blanks:*

A. _____

B. _____

CHECK POINT:

Thorax, Veins, Anterior View

7. Name the vessels that carry oxygen-rich blood to the heart.
8. Where do they terminate?
9. Name the drainage of the inferior vena cava. Where does it terminate?

• *Click on* **TAG 6**, *and the following image will appear:*

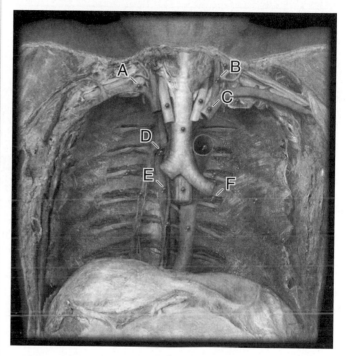

• *Mouse-over the blue pins on the screen to find the information necessary to fill in the following blanks:*

A. _____

B. _____

C. _____

D. _____

E. _____

F. _____

Thorax, Veins, Anterior View

10. Name the venous system that forms a collateral pathway between the superior and inferior vena cavae.

HEADS UP! ———————

View the radiographic images for the cardiovascular system.

- *Click on the* **IMAGING** *button at the bottom of the screen.*
- *Click on the* **Select topic** *menu, and select* **Thorax**.
- *Click the flashing* **GO** *button.*
- *Click on the* **Select structure** *menu in the* **STRUCTURE LIST** *box and then on the individual structures to highlight them in the* **IMAGE AREA**.

IN REVIEW

What Have I Learned?

The following questions cover the material that you have just learned—the blood vessels of the thorax. Use the information in the **STRUCTURE INFORMATION** window for these vessels to answer these questions:

1. What is the name for the fibrous sac that encloses the heart?

2. Name the lymphatic organ that is large in children but atrophies during adolescence.

3. Name the bilobed endocrine gland located lateral to the trachea and larynx.

4. How do large arteries supply blood to body structures?

5. Name the large vessel that conveys oxygen-poor blood from the right ventricle of the heart.

6. Name the two branches of the above blood vessel that convey oxygen-poor blood to the lungs.

7. Name the blunt tip of the left ventricle.

8. What is the carotid sheath? What structures are found within it?

9. What is the serous pericardium?

10. Name the structure that served to shunt blood in the fetus from the pulmonary artery to the aorta, bypassing the lungs. What is it called in the adult?

11. Name all of the different sections of the aorta, and describe where they are found.

12. Name the two major vessels that return oxygen-poor blood to the heart. What are the drainages for each? Where do they terminate?

The Head and Neck

EXERCISE 5.6: Cardiovascular System - Head and Neck, Vasculature, Lateral View

- *Insert* Anatomy & Physiology | Revealed® **Cardiovascular, Lymphatic, and Respiratory Systems** *CD, or, if you are already in the* **Dissection** *section, click the* **CHANGE VIEW** *button at the top of the screen, and skip the next two steps.*

- *From the* **Select system** *menu, select* **Cardiovascular**.

- *In the* **Home screen**, *select the* **Dissection** *button in the left portion of the screen. You may click either on the* **Dissection** *button or on the word.*

- *In the* **SELECT A VIEW** *window that appears, click on the* **Select topic** *button.*

- *Choose* **Head and neck** *from the menu.*

- **Vasculature - lateral** *will appear in the* **Select view** *menu.*

- *The* **GO** *button will flash green. Click on it.*

- *Click on* **TAG 1**, *and the following screen will appear:*

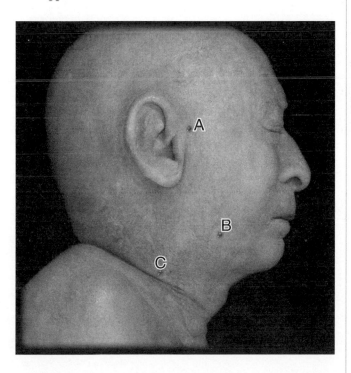

- *Mouse-over the blue pins on the screen to find the information necessary to fill in the following blanks:*

A. _____

B. _____

C. _____

CHECK POINT:

Head and Neck, Vasculature, Lateral View

1. What blood vessel is responsible for the superficial temporal pulse?
2. What blood vessel is responsible for the facial pulse?
3. Between which two structures of the neck would you find the carotid pulse?

- *Click on* **TAG 2**, *and the following image will appear:*

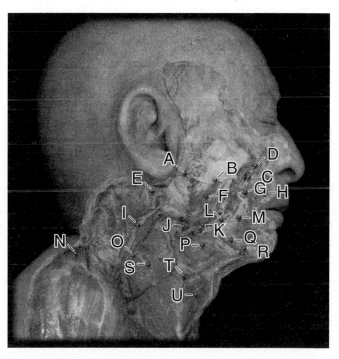

- *Mouse-over the blue pins on the screen to find the information necessary to fill in the following blanks:*

A. _____

B. _____

C. _____

D. _____

E. _____

F. _____

G. _____

H. _____

I. _____

J. _____

K. _____

L. _____

M. _____

N. _____

O. _____

P. _____

Q. _____

R. _____

S. _____

T. _____

U. _____

CHECK POINT:

Head and neck, Vasculature, Lateral View

4. Name a subcutaneous vein that drains the scalp and the face and terminates in the subclavian vein.
5. Name a vein, the size of which is inversely proportional to the vein in question 4.
6. Name an artery that supplies the muscles, glands, and mucous membranes of the lower lip.

• *Click on* **TAG 3**, *and the following image will appear:*

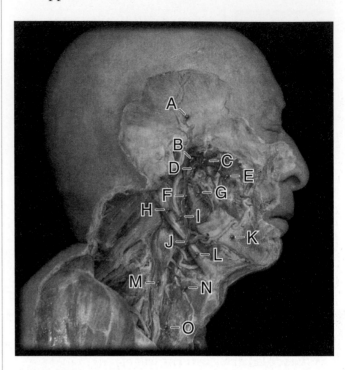

• *Mouse-over the blue pins on the screen to find the information necessary to fill in the following blanks:*

A. _____

B. _____

C. _____

D. _____

E. _____

F. _____

G. _____

H. _____

I. _____

J. _____

K. _____

L. _____

M. _____

N. _____

O. _____

CHECK POINT:

Head and neck, Vasculature, Lateral View

7. Name the large vein that drains the cranial cavity, including the brain, as well as the face and neck.
8. Name a vein that drains the superior molar and premolar teeth and the mucous membrane of the maxillary sinus.
9. Name a vein that drains the deep face and temporal region and descends within the parotid gland.

- *Click on* **TAG 4**, *and the following image will appear:*

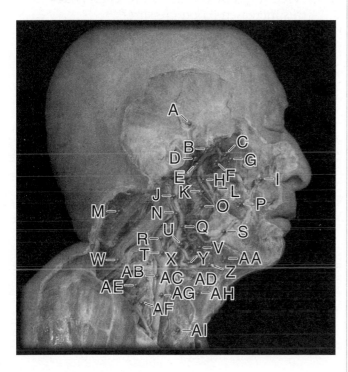

- *Mouse-over the blue pins on the screen to find the information necessary to fill in the following blanks:*

A. _____

B. _____

C. _____

D. _____

E. _____

F. _____

G. _____

H. _____

I. _____

J. _____

K. _____

L. _____

M. _____

N. _____

O. _____

P. _____

Q. _____

R. _____

S. _____

T. _____

U. _____

V. _____

W. _____

X. _____

Y. _____

Z. _____

AA. _____

AB. _____

AC. _____

AD. _____

AE. _____

AF. _____

AG. _____

AH. _____

AI. _____

CHECK POINT:

Head and neck, Vasculature, Lateral View

10. Name the major artery of the neck where the left one originates from the aortic arch and the right one originates from the brachiocephalic trunk.
11. Name the branch of the artery in question 10 that distributes to the exterior of the head (except the orbit), the face, meninges, and neck structures.
12. Name the branch of the artery in question 10 that enters the cranial cavity and terminates in the cerebral arterial circle.

• *Click on* **TAG 5**, and the following image will appear:

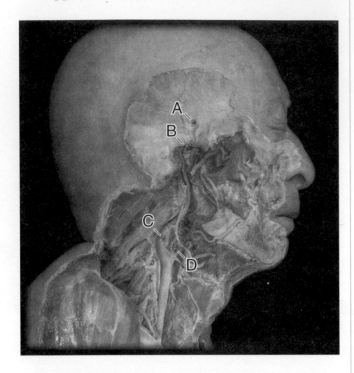

• *Mouse-over the blue pins on the screen to find the information necessary to fill in the following blanks:*

A. _____

B. _____

C. _____

D. _____

C H E C K   P O I N T :

Head and neck, Vasculature, Lateral View

13. Name the structure of the skull that contains the internal carotid artery and the carotid plexus of sympathetic nerves.

Animation: Baroreceptor Reflex

• *Click on the* **ANIMATIONS** *button at the bottom of the screen.*

• *In the* **Select topic** *menu, select* **Physiology**.

• *From the* **Select animation** *menu, select* **Baroreceptor reflex**.

• *Click the* **Play** *button, and the animation will run in the* **IMAGE AREA**.

• *After viewing the animation, answer the following questions:*

1. Where are baroreceptors located?

2. What is their function?

3. How do the arteries with baroreceptors, and all arteries for that matter, respond to increased blood pressure?

4. What response do the baroreceptors have to this increased blood pressure?

5. Where are these action potentials conducted?

6. What nerves conduct these impulses?

7. What is the parasympathetic response to and result of this stimulation?

8. What is the sympathetic response and result?

9. What about sympathetic stimulation of the blood vessels?

10. What physical events combine to bring elevated blood pressure back toward normal?

Animation: Chemoreceptor Reflex

• *Return to the animation menu at the top left of the screen, and from the* **Select animation** *menu, select* **Chemoreceptor reflex**.

• *Click the* **Play** *button, and the animation will run in the* **IMAGE AREA**.

• *After viewing the animation, answer the following questions:*

1. Where are chemoreceptors located?

2. What is their function?

3. Where are the impulses from the chemoreceptors conducted?

4. What nerves conduct these impulses?

5. What three events decrease parasympathetic stimulation of the heart?

6. What effect does this decreased stimulation have on the physiology of the heart?

7. What sympathetic response occurs during the three events listed in question 5?

8. What sympathetic response occurs in the blood vessels?

9. If the chemoreceptors are stimulated by decreased blood oxygen, what physical changes occur as a result of the changes in autonomic stimulation?

10. What would you deduce is the effect on blood pressure as a result of the answer to question 9?

HEADS UP!

View the radiographic images for the cardiovascular system.

- *Click on the* **IMAGING** *button at the bottom of the screen.*
- *Click on* **CHANGE REGION.** *Select* **Head and neck**.
- *Click the flashing* **GO** *button.*
- *Click on the* **Select view** *menu and select each view.*
- *Click on the structures in the* **Select structure** *menu located in the* **STRUCTURE LIST** *box to highlight them in the* **IMAGE AREA**.

IN REVIEW

What Have I Learned?

The following questions cover the material that you have just learned—the blood vessels of the head and neck. Use the information in the **STRUCTURE INFORMATION** window for these vessels to answer the following questions:

1. Name the gland that produces 25 percent of your saliva.

2. List three structures that pass through this gland.

3. Where does the duct from this gland empty?

4. Name two veins of the head and neck that lack valves.

5. Name a superficial vein that drains the temporal region.

6. List the areas drained by the maxillary veins.

7. What is an anastomosis? Give an example of one in the head and neck region.

The Brain

EXERCISE 5.7:   Cardiovascular System— Brain, Arteries, Inferior View

- *Insert* Anatomy & Physiology | Revealed® **Cardiovascular, Lymphatic, and Respiratory Systems** *CD, or, if you are already in the* **Dissection** *section, click the* **CHANGE VIEW** *button at the top of the screen, and skip the next two steps.*

- *From the* **Select system** *menu, select* **Cardiovascular**.

- *In the* **Home screen**, *select the* **Dissection** *button in the left portion of the screen. You may click either on the* **Dissection** *button or on the word.*

- *In the* **SELECT A VIEW** *window that appears, click on the* **Select topic** *button.*

- *Choose* **Brain** *from the menu.*

- *Click the* **Select view** *menu, and choose* **Arteries - inferior**.

- *The* **GO** *button will flash green. Click on it.*

- *Click on* **TAG 1**, *and the following image will appear:*

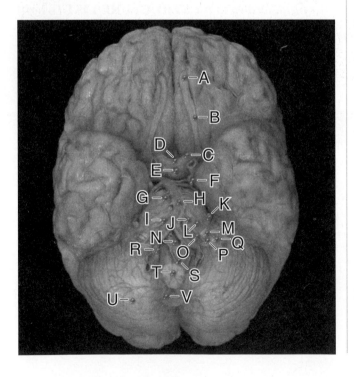

- *Mouse-over the blue pins on the screen to find the information necessary to fill in the following blanks:*

A. _____

B. _____

C. _____

D. _____

E. _____

F. _____

G. _____

H. _____

I. _____

J. _____

K. _____

L. _____

M. _____

N. _____

O. _____

P. _____

Q. _____

R. _____

S. _____

T. _____

U. _____

V. _____

C H E C K   P O I N T :

Brain, Arteries, Inferior View

1. The central branches of which artery supply the anterior two-thirds of the spinal cord?
2. Name the artery that distributes to the posterior cerebellum and lateral medulla oblongata.
3. Name the artery that distributes to the inferior cerebellum, pons, and medulla oblongata.

Animation: Blood Flow Through Brain

- *Click on the **ANIMATIONS** button at the bottom of the screen.*

- *In the **Select topic** menu, select **Anatomy**.*

- *From the **Select animation** menu, select **Blood flow through brain**.*

- *Click the **Play** button, and the animation will run in the **IMAGE AREA**.*

- *After viewing the animation, answer the following questions:*

1. Name the three arteries through which blood flows to the brain.

2. Which two arteries supply blood to 80 percent of the cerebrum? These are the terminal branches of which arteries?

3. Which arteries enter the cranial cavity through the foramen magnum? They unite to form which artery?

4. Which arteries supply the occipital and temporal lobes of the cerebrum?

5. Which arteries form an anastomosis at the base of the brain? What is the name of this anastomosis?

6. What is the function of the anastomosis?

7. Blood is drained from the brain through small veins that empty into vessel channels called _____.

8. From where does blood flow to enter the confluence of sinuses?

9. From the confluence of sinuses, where does blood flow before leaving the skull?

10. The blood leaves the skull via the _____.

- *Click on the **DISSECTION button** at the bottom of the screen, and continue by clicking on **TAG 2,** and the following image will appear:*

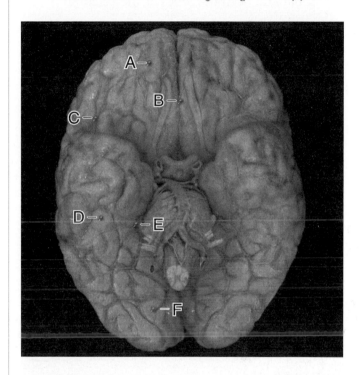

- *Mouse-over the blue pins on the screen to find the information necessary to fill in the following blanks:*

A. _____

B. _____

C. _____

D. _____

E. _____

F. _____

Brain, Arteries, Inferior View

4. Name the artery that distributes to the medial aspect of the frontal lobes of the cerebral cortex.
5. Name the artery that distributes to the lateral aspects of the frontal, parietal, and temporal lobes.
6. Name the artery that distributes to the temporal and occipital lobes of the cerebral cortex.

- *Click on* **TAG 3**, *and the following image will appear:*

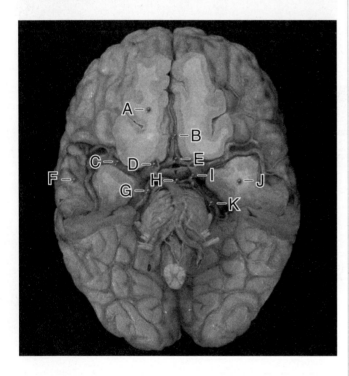

- *Mouse-over the blue pins on the screen to find the information necessary to fill in the following blanks:*

A. _____

B. _____

C. _____

D. _____

E. _____

F. _____

G. _____

H. _____

I. _____

J. _____

K. _____

Brain, Arteries, Inferior View

7. Name the arterial anastomosis of the brain complete in only 20 percent of individuals.
8. Name the paired arteries that course along each lateral sulcus of the cerebral hemisphere.
9. Name the cerebral lobe not visible from the surface, also known as the isle of Reil.

EXERCISE 5.8:   Cardiovascular System— Brain, Veins, Lateral View

- *Insert* Anatomy & Physiology | Revealed® **Cardiovascular, Lymphatic, and Respiratory Systems** *CD, or, if you are already in the* **Dissection** *section, click the* **CHANGE VIEW** *button at the top of the screen, and skip the next two steps.*

- *From the* **Select system** *menu, select* **Cardiovascular**.

- *In the* **Home screen**, *select the* **Dissection** *button in the left portion of the screen. You may click either on the* **Dissection** *button or on the word.*

- *In the* **SELECT A VIEW** *window that appears, click on the* **Select topic** *button.*

- *Choose* **Brain** *from the menu, if it is not already selected.*

- *Click the* **Select view** *menu and choose* **Veins - lateral**.

- *The* **GO** *button will flash green. Click on it.*

- *Click on* **TAG 3**, *and the following image will appear:*

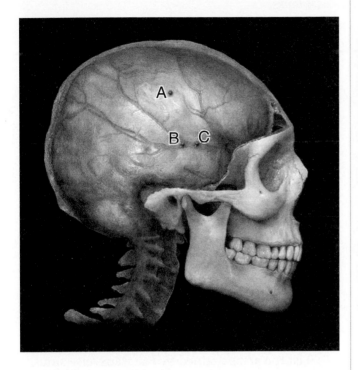

- *Mouse-over the blue pins on the screen to find the information necessary to fill in the following blanks:*

A. _____

B. _____

C. _____

CHECK POINT:

Brain, Veins, Lateral View

1. Name the most exterior of the meninges. Where is it located? What is its function?
2. Name the artery whose distribution includes the dura mater, skull, trigeminal, and facial ganglia.
3. Name the vein that drains the dura mater and bones of the anterior and middle cranial fossae.

- *Click on* **TAG 4**, *and the following image will appear:*

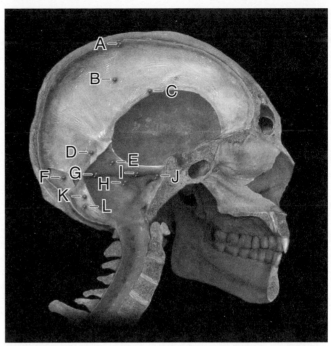

- *Mouse-over the blue pins on the screen to find the information necessary to fill in the following blanks:*

A. _____

B. _____

C. _____

D. _____

E. _____

F. _____

G. _____

H. _____

I. _____

J. _____

K. _____

L. _____

CHECK POINT:

Brain, Veins, Lateral View

4. What are dural venous sinuses?
5. Name the structure located in the internal occipital protuberance that drains the superior sagittal, straight, and occipital sinuses.
6. Name the sinus that drains the cerebellum.

- *Click on* **TAG 5**, and the following image will appear:

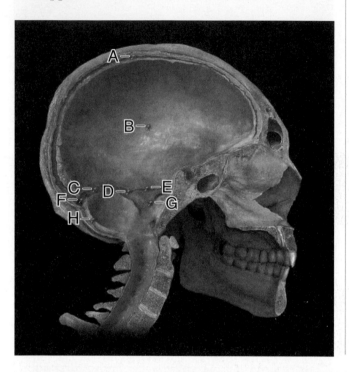

- *Mouse-over the blue pins on the screen to find the information necessary to fill in the following blanks:*

A. _____

B. _____

C. _____

D. _____

E. _____

F. _____

G. _____

H. _____

CHECK POINT:

Brain, Veins, Lateral View

7. Name the s-shaped continuation of the transverse sinus.
8. Name a paired dural venous sinus that terminates at the jugular foramen.
9. List the areas drained by the transverse sinus.

IN REVIEW

What Have I Learned?

The following questions cover the material that you have just learned—the blood vessels of the brain. Use the information in the **STRUCTURE INFORMATION** window for these vessels to answer the following questions:

1. Name the large crescent-shaped fold of dura mater that separates the right and left cerebral hemispheres.

2. Name the small crescent-shaped fold of dura mater that separates the right and left cerebellar hemispheres.

3. Name the dural venous sinus that contains arachnoid granulations. What is the function of these granulations?

4. Name the dural venous sinus that drains the medial cerebral hemispheres.

5. Name the horizontal crescent-shaped fold of the dura mater that separates the cerebral hemispheres and the cerebellum.

The Shoulder

EXERCISE 5.9:   The Shoulder, Arteries, Anterior View

- *Insert* Anatomy & Physiology | Revealed® **Cardiovascular, Lymphatic, and Respiratory Systems** *CD, or, if you are already in the* **Dissection** *section, click the* **CHANGE VIEW** *button at the top of the screen, and skip the next two steps.*

- *From the* **Select system** *menu, select* **Cardiovascular**.

- *In the* **Home screen**, *select the* **Dissection** *button in the left portion of the screen. You may click either on the* **Dissection** *button or on the word.*

- *In the* **SELECT A VIEW** *window that appears, click on the* **Select topic** *button.*

- *Choose* **Shoulder** *from the menu.*

- *Click the* **Select view** *menu and choose* **Arteries - anterior**.

- *The* **GO** *button will flash green. Click on it.*

- *Click on* **TAG 2**, *and the following image will appear:*

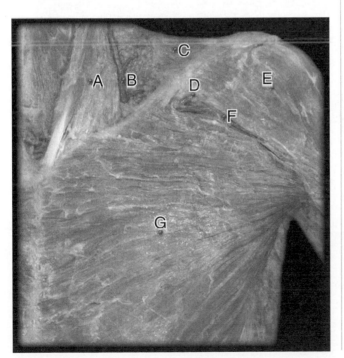

- *Mouse-over the blue pins on the screen to find the information necessary to fill in the following blanks:*

A. _____

B. _____

C. _____

D. _____

E. _____

F._____

G. _____

- *Click on* **TAG 3**, *and the following image will appear:*

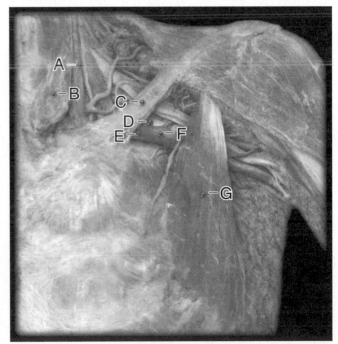

• *Mouse-over the blue pins on the screen to find the information necessary to fill in the following blanks:*

A. _____

B. _____

C. _____

D. _____

E. _____

F. _____

G. _____

CHECK POINT:

Shoulder, Arteries, Anterior View

4. Name the large vein within the carotid sheath that is a continuation of the sigmoid dural sinus of the cranial cavity.
5. Which two veins unite to form the brachio-cephalic vein?
6. Where does the subclavian vein become the axillary vein?

• *Click on* **TAG 4**, *and the following image will appear:*

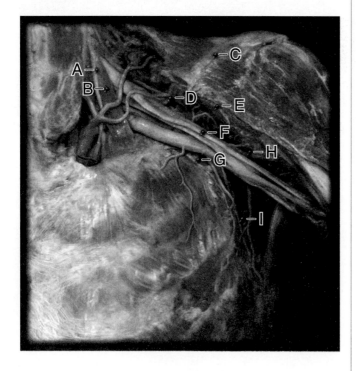

• *Mouse-over the blue pins on the screen to find the information necessary to fill in the following blanks:*

A. _____

B. _____

C. _____

D. _____

E. _____

F. _____

G. _____

H. _____

I. _____

• *Click on* **TAG 5**, *and the following image will appear:*

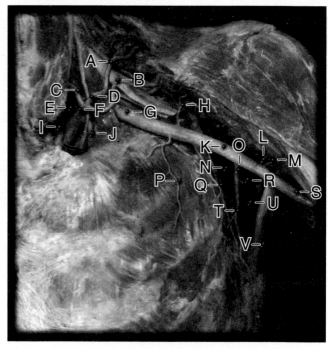

• *Mouse-over the blue pins on the screen to find the information necessary to fill in the following blanks:*

A. _____

B. _____

C. _____

D. _____

E. _____

F. _____

G._____

H._____

I._____

J._____

K._____

L._____

M._____

N._____

O._____

P._____

Q._____

R._____

S._____

T._____

U._____

V._____

CHECK POINT:

Shoulder, Arteries, Anterior View

7. What is the definition of a "trunk"? (This one will require inference from the information available in the **STRUCTURE INFORMATION** window.)
8. Name a trunk that ascends in the lower neck.
9. Where does the subclavian artery become the axial artery? Where does the axillary artery become the brachial artery?

EXERCISE 5.10:   Cardiovascular System—
Shoulder, Veins,
Anterior View

- *Insert* Anatomy & Physiology | Revealed® **Cardiovascular, Lymphatic, and Respiratory Systems** *CD, or, if you are already in the* **Dissection** *section, click the* **CHANGE VIEW** *button at the top of the screen, and skip the next two steps.*

- *From the* **Select system** *menu, select* **Cardiovascular**.

- *In the* **Home screen***, select the* **Dissection** *button in the left portion of the screen. You may click either on the* **Dissection** *button or on the word.*

- *In the* **SELECT A VIEW** *window that appears, click on the* **Select topic** *button.*

- *Choose* **Shoulder** *from the menu, if it is not already selected.*

- *Click the* **Select view** *menu and choose* **Veins - anterior**.

- *The* **GO** *button will flash green. Click on it.*

- *Click on* **TAG 4***, and the following screen will appear:*

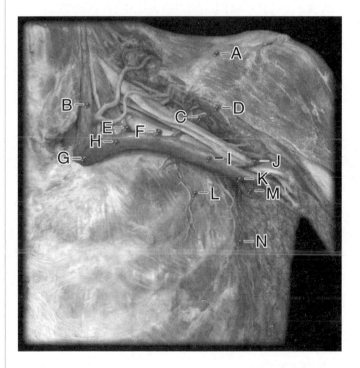

- *Mouse-over the blue pins on the screen to find the information necessary to fill in the following blanks:*

A._____

B._____

C._____

D._____

E._____

F._____

G._____

H._____

I._____

J._____

K._____

L._____

M._____

N._____

CHECK POINT:

Shoulder, Veins, Anterior View

1. List the areas drained by the left brachio-cephalic vein. What is the other name for this vein in Latin and in English?
2. Where does the vein in question 1 terminate?
3. List the areas drained by the axillary vein. What are its tributaries?

IN REVIEW

What Have I Learned?

The following questions cover the material that you have just learned—the blood vessels of the shoulder. Use the information in the **STRUCTURE INFORMATION** window for these structures to answer the following questions:

1. What is the deltopectoral triangle? What is another name for it?

2. Name three arteries that distribute to the muscles of the scapula.

3. Name the paired arteries that ascend through the neck via the transverse foramina in the cervical vertebrae.

4. Where do these paired arteries enter the cranial cavity?

The Shoulder and Arm

EXERCISE 5.11: Cardiovascular System - Shoulder and Arm, Arteries, Anterior View

- *Insert* Anatomy & Physiology | Revealed® **Cardiovascular, Lymphatic, and Respiratory Systems** *CD, or, if you are already in the* **Dissection** *section, click the* **CHANGE VIEW** *button at the top of the screen, and skip the next two steps.*

- *From the* **Select system** *menu, select* **Cardiovascular**.

- *In the* **Home screen**, *select the* **Dissection** *button in the left portion of the screen. You may click either on the* **Dissection** *button or on the word.*

- *In the* **SELECT A VIEW** *window that appears, click on the* **Select topic** *button.*

- *Choose* **Shoulder and arm** *from the menu.*

- *Click the* **Select view** *menu and choose* **Arteries - anterior**.

- *The* **GO** *button will flash green. Click on it.*

• *Click on* **TAG 1**, *and the following image will appear:*

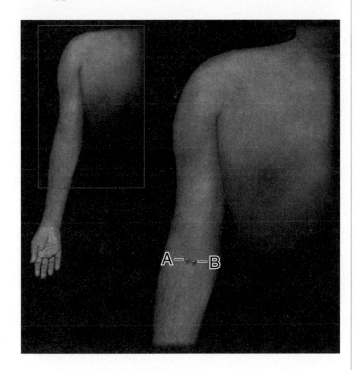

• *Mouse-over the blue pins on the screen to find the information necessary to fill in the following blanks:*

A. _____

B. _____

CHECK POINT:

Shoulder and Arm, Arteries, Anterior View

1. What is the cubital fossa?
2. What structures are found in the cubital fossa?
3. Where is the location for the stethoscope when taking blood pressure with a pressure cuff?

• *Click on* **TAG 3**, *and the following image will appear:*

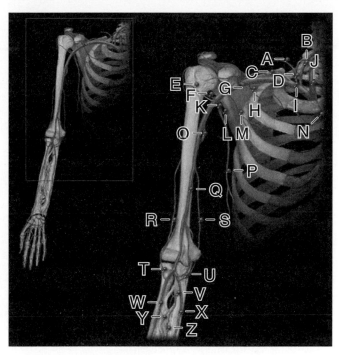

• *Mouse-over the blue pins on the screen to find the information necessary to fill in the following blanks:*

A. _____

B. _____

C. _____

D. _____

E. _____

F._____

G. _____

H. _____

I. _____

J._____

K. _____

L. _____

M. _____

N. _____

O. _____

P. _____

Q._____

R. _____

S. _____

T. _____

U. _____

V. _____

W. _____

X. _____

Y. _____

Z. _____

C H E C K   P O I N T :

Shoulder and Arm, Arteries, Anterior View

4. Name the major artery of the arm.
5. What two arteries branch from this artery at the anterior elbow?
6. What artery is also known as the *profunda brachii* artery? How does *profunda brachii* translate?

EXERCISE 5.12:   Cardiovascular System - Shoulder and Arm, Veins, Anterior View

- *Insert* Anatomy & Physiology | Revealed® **Cardiovascular, Lymphatic, and Respiratory Systems** *CD, or, if you are already in the* **Dissection** *section, click the* **CHANGE VIEW** *button at the top of the screen, and skip the next two steps.*

- *From the* **Select system** *menu, select* **Cardiovascular**.

- *In the* **Home screen**, *select the* **Dissection** *button in the left portion of the screen. You may click either on the* **Dissection** *button or on the word.*

- *In the* **SELECT A VIEW** *window that appears, click on the* **Select topic** *button.*

- *Choose* **Shoulder and arm** *from the menu, if it is not already selected.*

- *Click the* **Select view** *menu, and choose* **Veins - anterior**.

- *The* **GO** *button will flash green. Click on it.*

- *Click on* **TAG 1**, *and the following image will appear:*

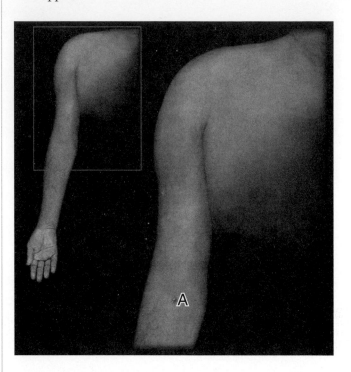

- *Mouse-over the blue pin on the screen to find the information necessary to fill in the following blank:*

A. _____

- *Click on* **TAG 2**, *and the following image will appear:*

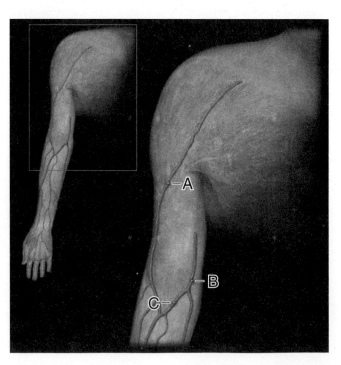

- *Mouse-over the blue pins on the screen to find the information necessary to fill in the following blanks:*

A. _____

B. _____

C. _____

CHECK POINT:

Shoulder and Arm, Veins, Anterior View

1. Which vein of the arm is frequently used for venipuncture? What is venipuncture?
2. What structures are drained by the basilica vein?
3. What structures are drained by the cephalic vein?

- *Click on **TAG 4**, and the following image will appear:*

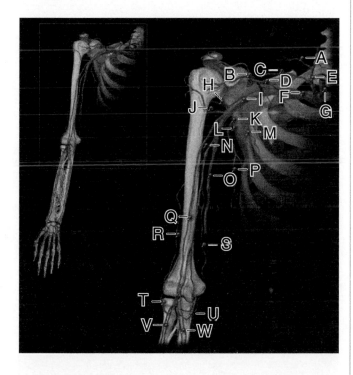

- *Mouse-over the blue pins on the screen to find the information necessary to fill in the following blanks:*

A. _____

B. _____

C. _____

D. _____

E. _____

F. _____

G. _____

H. _____

I. _____

J. _____

K. _____

L. _____

M. _____

N. _____

O. _____

P. _____

Q. _____

R. _____

S. _____

T. _____

U. _____

V. _____

W. _____

CHECK POINT:

Shoulder and Arm, Veins, Anterior View

1. Name the paired veins that ascend along the brachial artery.
2. Name the structures drained by the axillary vein. What are its major tributaries?
3. Name the large vein that terminates in the axillary artery.

IN REVIEW

What Have I Learned?

The following questions cover the material that you have just learned—the blood vessels of the shoulder and arm. Use the information in the **STRUCTURE INFORMATION** window for these structures to answer the following questions:

1. Name the two arteries that arise from the brachiocephalic trunk.

2. Which of the two ascends into the neck?

3. Name the two branches at the terminus of this artery.

4. Name the arteries that form an anastomosis around the elbow.

5. Name two arteries that form an anastomosis around the scapula.

6. Name the structures drained by the external jugular vein. What are its major tributaries?

7. Name the structures drained by the internal jugular vein. What are its major tributaries?

8. Name two locations where venous anastomoses occur in the shoulder and arm.

9. How do these venous anastomoses compare to arterial anastomoses?

10. Why?

The Forearm and Hand

EXERCISE 5.13: Cardiovascular System - Forearm and Hand, Arteries, Anterior View

- *Insert* Anatomy & Physiology | Revealed® **Cardiovascular, Lymphatic, and Respiratory Systems** *CD, or, if you are already in the* **Dissection** *section, click the* **CHANGE VIEW** *button at the top of the screen, and skip the next two steps.*

- *From the* **Select system** *menu, select* **Cardiovascular**.

- *In the* **Home screen**, *select the* **Dissection** *button in the left portion of the screen. You may click either on the* **Dissection** *button or on the word.*

- *In the* **SELECT A VIEW** *window that appears, click on the* **Select topic** *button.*

- *Choose* **Forearm and hand** *from the menu.*

- *Click the* **Select view** *menu, and choose* **Arteries - anterior**.

- *The* **GO** *button will flash green. Click on it.*

- *Click on* **TAG 1**, *and the following image will appear:*

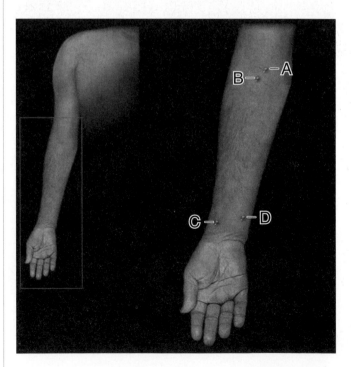

- *Mouse-over the blue pins on the screen to find the information necessary to fill in the following blanks:*

A. _____

B. _____

C. _____

D. _____

Forearm and Hand, Arteries, Anterior View

1. Name the location commonly used to take a pulse at the wrist.
2. Name another location for taking a pulse at the wrist.
3. Name another location for taking a pulse at the elbow.

- *Click on* **TAG 2**, *and the following image will appear:*

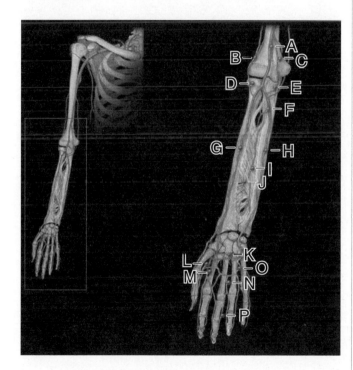

- *Mouse-over the blue pins on the screen to find the information necessary to fill in the following blanks:*

A. _____

B. _____

C. _____

D. _____

E. _____

F. _____

G. _____

H. _____

I. _____

J. _____

K. _____

L. _____

M. _____

N. _____

O. _____

P. _____

Forearm and Hand, Arteries, Anterior View

4. Name the two arteries that form an anastomosis through the superficial and deep palmar arch.
5. Name the artery that descends along the anterior aspect of the interosseous membrane of the forearm.
6. Name the artery commonly used for taking a pulse at the wrist.

EXERCISE 5.14:   Cardiovascular System - Forearm and Hand, Veins, Anterior View

- *Insert* Anatomy & Physiology | Revealed® **Cardiovascular, Lymphatic, and Respiratory Systems** *CD, or, if you are already in the* **Dissection** *section, click the* **CHANGE VIEW** *button at the top of the screen, and skip the next two steps.*

- *From the **Select system** menu, select **Cardiovascular**.*

- *In the **Home screen**, select the **Dissection** button in the left portion of the screen. You may click either on the **Dissection** button or on the word.*

- *In the **SELECT A VIEW** window that appears, click on the **Select topic** button.*

- *Choose **Forearm and hand** from the menu, if it is not already selected.*

- *Click the **Select view** menu and choose **Veins - anterior**.*

- *The **GO** button will flash green. Click on it.*

- *Click on **TAG 2**, and the following image will appear:*

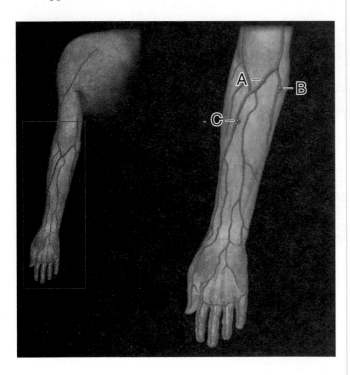

- *Mouse-over the blue pins on the screen to find the information necessary to fill in the following blanks:*

A. _____

B. _____

C. _____

CHECK POINT:

Forearm and Hand, Veins, Anterior View

1. Name the two veins that extend from the dorsal hand to the axillary vein.
2. Name the vein that has an oblique subcutaneous path between these two veins over the cubital fossa.

- *Click on **TAG 3**, and the following image will appear:*

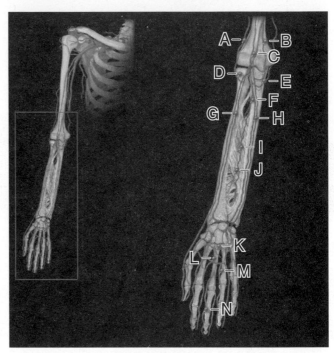

- *Mouse-over the blue pins on the screen to find the information necessary to fill in the following blanks:*

A. _____

B. _____

C. _____

D. _____

E. _____

F. _____

G. _____

H. _____

I. _____

J. _____

K. _____

L. _____

M. _____

N. _____

- *Click on the* **IMAGING** *button at the bottom of the screen.*
- *Click on the* **Select a topic** *menu, and select* **Hand**.
- *Click the flashing* **GO** *button.*
- *Click on the* **Select structure** *menu in the* **STRUCTURE LIST** *box and then on the individual structures to highlight them in the* **IMAGE AREA**.

CHECK POINT:

Forearm and Hand, Veins, Anterior View

3. Name the paired veins that ascend along the lateral aspect of the forearm.
4. Name the paired veins that ascend along the medial aspect of the forearm.
5. Name a set of paired veins of the forearm that are not always present.

IN REVIEW

What Have I Learned?

The following questions cover the material that you have just learned—the blood vessels of the forearm and hand. Use the information in the **STRUCTURE INFORMATION** window for these vessels to answer the following questions:

1. Describe the pathway of arterial blood flow from the arm through the elbow and lateral forearm to the palm of the hand.

2. Describe the pathway of arterial blood flow from the arm through the elbow and medial forearm to the palm of the hand.

3. Describe the pathway of arterial blood flow from the palm of the hand to the middle finger.

4. Describe the *superficial* pathway of venous blood drainage from the anterior forearm to the elbow.

5. Describe the *deep* pathway of venous blood drainage from the middle finger to the anterior elbow and forearm.

The Abdomen

EXERCISE 5.15: Cardiovascular System - Abdomen, Celiac Trunk, Anterior View

- *Insert* Anatomy & Physiology | Revealed® **Cardiovascular, Lymphatic, and Respiratory Systems** *CD, or, if you are already in the* **Dissection** *section, click the* **CHANGE**

VIEW *button at the top of the screen, and skip the next two steps.*

- *From the* **Select system** *menu, select* **Cardiovascular**.
- *In the* **Home screen**, *select the* **Dissection** *button in the left portion of the screen. You may click either on the* **Dissection** *button or on the word.*

• *In the* **SELECT A VIEW** *window that appears, click on the* **Select topic** *button.*

• *Choose* **Abdomen** *from the menu.*

• *Click the* **Select view** *menu, and choose* **Celiac trunk - anterior**.

• *The* **GO** *button will flash green. Click on it.*

• *Click on* **TAG 1**, *and the following image will appear:*

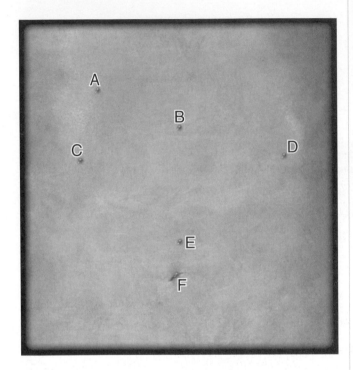

• *Mouse-over the blue pins on the screen to find the information necessary to fill in the following blanks:*

A. _____

B. _____

C. _____

D. _____

E. _____

F._____

CHECK POINT:

Abdomen, Celiac Trunk, Anterior View

1. Name the upper median abdominal region.
2. Name the two abdominal regions lateral to question 1.
3. Name the median abdominal region.

• *Click on* **TAG 2**, *and the following image will appear:*

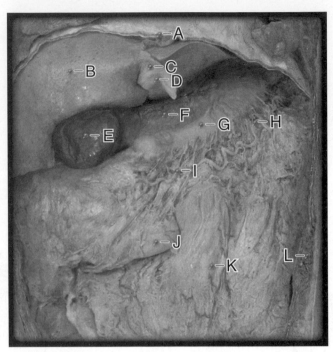

• *Mouse-over the blue pins on the screen to find the information necessary to fill in the following blanks:*

A. _____

B. _____

C. _____

D. _____

E. _____

F._____

G. _____

H._____

I._____

J._____

K. _____

L. _____

CHECK POINT:

Abdomen, Celiac Trunk, Anterior View

4. Name the structure that is the remnant of the umbilical vein of the fetus.
5. Name the anastomosis that has distribution to the stomach and the greater omentum.
6. Once again, what is an anastomosis?

- *Click on* **TAG 3**, *and the following image will appear:*

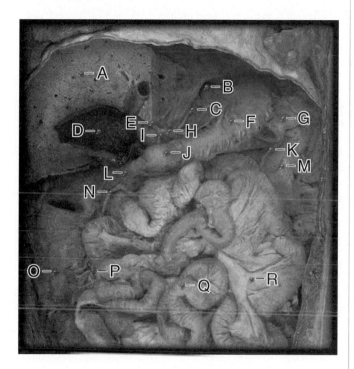

- *Mouse-over the blue pins on the screen to find the information necessary to fill in the following blanks:*

A. _____

B. _____

C. _____

D. _____

E. _____

F. _____

G. _____

H. _____

I. _____

J. _____

K. _____

L. _____

M. _____

N. _____

O. _____

P. _____

Q. _____

R. _____

CHECK POINT:

Abdomen, Celiac Trunk, Anterior View

7. Name two arteries that originate at the celiac trunk.
8. What is the smallest branch of the celiac artery? What is its distribution?
9. Name the large, highly vascular, accessory digestive organ. What are its four lobes?

- *Click on* **TAG 4**, *and the following image will appear:*

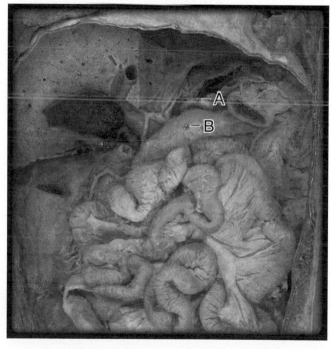

- *Mouse-over the blue pins on the screen to find the information necessary to fill in the following blanks:*

A. _____

B. _____

CHECK POINT:

Abdomen, Celiac Trunk, Anterior View

10. Name the artery that is the largest branch of the celiac trunk.
11. List the branches of this artery.
12. What is meant by serpentine?

- *Click on* **TAG 5,** and the following image will appear:

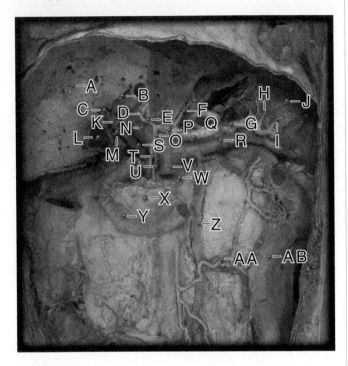

- *Mouse-over the blue pins on the screen to find the information necessary to fill in the following blanks:*

A. _____

B. _____

C. _____

D. _____

E. _____

F. _____

G. _____

H. _____

I. _____

J. _____

K. _____

L. _____

M. _____

N. _____

O. _____

P. _____

Q. _____

R. _____

S. _____

T. _____

U. _____

V. _____

W. _____

X. _____

Y. _____

Z. _____

AA. _____

AB. _____

CHECK POINT:

Abdomen, Celiac Trunk, Anterior View

13. Name the unpaired anterior artery immediately inferior to the celiac trunk.
14. What is the distribution of this artery?
15. Name three arteries that originate at the celiac trunk.

Animation: Hepatic Portal System

- *Click on the* **ANIMATIONS** *button at the bottom of the screen.*

- *In the* **Select topic** *menu, select* **Anatomy.**

- *From the* **Select animation** *menu, select* **Hepatic portal system.**

- *Click the* **Play** *button, and the animation will run in the* **IMAGE AREA.**

- *After viewing the animation, answer the following questions:*

1. What is the hepatic portal system?

2. What does this do for the liver?

3. The hepatic portal system originates from which organs?

4. Which vein is the largest in the hepatic portal system?

5. Which two veins unite to form it?

6. Which areas are drained by the splenic vein?

7. Which areas are drained by the superior mesenteric vein?

8. Blood from which structures drains directly into the hepatic portal vein?

9. What happens to the hepatic portal vein upon entering the liver?

10. What are hepatic sinusoids?

11. What occurs there?

12. What occurs in central veins of the liver?

13. What veins are formed by the union of thousands of these central veins?

14. Upon exiting the liver, these veins _____
_____.

• *Click on the* **DISSECTION** *button at the bottom of the screen to resume your study of the structures of the abdomen. Click on* **TAG 6**, *and the following image will appear:*

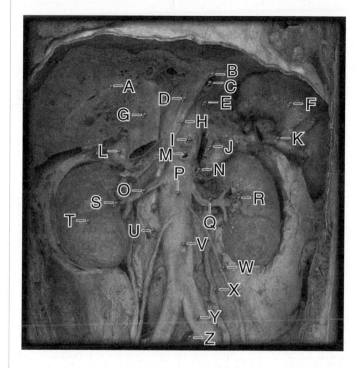

• *Mouse-over the blue pins on the screen to find the information necessary to fill in the following blanks:*

A. _____

B. _____

C. _____

D. _____

E. _____

F. _____

G. _____

H. _____

I. _____

J. _____

K. _____

L. _____

M. _____

N. _____

O. _____

P. _____

Q. _____

R. _____

S. _____

T. _____

U. _____

V. _____

W. _____

X. _____

Y. _____

Z. _____

CHECK POINT:

Abdomen, Celiac Trunk, Anterior View

16. Name the three arteries that branch off of the anterior abdominal aorta, from superior to inferior.
17. Name the large paired arteries that branch off from the abdominal aorta laterally to the kidneys.
18. Name the small paired arteries that branch off from the abdominal aorta laterally in the vicinity of the kidneys. What is their name in the male? In the female?

EXERCISE 5.16: Cardiovascular System - Abdomen, Mesenteric Arteries, Anterior View

- *Insert* Anatomy & Physiology | Revealed® **Cardiovascular, Lymphatic, and Respiratory Systems** *CD, or, if you are already in the* **Dissection** *section, click the* **CHANGE VIEW** *button at the top of the screen, and skip the next two steps.*

- *From the* **Select system** *menu, select* **Cardiovascular**.

- *In the* **Home screen**, *select the* **Dissection** *button in the left portion of the screen. You may click either on the* **Dissection** *button or on the word.*

- *In the* **SELECT A VIEW** *window that appears, click on the* **Select topic** *button.*

- Choose **Abdomen** *from the menu, if it is not already selected.*

- *Click the* **Select view** *menu and choose* **Mesenteric arteries - anterior**.

- *The* **GO** *button will flash green. Click on it.*
- *Click on* **TAG 1**, *and the following image will appear:*

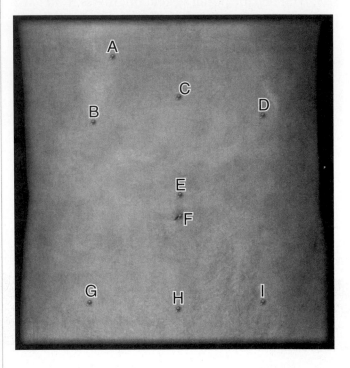

- *Mouse-over the blue pins on the screen to find the information necessary to fill in the following blanks:*

A. _____

B. _____

C. _____

D. _____

E. _____

F. _____

G. _____

H. _____

I. _____

CHECK POINT:

Abdomen, Mesenteric Arteries, Anterior View

1. Name the lower medial abdominal region.
2. What regions flank this lower medial region?
3. What regions flank the umbilical region?

• *Click on* **TAG 4**, *and the following image will appear:*

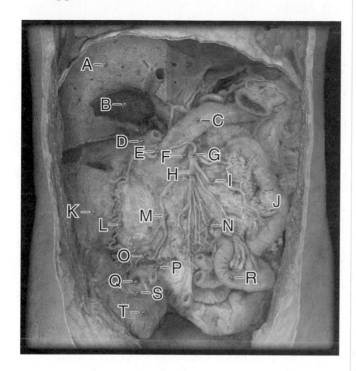

• *Mouse-over the blue pins on the screen to find the information necessary to fill in the following blanks:*

A. _____

B. _____

C. _____

D. _____

E. _____

F. _____

G. _____

H. _____

I. _____

J. _____

K. _____

L. _____

M. _____

N. _____

O. _____

P. _____

Q. _____

R. _____

S. _____

T. _____

Abdomen, Mesenteric Arteries, Anterior View

4. Name the abdominal artery that courses inferiorly to enter the mesentery of the small intestine.
5. What is the termination of this artery?
6. What arteries are also known as *vasa recta*? What is their distribution?

• *Click on* **TAG 5**, *and the following image will appear:*

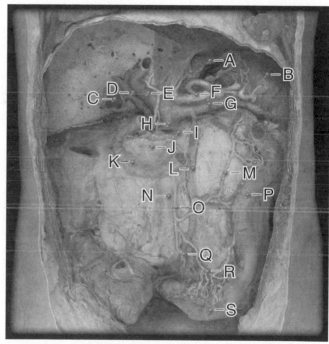

• *Mouse-over the blue pins on the screen to find the information necessary to fill in the following blanks:*

A. _____

B. _____

C. _____

D. _____

E. _____

F. _____

G. _____

H. _____

I._____

J._____

K. _____

L. _____

M. _____

N._____

O._____

P. _____

Q._____

R. _____

S. _____

CHECK POINT:

Abdomen, Mesenteric Arteries, Anterior View

1. What is the distribution of the inferior mesenteric artery?
2. Name the artery with the distribution to the lower part of the descending colon and the sigmoid colon.
3. Name the artery that is the direct continuation of the inferior mesenteric artery and distributes to the rectum.

• *Click on* **TAG 6**, *and the following image will appear:*

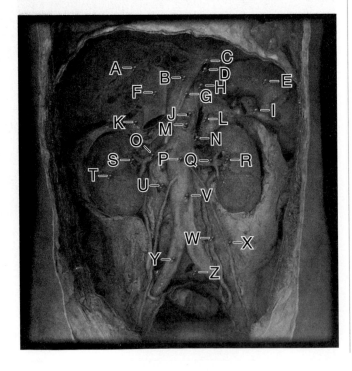

• *Mouse-over the blue pins on the screen to find the information necessary to fill in the following blanks:*

A. _____

B. _____

C. _____

D. _____

E. _____

F._____

G._____

H._____

I._____

J._____

K. _____

L. _____

M. _____

N._____

O._____

P. _____

Q._____

R. _____

S. _____

T. _____

U. _____

V. _____

W. _____

X._____

Y. _____

Z. _____

CHECK POINT:

Abdomen, Mesenteric Arteries, Anterior View

10. List the drainages of the common iliac veins. Where do they terminate?
11. What is the origin and termination of the common iliac arteries?
13. Name the vein that drains the spleen, pancreas, and the fundus and greater curvature of the stomach.

EXERCISE 5.17: Cardiovascular System - Abdomen, Veins, Anterior View

- *Insert* Anatomy & Physiology | Revealed® **Cardiovascular, Lymphatic, and Respiratory Systems** *CD, or, if you are already in the* **Dissection** *section, click the* **CHANGE VIEW** *button at the top of the screen, and skip the next two steps.*

- *From the* **Select system** *menu, select* **Cardiovascular**.

- *In the* **Home screen**, *select the* **Dissection** *button in the left portion of the screen. You may click either on the* **Dissection** *button or on the word.*

- *In the* **SELECT A VIEW** *window that appears, click on the* **Select topic** *button.*

- *Choose* **Abdomen** *from the menu, if it is not already selected.*

- *Click the* **Select view** *menu and choose* **Veins - anterior**.

- *The* **GO** *button will flash green. Click on it.*

- *Click on* **TAG 2**, *and the following image will appear:*

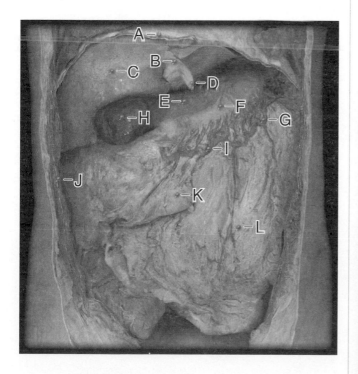

- *Mouse-over the blue pins on the screen to find the information necessary to fill in the following blanks:*

A. _____

B. _____

C. _____

D. _____

E. _____

F. _____

G. _____

H. _____

I. _____

J. _____

K. _____

L. _____

CHECK POINT:

Abdomen, Veins, Anterior View

1. Name the veins that drain the stomach and greater omentum.

- *Click on* **TAG 4**, *and the following image will appear:*

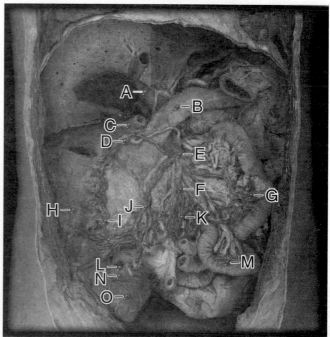

- *Mouse-over the blue pins on the screen to find the information necessary to fill in the following blanks:*

A. _____

B. _____

C. _____

D. _____

E. _____

F. _____

G. _____

H. _____

I. _____

J. _____

K. _____

L. _____

M. _____

N. _____

O. _____

CHECK POINT:

Abdomen, Veins, Anterior View

2. Name the vein formed by 15–20 intestinal veins that courses superiorly in the mesentery of the small intestine.
3. List the structures drained by the hepatic portal vein.
4. What do marginal venous arcades represent?

- *Click on **TAG 5**, and the following image will appear:*

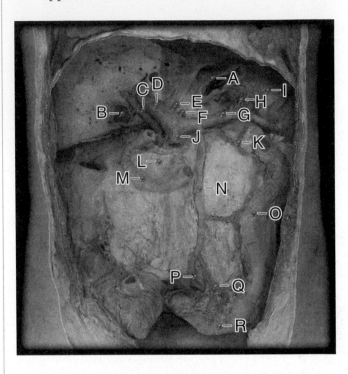

- *Mouse-over the blue pins on the screen to find the information necessary to fill in the following blanks:*

A. _____

B. _____

C. _____

D. _____

E. _____

F. _____

G. _____

H. _____

I. _____

J. _____

K. _____

L. _____

M. _____

N. _____

O. _____

P. _____

Q. _____

R. _____

CHECK POINT:

Abdomen, Veins, Anterior View

5. Name the vein that ascends from the true pelvis to become the inferior mesenteric vein in the false pelvis.

• *Click on* **TAG 6**, *and the following image will appear:*

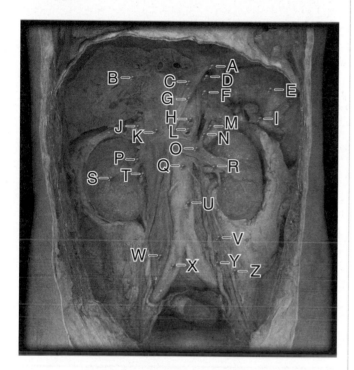

• *Mouse-over the blue pins on the screen to find the information necessary to fill in the following blanks:*

A. _____

B. _____

C. _____

D. _____

E. _____

F. _____

G. _____

H. _____

I. _____

J. _____

K. _____

L. _____

M. _____

N. _____

O. _____

P. _____

Q. _____

R. _____

S. _____

T. _____

U. _____

V. _____

W. _____

X. _____

Y. _____

Z. _____

CHECK POINT:

Abdomen, Veins, Anterior View

6. Which is normally on the right side of the body, the abdominal aorta or the inferior vena cava?
7. Which are normally anterior to the others, the renal arteries or the renal veins?
8. What are the two terminations of the gonadal veins? (Be sure to note which one terminates where.)

IN REVIEW

What Have I Learned?

The following questions cover the material that you have just learned—blood vessles of the abdomen. Use the information in the **STRUCTURE INFORMATION** window for these structures to answer the following questions:

1. Using the following figure, label the nine abdominal regions.

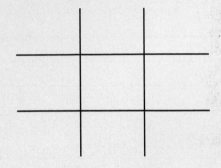

2. What is the umbilicus? (Not the region, the structure.)

3. Name the structure that provides attachment for the diaphragm.

4. Name the vein that carries absorbed products of digestion to the liver.

5. What two veins unite to form the hepatic portal vein?

6. List the organs drained by the splenic vein.

7. Name the opening that transmits the aorta from the thoracic to the abdominal cavity.

8. Name the opening that transmits the esophagus from the thoracic to the abdominal cavity.

9. Name the paired arteries at the terminus of the abdominal aorta.

10. Name the vein that drains the diaphragm and the left suprarenal gland.

The Pelvis

EXERCISE 5.18: Cardiovascular System - Pelvis - Female, Vasculature, Anterior View

- *Insert* Anatomy & Physiology | Revealed® **Cardiovascular, Lymphatic, and Respiratory Systems** *CD, or, if you are already in the* **Dissection** *section, click the* **CHANGE VIEW** *button at the top of the screen, and skip the next two steps.*

- *From the* **Select system** *menu, select* **Cardiovascular**.

- *In the* **Home screen**, *select the* **Dissection** *button in the left portion of the screen. You may click either on the* **Dissection** *button or on the word.*

- *In the* **SELECT A VIEW** *window that appears, click on the* **Select topic** *button.*

- *Choose* **Pelvis - female** *from the menu.*

- **Vasculature - anterior** *appears in the* **Select view** *menu.*

- *The* **GO** *button will flash green. Click on it.*

- *After viewing the first two **TAGs** , click on **TAG** 3 and the following image will appear:*

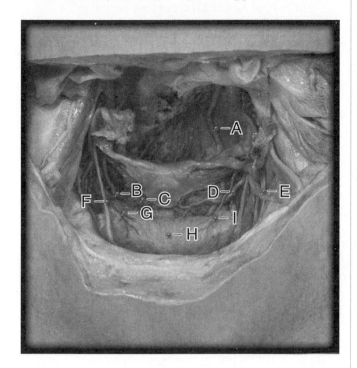

- *Mouse-over the blue pins on the screen to find the information necessary to fill in the following blanks:*

A. _____

B. _____

C. _____

D. _____

E. _____

F._____

G. _____

H. _____

I._____

CHECK POINT:

Pelvis - Female, Vasculature, Anterior View

1. Name the artery that carries blood between the placenta and fetus.
2. Name the artery that reflects along the side of the uterus, distributing to the uterus and vagina.
3. Name the vein that drains the uterus.

- *Click on **TAG 4**, and the following image will appear:*

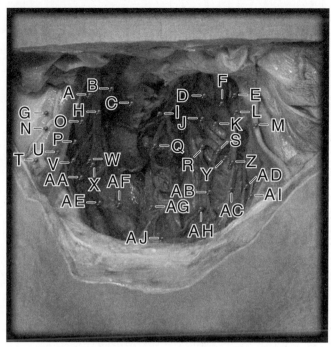

- *Mouse-over the blue pins on the screen to find the information necessary to fill in the following blanks:*

A. _____

B. _____

C. _____

D. _____

E. _____

F._____

G. _____

H. _____

I._____

J._____

K. _____

L. _____

M. _____

N._____

O._____

P. _____

Q._____

R. _____

S. _____

T. _____

U. _____

V. _____

W. _____

X. _____

Y. _____

Z. _____

AA. _____

AB. _____

AC. _____

AD. _____

AE. _____

AF. _____

AG. _____

AH. _____

AI. _____

AJ. _____

CHECK POINT:

Pelvis - Female, Vasculature, Anterior View

4. Name the vein that drains the pelvis and gluteal region and terminates in the common iliac vein.
5. Name two veins that are tributaries to the vein in question 4.
6. Name the artery that originates at the common iliac artery and is continuous with the femoral artery.

EXERCISE 5.19: Cardiovascular System - Pelvis - Male, Vasculature, Anterior View

- *Insert* Anatomy & Physiology | Revealed® **Cardiovascular, Lymphatic, and Respiratory Systems** CD, or, if you are already in the **Dissection** section, click the **CHANGE VIEW** button at the top of the screen, and skip the next two steps.

- *From the* **Select system** *menu, select* **Cardiovascular**.

- *In the* **Home screen**, *select the* **Dissection** *button in the left portion of the screen. You may click either on the* **Dissection** *button or on the word.*

- *In the* **SELECT A VIEW** *window that appears, click on the* **Select topic** *button.*

- *Choose* **Pelvis - male** *from the menu.*

- **Vasculature - anterior** *appears in the* **Select view** *menu.*

- *The* **GO** *button will flash green. Click on it.*

- *Click on* **TAG 2**, *and the following image will appear:*

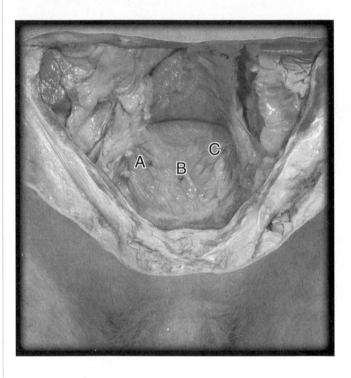

- *Mouse-over the blue pins on the screen to find the information necessary to fill in the following blanks:*

A. _____

B. _____

C. _____

CHECK POINT:

Pelvis - Male, Vasculature, Anterior View

1. Name an artery that distributes to the urinary bladder.
2. Name a vein that drains the urinary bladder.
3. The volume of which organ effects the position of the surrounding organs?

• *Click on* **TAG 4**, *and the following image will appear:*

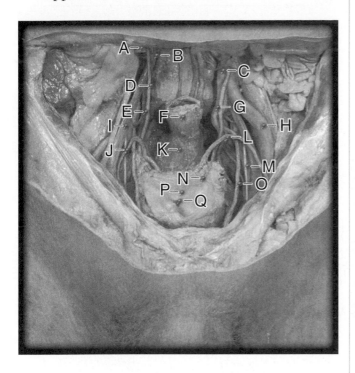

Pelvis - Male, Vasculature, Anterior View

4. Name an artery that passes medially to reach the superior surface of the urinary bladder.

• *Click on* **TAG 5**, *and the following image will appear:*

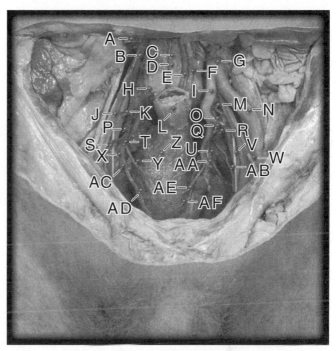

• *Mouse-over the blue pins on the screen to find the information necessary to fill in the following blanks:*

A. _____

B. _____

C. _____

D. _____

E. _____

F. _____

G. _____

H. _____

I. _____

J. _____

K. _____

L. _____

M. _____

N. _____

O. _____

P. _____

Q. _____

• *Mouse-over the blue pins on the screen to find the information necessary to fill in the following blanks:*

A. _____

B. _____

C. _____

D. _____

E. _____

F. _____

G. _____

H. _____

I. _____

J. _____

K. _____

L. _____

M. _____

N. _____

O. _____

P. _____

Q. _____

R. _____

S. _____

T. _____

U. _____

V. _____

W. _____

X. _____

Y. _____

Z. _____

AA. _____

AB. _____

AC. _____

AD. _____

AE. _____

AF. _____

CHECK POINT:

Pelvis - Male, Vasculature, Anterior View

5. Name a vein that passes laterally from the inferior surface of the urinary bladder and drains the urinary bladder and the prostate gland. What is this vein known as in the female?

6. List the structures drained by the median sacral vein.

7. List the structures distributed by the ilio-lumbar artery.

IN REVIEW

What Have I Learned?

The following questions cover the material that you have just learned—the vasculature of the female and male pelvis. Use the information in the **STRUCTURE INFORMATION** window for this region to answer the following questions:

1. Name the thick-walled, pear-shaped hollow muscular organ that is the site of implantation of the zygote.

2. What is a zygote?

3. Name the three parts of the organs in question 1.

4. Name the paired female gonads.

5. What is their function?

6. Name the site of fertilization of the egg.

7. Name the artery that descends into the pelvis and passes posteriorly toward the superior margin of the greater sciatic notch.

8. Name an artery that distributes to the anal canal and external genitalia of both sexes.

9. Name an artery whose distribution includes the rectum and vagina in females and the rectum, prostate, and seminal vesicle in males.

10. Name a pelvic vein that drains the muscles and skin of the medial thigh and terminates at the internal iliac vein.

The Hip and Thigh

EXERCISE 5.20: Cardiovascular System - Hip and Thigh, Arteries, Anterior View

- *Insert* Anatomy & Physiology | Revealed® **Cardiovascular, Lymphatic, and Respiratory Systems** *CD, or, if you are already in the* **Dissection** *section, click the* **CHANGE VIEW** *button at the top of the screen, and skip the next two steps.*

- *From the* **Select system** *menu, select* **Cardiovascular**.

- *In the* **Home screen**, *select the* **Dissection** *button in the left portion of the screen. You may click either on the* **Dissection** *button or on the word.*

- *In the* **SELECT A VIEW** *window that appears, click on the* **Select topic** *button.*

- *Choose* **Hip and thigh** *from the menu.*

- *Click the* **Select view** *menu and choose* **Arteries - anterior**.

- *The* **GO** *button will flash green. Click on it.*

- *Click on* **TAG 1**, *and the following image will appear:*

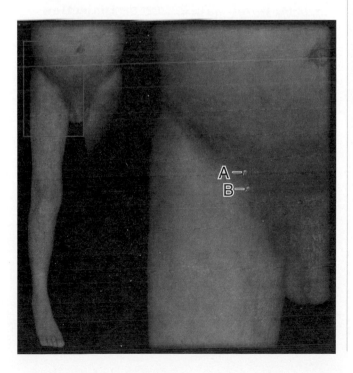

- *Mouse-over the blue pins on the screen to find the information necessary to fill in the following blanks:*

A. _____

B. _____

CHECK POINT:

Hip and Thigh, Arteries, Anterior View

1. Name the location for checking the pulse at the groin.
2. What emergency function is applicable at this location?
3. What structures are contained within the femoral triangle?

- *Click on* **TAG 2**, *and the following image will appear:*

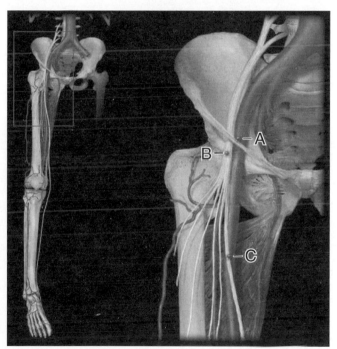

- *Mouse-over the blue pins on the screen to find the information necessary to fill in the following blanks:*

A. _____

B. _____

C. _____

M. _____

N. _____

O. _____

C H E C K   P O I N T :

Hip and Thigh, Arteries, Anterior View

4. What structure forms the floor of the inguinal canal?
5. Name the nerve and its branches that enter the thigh posterior to the inguinal ligament.
6. Which branch is the largest?

• *Click on* **TAG 3**, *and the following image will appear:*

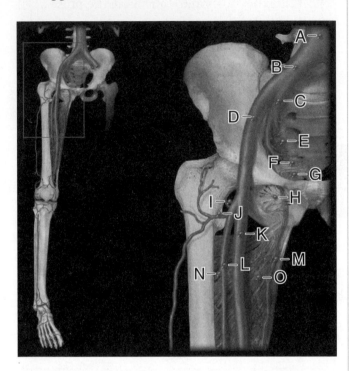

• *Mouse-over the blue pins on the screen to find the information necessary to fill in the following blanks:*

A. _____

B. _____

C. _____

D. _____

E. _____

F. _____

G. _____

H. _____

I. _____

J. _____

K. _____

L. _____

C H E C K   P O I N T :

Hip and Thigh, Arteries, Anterior View

7. List the sequence of the continuous major arteries from the abdominal aorta through the thigh.
8. Name the largest branch of the femoral artery. What is its distribution? What are two other names for this artery?
9. List the distribution of the obturator artery.

EXERCISE 5.21: Cardiovascular System - Hip and Thigh, Arteries, Posterior View

• *Insert* Anatomy & Physiology | Revealed® **Cardiovascular, Lymphatic, and Respiratory Systems** CD, *or, if you are already in the* **Dissection** *section, click the* **CHANGE VIEW** *button at the top of the screen, and skip the next two steps.*

• *From the* **Select system** *menu, select* **Cardiovascular**.

• *In the* **Home screen**, *select the* **Dissection** *button in the left portion of the screen. You may click either on the* **Dissection** *button or on the word.*

• In the **SELECT A VIEW** *window that appears, click on the* **Select topic** *button.*

• Choose **Hip and thigh** *from the menu, if it is not already selected.*

• Click the **Select view** *menu and choose* **Arteries - posterior**.

• *The* **GO** *button will flash green. Click on it.*

• *Click on* **TAG 2**, *and the following image will appear:*

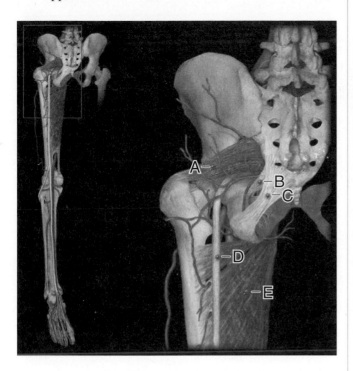

• *Mouse-over the blue pins on the screen to find the information necessary to fill in the following blanks:*

A. _____

B. _____

C. _____

D. _____

E. _____

CHECK POINT:

Hip and Thigh, Arteries, Posterior View

1. Name the largest nerve in the body.
2. Name the two nerves that form that nerve.
3. List the sensory and motor innervations of this nerve.

• *Click on* **TAG 3**, *and the following image will appear:*

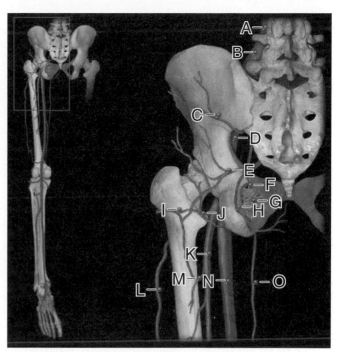

• *Mouse-over the blue pins on the screen to find the information necessary to fill in the following blanks:*

A. _____

B. _____

C. _____

D. _____

E. _____

F. _____

G. _____

H. _____

I. _____

J. _____

K. _____

L. _____

M. _____

N. _____

O. _____

CHECK POINT:

Hip and Thigh, Arteries, Posterior View

4. Name the largest branch of the internal iliac artery. What is its distribution?
5. List the distribution of the abdominal aorta.
6. What artery is usually crossed by the ureter and gonadal vessels at its origin?

EXERCISE 5.22:   Cardiovascular System - Hip and Thigh, Veins, Anterior View

- *Insert* Anatomy & Physiology | Revealed® **Cardiovascular, Lymphatic, and Respiratory Systems** *CD, or, if you are already in the* **Dissection** *section, click the* **CHANGE VIEW** *button at the top of the screen, and skip the next two steps.*

- *From the* **Select system** *menu, select* **Cardiovascular**.

- *In the* **Home screen**, *select the* **Dissection** *button in the left portion of the screen. You may click either on the* **Dissection** *button or on the word.*

- *In the* **SELECT A VIEW** *window that appears, click on the* **Select topic** *button.*

- *Choose* **Hip and thigh** *from the menu, if it is not already selected.*

- *Click the* **Select view** *menu and choose* **Veins - anterior**.

- *The* **GO** *button will flash green. Click on it.*

- *Click on* **TAG 2**, *and the following image will appear:*

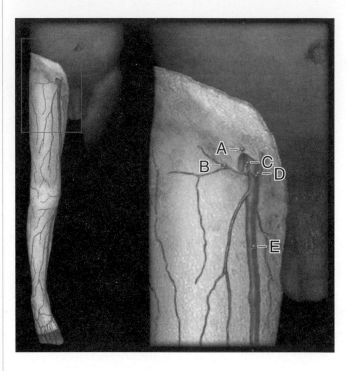

- *Mouse-over the blue pins on the screen to find the information necessary to fill in the following blanks:*

A. _____

B. _____

C. _____

D. _____

E. _____

CHECK POINT:

Hip and Thigh, Veins, Anterior View

1. Name the common source of vessel tissue for coronary bypass surgery.
2. Defective valves in this vein may cause _____.
3. What is another name for this vein?

- *Click on* **TAG 4**, *and the following image will appear:*

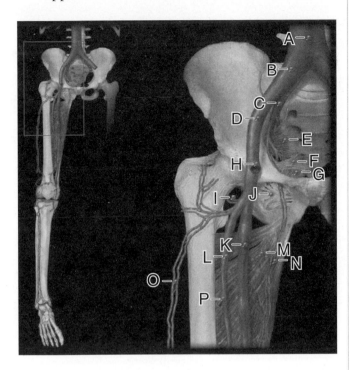

- *Mouse-over the blue pins on the screen to find the information necessary to fill in the following blanks:*

A. _____

B. _____

C. _____

D. _____

E. _____

F. _____

G. _____

H. _____

I. _____

J. _____

K. _____

L. _____

M. _____

N. _____

O. _____

P. _____

EXERCISE 5.23:   Cardiovascular System - Hip and Thigh, Veins, Posterior View

- *Insert* Anatomy & Physiology | Revealed® **Cardiovascular, Lymphatic, and Respiratory Systems** CD, *or, if you are already in the* **Dissection** *section, click the* **CHANGE VIEW** *button at the top of the screen, and skip the next two steps.*

- *From the* **Select system** *menu, select* **Cardiovascular.**

- *In the* **Home screen**, *select the* **Dissection** *button in the left portion of the screen. You may click either on the* **Dissection** *button or on the word.*

- In the **SELECT A VIEW** *window that appears, click on the* **Select topic** *button.*

- Choose **Hip and thigh** *from the menu, if it is not already selected.*

- Click the **Select view** *menu and choose* **Veins - posterior**.

- *The* **GO** *button will flash green. Click on it.*

• *Click on **TAG 2**, and the following image will appear:*

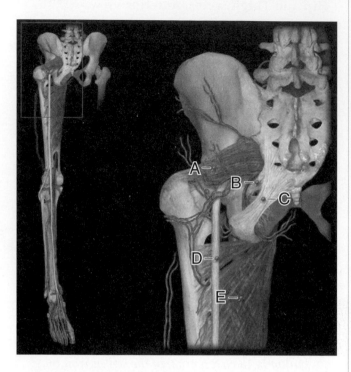

• *Mouse-over the blue pins on the screen to find the information necessary to fill in the following blanks:*

A. _____

B. _____

C. _____

D. _____

E. _____

• *Click on **TAG 3**, and the following image will appear:*

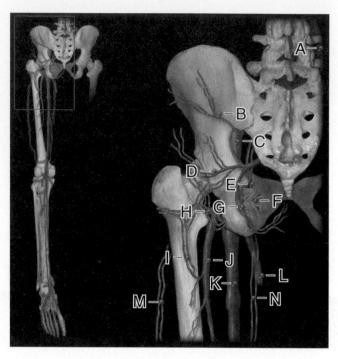

• *Mouse-over the blue pins on the screen to find the information necessary to fill in the following blanks:*

A. _____

B. _____

C. _____

D. _____

E. _____

F._____

G. _____

H._____

I._____

J._____

K. _____

L. _____

M. _____

N._____

CHAPTER 5   The Cardiovascular System

Hip and Thigh, Veins, Posterior View

1. Name the medial vein of the leg that spans from the medial malleolus to the femoral vein just inferior to the inguinal ligament.
2. List the drainages of the deep vein of the thigh.
3. Name the largest vein in the body.

IN REVIEW

What Have I Learned?

The following questions cover the material that you have just learned—the blood vessels of the hip and thigh. Use the information in the **STRUCTURE INFORMATION** window for these structures to answer the following questions:

1. Name the four arteries that contribute to the cruciate anastomosis of the hip joint.

2. Name the three veins that contribute to the anastomosis of the hip joint.

3. What is the largest vein in the body? Did you know that it is also the thickest *vessel* in your body? Which do you suppose is the longest vein in your body?

4. Name the triangular region of the groin that contains the femoral nerve, artery, and vein.

5. Name the opening in the deep fascia through which the great saphenous vein passes.

6. Name the subcutaneous vein that runs parallel to the inguinal ligament.

7. Name the deep vertical space that separates the left and right buttock.

8. What structure represents the superior limit of the posterior thigh?

The Knee

EXERCISE 5.24: Cardiovascular System - Knee, Arteries, Anterior View

- *Insert* Anatomy & Physiology | Revealed® **Cardiovascular, Lymphatic, and Respiratory Systems** *CD, or, if you are already in the* **Dissection** *section, click the* **CHANGE VIEW** *button at the top of the screen, and skip the next two steps.*

- *From the* **Select system** *menu, select* **Cardiovascular.**

- *In the* **Home screen**, *select the* **Dissection** *button in the left portion of the screen. You may click either on the* **Dissection** *button or on the word.*

- In the **SELECT A VIEW** *window that appears, click on the* **Select topic** *button.*

- Choose **Knee** *from the menu.*

- Click the **Select view** *menu and choose* **Arteries - anterior**.

- *The* **GO** *button will flash green. Click on it.*

- *Click on* **TAG 1**, *and the following image will appear:*

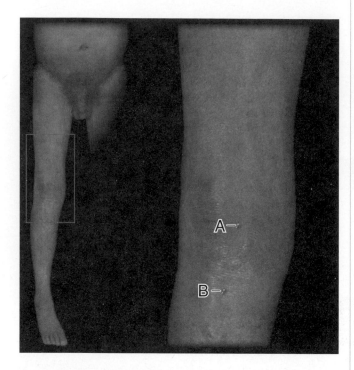

- *Mouse-over the blue pins on the screen to find the information necessary to fill in the following blanks:*

A. _____

B. _____

CHECK POINT:

Knee, Arteries, Anterior View

1. Name the sesamoid bone embedded in the tendon of the quadriceps femoris muscles.
2. Name the bony elevation on the proximal shaft of the tibia.

- *Click on* **TAG 2**, and the following image will appear:

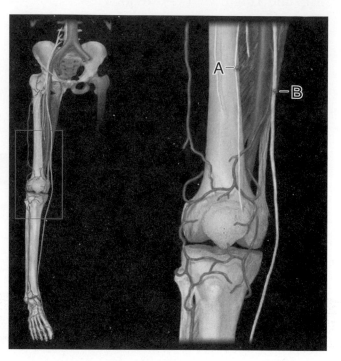

- *Mouse-over the blue pins on the screen to find the information necessary to fill in the following blanks:*

A. _____

B. _____

- *Click on* **TAG 3**, and the following image will appear:

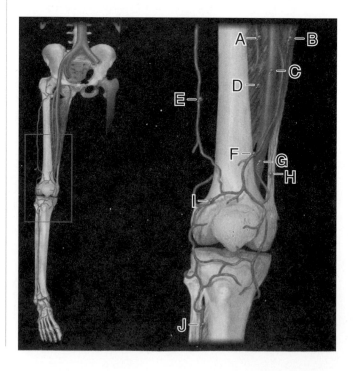

- *Mouse-over the blue pins on the screen to find the information necessary to fill in the following blanks:*

A. _____

B. _____

C. _____

D. _____

E. _____

F._____

G. _____

H. _____

I._____

J._____

C H E C K   P O I N T :

Knee, Arteries, Anterior View

3. List the areas distributed by the femoral artery.
4. Name a deep artery of the thigh that distributes to the hip joint and the quadriceps femoris muscle.
5. Name an artery that distributes to the muscles and skin of the medial thigh.

EXERCISE 5.25: Cardiovascular System - Knee, Arteries, Posterior View

- *Insert* Anatomy & Physiology | Revealed® **Cardiovascular, Lymphatic, and Respiratory Systems** *CD, or, if you are already in the* **Dissection** *section, click the* **CHANGE VIEW** *button at the top of the screen, and skip the next two steps.*

- *From the* **Select system** *menu, select* **Cardiovascular.**

- *In the* **Home screen,** *select the* **Dissection** *button in the left portion of the screen. You may click either on the* **Dissection** *button or on the word.*

- *In the* **SELECT A VIEW** *window that appears, click on the* **Select topic** *button.*

- Choose **Knee** *from the menu, if it is not already selected.*

- Click the **Select view** *menu and choose* **Arteries - posterior**.

- *The* **GO** *button will flash green. Click on it.*

- *Click on* **TAG 1**, *and the following image will appear:*

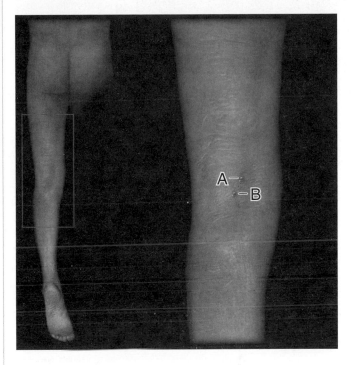

- *Mouse-over the blue pins on the screen to find the information necessary to fill in the following blanks:*

A. _____

B. _____

C H E C K   P O I N T :

Knees, Arteries, Posterior View

1. Name the diamond-shaped area of the posterior knee.
2. Name the artery that provides a pulse in this area.
3. What special requirements must be met to appreciate this pulse?

• *Click on* **TAG 2**, *and the following image will appear:*

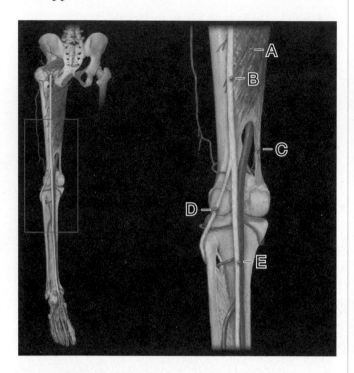

• *Click on* **TAG 3**, *and the following image will appear:*

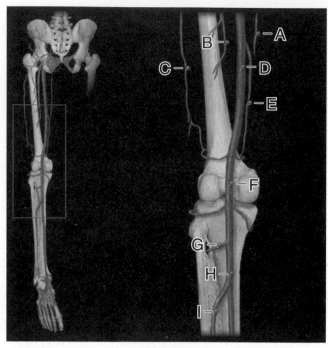

• *Mouse-over the blue pins on the screen to find the information necessary to fill in the following blanks:*

A. _____

B. _____

C. _____

D. _____

E. _____

• *Mouse-over the blue pins on the screen to find the information necessary to fill in the following blanks:*

A. _____

B. _____

C. _____

D. _____

E. _____

F. _____

G. _____

H. _____

I. _____

C H E C K   P O I N T :

Knee, Arteries, Posterior View

4. Name the large nerve of the posterior hip and thigh. Name the two nerves that merge to form this nerve.
5. What effect will a herniated intervertebral disk in the lower lumbar region have along the distribution of this nerve?
6. What is the adductor hiatus?

C H E C K   P O I N T :

Knee, Arteries, Posterior View

7. What artery does the femoral artery become in the vicinity of the knee?
8. Where exactly does this name change occur?
9. Name an artery that distributes to the anterior (extensor) compartment of the leg.

EXERCISE 5.26: Cardiovascular System -
Knee, Veins, Anterior View

- *Insert* Anatomy & Physiology | Revealed®
Cardiovascular, Lymphatic and Respiratory Systems *CD, or, if you are already in the* **Dissection** *section, click the* **CHANGE VIEW** *button at the top of the screen, and skip the next two steps.*

- *From the* **Select system** *menu, select* **Cardiovascular.**

- *In the* **Home screen**, *select the* **Dissection** *button in the left portion of the screen. You may click either on the* **Dissection** *button or on the word.*

- *In the* **SELECT A VIEW** *window that appears, click on the* **Select topic** *button.*

- Choose **Knee** *from the menu, if it is not already selected.*

- *Click the* **Select view** *menu and choose* **Veins - anterior**.

- *The* **GO** *button will flash green. Click on it.*

- *Click on* **TAG 2**, *and the following image will appear:*

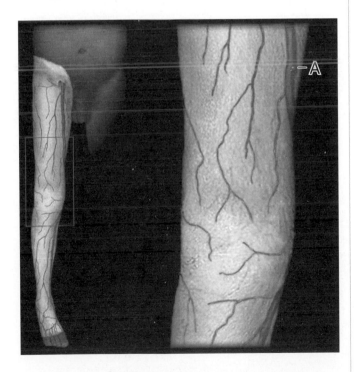

- *Mouse-over the blue pin on the screen to find the information necessary to fill in the following blank:*

A. _____

Knee, Veins, Anterior View

1. Name the large subcutaneous vein of the median leg and thigh.
2. What is its drainage?

- *Click on* **TAG 4**, *and the following image will appear:*

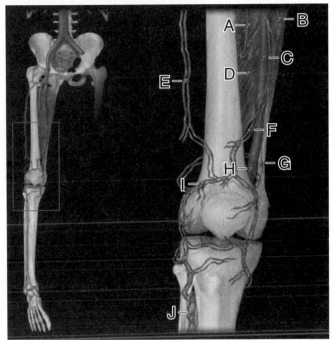

- *Mouse-over the blue pins on the screen to find the information necessary to fill in the following blanks:*

A. _____
B. _____
C. _____
D. _____
E. _____
F. _____
G. _____
H. _____
I. _____
J. _____

CHECK POINT:

Knee, Veins, Anterior View

3. List the structures drained by the femoral vein.
4. Name the veins that drain the anterior (extensor) compartment of the leg.
5. Name the venous anastomosis that forms a network around the knee. What veins contribute to this anastomosis?

EXERCISE 5.27: Cardiovascular System - Knee, Veins, Posterior View

- *Insert* Anatomy & Physiology | Revealed® **Cardiovascular, Lymphatic, and Respiratory Systems** *CD, or, if you are already in the* **Dissection** *section, click the* **CHANGE VIEW** *button at the top of the screen, and skip the next two steps.*

- *From the* **Select system** *menu, select* **Cardiovascular**.

- *In the* **Home screen**, *select the* **Dissection** *button in the left portion of the screen. You may click either on the* **Dissection** *button or on the word.*

- *In the* **SELECT A VIEW** *window that appears, click on the* **Select topic** *button.*

- *Choose* **Knee** *from the menu, if it is not already selected.*

- *Click the* **Select view** *menu and choose* **Veins - posterior**.

- *The* **GO** *button will flash green. Click on it.*

- *Click on* **TAG 2**, *and the following image will appear:*

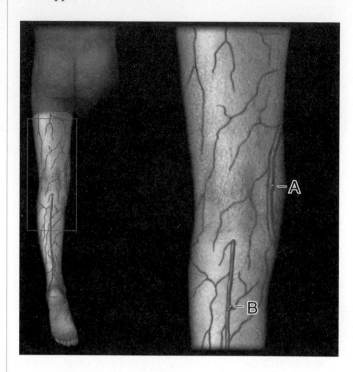

- *Mouse-over the blue pins on the screen to find the information necessary to fill in the following blanks:*

A. _____

B. _____

CHECK POINT:

Knee, Veins, Posterior View

1. Name the vein providing drainage for the dorsum of the foot and subcutaneous posterior leg.
2. Name the vein providing drainage for the dorsum of the foot and subcutaneous medial leg and thigh.

- *Click on* **TAG 4**, *and the following image will appear:*

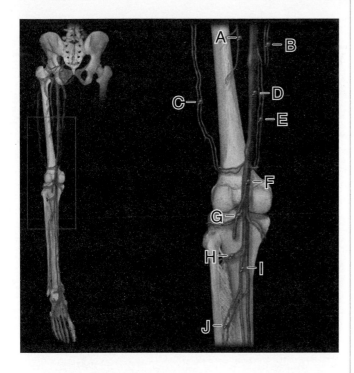

- *Mouse-over the blue pins on the screen to find the information necessary to fill in the following blanks:*

A. _____

B. _____

C. _____

D. _____

E. _____

F. _____

G. _____

H. _____

I. _____

J. _____

CHECK POINT:

Knee, Veins, Posterior View

3. Name the vein that drains the knee joint and the surrounding structures.
4. Name the paired veins that drain the posterior muscles of the leg.
5. List the drainage of the fibular veins.

HEADS UP!

View the radiographic images for the cardiovascular system.

- *Click on the* **IMAGING** *button at the bottom of the screen.*
- *Click on the* **CHANGE REGION** *button.*
- *Click on the* **Select region** *menu, and select* **Knee**.
- *Click the flashing* **GO** *button.*
- *Click on the* **Select structure** *menu in the* **STRUCTURE LIST** *box and then on the individual structures to highlight them in the* **IMAGE AREA**.

IN REVIEW

What Have I Learned?

The following questions cover the material that you have just learned—the blood vessels of the knee. Use the information in the **STRUCTURE INFORMATION** window for these structures to answer the following questions:

1. Name two blood vessels that change their names at the posterior knee.

2. Where exactly does this name change occur?

3. What new names do these vessels acquire?

4. The terminal branches of which artery supplies everything inferior to the knee?

5. Which artery is the primary blood supply for the posterior leg muscles?

6. List the distribution of the fibular artery.

I N   R E V I E W

7. Name the artery with the distribution to the knee joint and the structures around it.

8. Many arteries of the limbs are accompanied by _____.

9. What is the meaning of the Latin word *genu?*

10. Name the paired veins that drain the quadriceps femoris muscles and the hip joint.

The Leg and Foot

EXERCISE 5.28: Cardiovascular System - Leg and Foot, Arteries, Anterior View

- *Insert* Anatomy & Physiology | Revealed® **Cardiovascular, Lymphatic, and Respiratory Systems** *CD, or, if you are already in the* **Dissection** *section, click the* **CHANGE VIEW** *button at the top of the screen, and skip the next two steps.*

- *From the* **Select system** *menu, select* **Cardiovascular**.

- *In the* **Home screen**, *select the* **Dissection** *button in the left portion of the screen. You may click either on the* **Dissection** *button or on the word.*

- *In the* **SELECT A VIEW** *window that appears, click on the* **Select topic** *button.*

- *Choose* **Leg and foot** *from the menu.*

- *Click the* **Select view** *menu and choose* **Arteries - anterior**.

- *The* **GO** *button will flash green. Click on it.*

- *Click on* **TAG 1**, *and the following image will appear:*

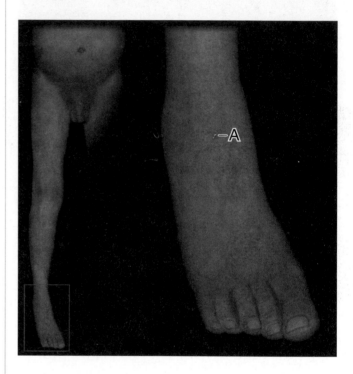

- *Mouse-over the blue pin on the screen to find the information necessary to fill in the following blank:*

A. _____

C H E C K   P O I N T :

Leg and Foot, Arteries, Anterior View

1. Name the artery used for a pedal pulse.
2. Where is this pulse located?
3. What is this pulse used to assess?

- *Click on* **TAG 2**, *and the following image will appear:*

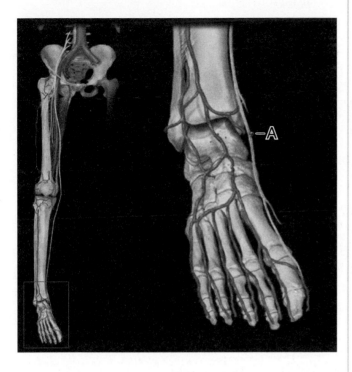

- *Mouse-over the blue pin on the screen to find the information necessary to fill in the following blank:*

A. _____

CHECK POINT:

Leg and Foot, Arteries, Anterior View

4. Name the largest branch of the femoral nerve.
5. Where is it located?
6. Where is the sensory innervation associated with this nerve?

- *Click on* **TAG 3**, *and the following image will appear:*

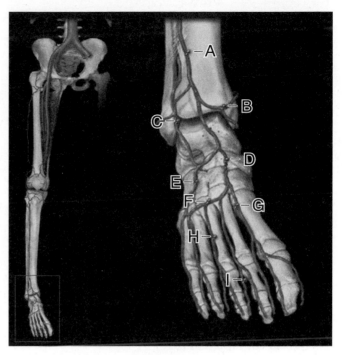

- *Mouse-over the blue pins on the screen to find the information necessary to fill in the following blanks:*

A. _____

B. _____

C. _____

D. _____

E. _____

F. _____

G. _____

H. _____

I. _____

CHECK POINT:

Leg and Foot, Arteries, Anterior View

7. Name the terminal branch of the popliteal artery that continues on the dorsum of the foot.
8. What is the name for this artery on the dorsum of the foot?
9. Name the arteries that distribute to the toes and their joints.

EXERCISE 5.29: Cardiovascular System - Leg and Foot, Arteries, Posterior View

- *Insert* Anatomy & Physiology | Revealed® **Cardiovascular, Lymphatic, and Respiratory Systems** *CD, or, if you are already in the* **Dissection** *section, click the* **CHANGE VIEW** *button at the top of the screen, and skip the next two steps.*

- *From the* **Select system** *menu, select* **Cardiovascular**.

- *In the* **Home screen**, *select the* **Dissection** *button in the left portion of the screen. You may click either on the* **Dissection** *button or on the word.*

- *In the* **SELECT A VIEW** *window that appears, click on the* **Select topic** *button.*

- *Choose* **Leg and foot** *from the menu, if it is not already selected.*

- *Click the* **Select view** *menu and choose* **Arteries - posterior**.

- *The* **GO** *button will flash green. Click on it.*

- *Click on* **TAG 2**, *and the following image will appear:*

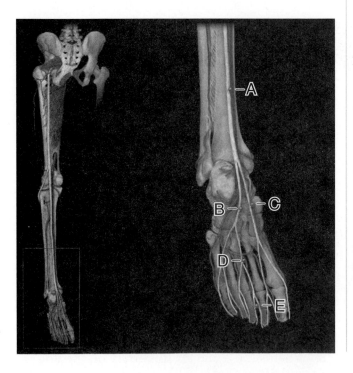

- *Mouse-over the blue pins on the screen to find the information necessary to fill in the following blanks:*

A. _____

B. _____

C. _____

D. _____

E. _____

CHECK POINT:

Leg and Foot, Arteries, Posterior View

1. Name the nerve that is both motor and sensory and innervates the muscles of the posterior leg.
2. Name the lateral terminal branch of this nerve. Where is its motor and sensory innervation?
3. Name the medial terminal branch of this nerve. Where is its motor and sensory innervation?

- *Click on* **TAG 3**, *and the following image will appear:*

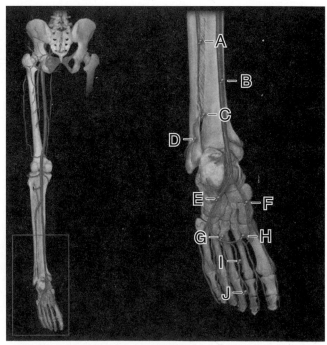

- *Mouse-over the blue pins on the screen to find the information necessary to fill in the following blanks:*

A. _____

B. _____

C. _____

D. _____

E. _____

F. _____

G. _____

H. _____

I. _____

J. _____

4. Name the arterial vessel that arches across the plantar surface at the bases of metatarsals 2–4.
5. Name the artery that distributes to the muscles and joints of the medial plantar foot.
6. Name the artery that distributes to the muscles and joints of the lateral plantar foot.

EXERCISE 5.30:   Cardiovascular System - Leg and Foot, Veins, Anterior View

- *Insert* Anatomy & Physiology | Revealed® **Cardiovascular, Lymphatic and Respiratory Systems** *CD, or, if you are already in the* **Dissection** *section, click the* **CHANGE VIEW** *button at the top of the screen, and skip the next two steps.*

- *From the* **Select system** *menu, select* **Cardiovascular**.

- *In the* **Home screen**, *select the* **Dissection** *button in the left portion of the screen. You may click either on the* **Dissection** *button or on the word.*

- *In the* **SELECT A VIEW** *window that appears, click on the* **Select topic** *button.*

- *Choose* **Leg and foot** *from the menu, if it is not already selected.*

- *Click the* **Select view** *menu and choose* **Veins - anterior**.

- *The* **GO** *button will flash green. Click on it.*

- *Click on* **TAG 2**, *and the following image will appear:*

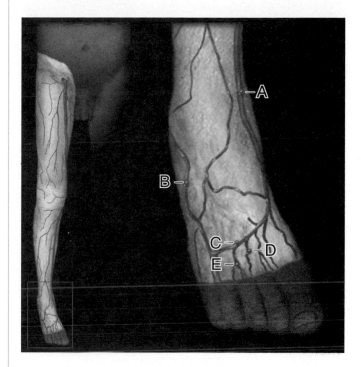

- *Mouse-over the blue pins on the screen to find the information necessary to fill in the following blanks:*

A. _____

B. _____

C. _____

D. _____

E. _____

1. Name the venous vessel that curves medial to lateral over the proximal ends of the metatarsals.
2. Name the veins that drain the toes, plus the distal dorsum and joints of the foot.
3. Name the veins that course along the medial and lateral sides of the digits and drains the toes.

• *Click on* **TAG 4**, *and the following image will appear:*

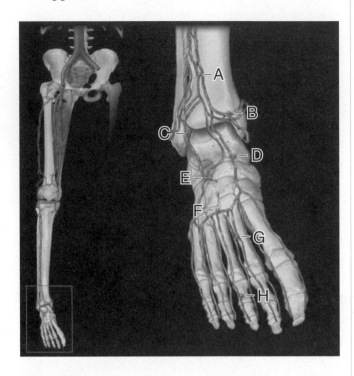

• *Mouse-over the blue pins on the screen to find the information necessary to fill in the following blanks:*

A. _____

B. _____

C. _____

D. _____

E. _____

F. _____

G. _____

H. _____

CHECK POINT:

Leg and Foot, Veins, Anterior View

4. Name the vein that drains the lateral tarsal bones.
5. Name the vein that drains the dorsum of the foot, including the toes. What is another name for this vein?
6. Name the veins that drain the toes and metatarsal region of the foot.

EXERCISE 5.31: Cardiovascular System - Leg and Foot, Veins, Posterior View

• *Insert* Anatomy & Physiology | Revealed® **Cardiovascular, Lymphatic, and Respiratory Systems** *CD, or, if you are already in the* **Dissection** *section, click the* **CHANGE VIEW** *button at the top of the screen, and skip the next two steps.*

• *From the* **Select system** *menu, select* **Cardiovascular**.

• *In the* **Home screen**, *select the* **Dissection** *button in the left portion of the screen. You may click either on the* **Dissection** *button or on the word.*

• *In the* **SELECT A VIEW** *window that appears, click on the* **Select topic** *button.*

• *Choose* **Leg and foot** *from the menu, if it is not already selected.*

• *Click the* **Select view** *menu and choose* **Veins - posterior**.

• *The* **GO** *button will flash green. Click on it.*

• *Click on* **TAG 2**, *and the following image will appear:*

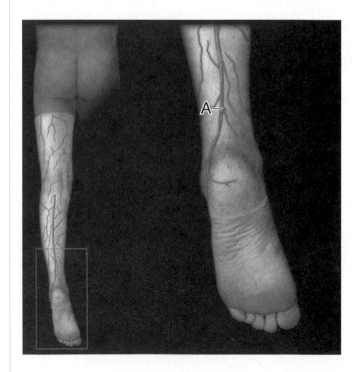

- *Mouse-over the blue pin on the screen to find the information necessary to fill in the following blank:*

A. _____

Leg and Foot, Veins, Posterior View

1. List the drainage and the course of the small saphenous vein.

- *Click on* **TAG 4**, *and the following image will appear:*

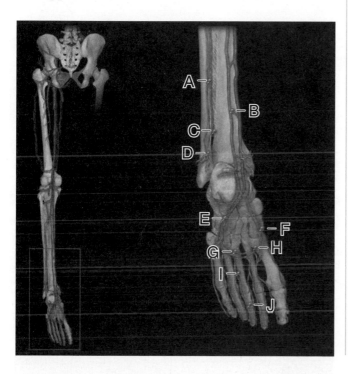

- *Mouse-over the blue pins on the screen to find the information necessary to fill in the following blanks:*

A. _____

B. _____

C. _____

D. _____

E. _____

F. _____

G. _____

H. _____

I. _____

J. _____

Leg and Foot, Veins, Posterior View

2. Name the venous vessels that arch across the plantar surface at the bases of metatarsals 2–4.
3. List the drainages of the medial plantar veins.
4. List the drainages of the lateral plantar veins.

IN REVIEW

What Have I Learned?

The following questions cover the material that you have just learned—the blood vessels of the leg and foot. Use the information in the **STRUCTURE INFORMATION** window for these structures to answer the following questions:

1. Name the arteries that pass directly along the metatarsals and distribute to the dorsum of the foot and its joints and toes.

2. Name the artery that distributes to the tarsal bones of the lateral foot.

3. Name the artery used for pedal pulse.

4. Name the artery that distributes to the ankle joint and the dorsum of the foot.

IN REVIEW

5. Name the artery that distributes to the lateral and medial aspects of the ankle joint.

6. Name the nerve located along the plantar aspect of the metatarsals supplying motor innervation to the intrinsic foot muscles. Where is its sensory innervation?

7. Name the nerve located on the medial and lateral sides of the digits and supplying sensory innervation to the skin and joints of the toes.

8. List the distribution of the plantar metatarsal arteries.

9. Name the arteries that course along the medial and lateral sides of the digits, distributing to the toes and their joints.

10. Name the vein that drains the posterior leg muscles as it ascends from the ankle through the posterior leg.

Elastic Artery - The Aorta

EXERCISE 5.32: Cardiovascular System - Elastic Artery (Aorta)— Histology

- *Insert* Anatomy & Physiology | Revealed® **Cardiovascular, Lymphatic, and Respiratory Systems** *CD, or, if you are already in the* **Dissection** *section, click the* **CHANGE VIEW** *button at the top of the screen, and skip the next two steps.*

- *From the* **Select system** *menu, select* **Cardiovascular**.

- *In the* **Home screen**, *select the* **Dissection** *button in the left portion of the screen. You may click either on the* **Dissection** *button or on the word.*

- *In the* **SELECT A VIEW** *window that appears, click on the* **Select topic** *button.*

- *Choose* **Elastic artery (aorta)** *from the menu.*

- *The* **Select view** *menu will then show* **Histology**.

- *The* **GO** *button will flash green. Click on it.*

- *Click on* **TAG 1**, *and the following image will appear:*

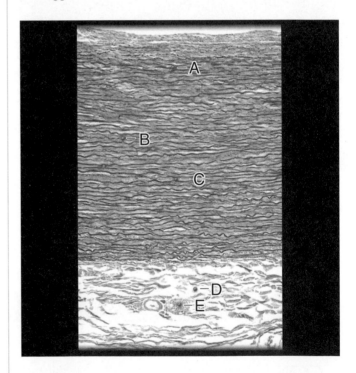

- *Mouse-over the blue pins on the screen to find the information necessary to fill in the following blanks:*

A. _____

B. _____

C. _____

D. _____

E. _____

CHECK POINT:

Elastic Artery (Aorta)—Histology

1. What is the name for the outermost layer of the vessel wall?
2. What is the name for the middle layer of the vessel wall?
3. What is the name for the innermost layer of the vessel wall?

IN REVIEW

What Have I Learned?

The following questions cover the material that you have just learned — the histology of an elastic artery. Use the information in the **STRUCTURE INFORMATION** window for these structures to answer the following questions:

1. In arteries, which tunic or layer is the thickest?

2. In veins, which tunic or layer is the thickest?

3. Which fibers are more abundant in the tunics of arteries than in the those of veins?

4. What is the function of the tunica externa of elastic arteries?

5. What is the function of the tunica media of elastic arteries?

6. What is the function of the tunica intima of elastic arteries?

7. Which of the three tunics contains the endothelium? Did you know that the endothelium of the blood vessels is continuous with the endocardium of the heart?

8. Which structures in the walls of medium and large arteries provide the elastic properties of those blood vessel walls?

9. Name the small blood vessels in the tunica externa.

10. How are the other two tunics supplied with oxygen and nutrients?

Large Vein - The Inferior Vena Cava

EXERCISE 5.33: Cardiovascular System - Large Vein (Inferior Vena Cava)—Histology

- *Insert* Anatomy & Physiology | Revealed® **Cardiovascular, Lymphatic, and Respiratory Systems** *CD, or, if you are already in the* **Dissection** *section, click the* **CHANGE VIEW** *button at the top of the screen, and skip the next two steps.*

- *From the* **Select system** *menu, select* **Cardiovascular**.

- *In the* **Home screen**, *select the* **Dissection** *button in the left portion of the screen. You may click either on the* **Dissection** *button or on the word.*

- *In the* **SELECT A VIEW** *window that appears, click on the* **Select topic** *button.*

- *Choose* **Large vein (inferior vena cava)** *from the menu.*

- *The* **Select view** *menu will then show* **Histology**.

- *The* **GO** *button will flash green. Click on it.*
- *Click on* **TAG 1**, *and the following screen will appear:*

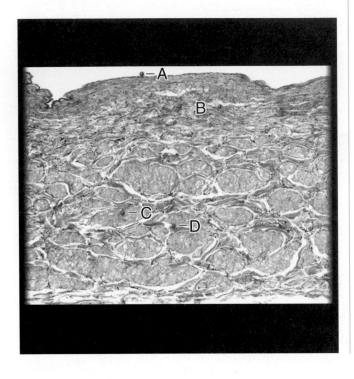

- *Mouse-over the blue pins on the screen to find the information necessary to fill in the following blanks:*

A. _____

B. _____

C. _____

D. _____

CHECK POINT:

Large Vein (Inferior Vena Cava)—Histology

1. What is the name for the outermost layer of the vessel wall?
2. What is the name for the middle layer of the vessel wall?
3. What is the name for the innermost layer of the vessel wall?

IN REVIEW

What Have I Learned?

The following questions cover the material that you have just learned—the histology of a large vein. Use the information in the **STRUCTURE INFORMATION** window for these structures to answer the following questions:

1. Where are the smooth muscle bundles located in veins?

2. What is their function?

3. How does the tunica intima of the inferior vena cava compare with that of the aorta?

4. How does the tunica media of the inferior vena cava compare with that of the aorta?

5. How does the tunica externa of the inferior vena cava compare with that of the aorta?

Muscular Artery and Medium-Sized Vein

EXERCISE 5.34: Cardiovascular System - Muscular Artery and Medium-Sized Vein— Histology

- *Insert* Anatomy & Physiology | Revealed® **Cardiovascular, Lymphatic, and Respiratory Systems** *CD, or, if you are already in the* **Dissection** *section, click the* **CHANGE VIEW** *button at the top of the screen, and skip the next two steps.*

- *From the* **Select system** *menu, select* **Cardiovascular**.

- *In the* **Home screen**, *select the* **Dissection** *button in the left portion of the screen. You may click either on the* **Dissection** *button or on the word.*

- *In the* **SELECT A VIEW** *window that appears, click on the* **Select topic** *button.*

- *Choose* **Muscular artery and medium-sized vein** *from the menu.*

- *The* **Select view** *menu will then show* **Histology**.

- *The* **GO** *button will flash green. Click on it.*

- *Click on* **TAG 1**, *and the following image will appear:*

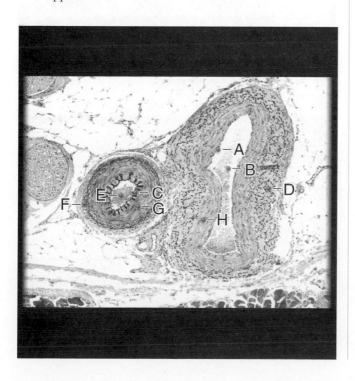

- *Mouse-over the blue pins on the screen to find the information necessary to fill in the following blanks:*

A. _____

B. _____

C. _____

D. _____

E. _____

F. _____

G. _____

H. _____

CHECK POINT:

Muscular Artery and Medium-Sized Vein, Histology

1. Where are muscular arteries located?
2. Where are medium-sized veins located?
3. What is the function of muscular arteries?

IN REVIEW

What Have I Learned?

The following questions cover the material that you have just learned—the histology of muscular arteries and medium-sized veins. Use the information in the **STRUCTURE INFORMATION** window for these vessels to answer the following questions:

1. What tissue predominates the tunica media of a muscular artery?

2. Give three examples of muscular arteries.

3. What is the function of medium-sized veins?

4. What tissues predominate the tunica media of these veins?

5. Give two examples of medium-sized veins.

Arteriole and Venule

EXERCISE 5.35: Cardiovascular System - Arteriole and Venule— Histology

- *Insert* Anatomy & Physiology | Revealed® **Cardiovascular, Lymphatic, and Respiratory Systems** *CD, or, if you are already in the* **Dissection** *section, click the* **CHANGE VIEW** *button at the top of the screen, and skip the next two steps.*

- *From the* **Select system** *menu, select* **Cardiovascular**.

- *In the* **Home screen**, *select the* **Dissection** *button in the left portion of the screen. You may click either on the* **Dissection** *button or on the word.*

- *In the* **SELECT A VIEW** *window that appears, click on the* **Select topic** *button.*

- *Choose* **Arteriole and venule** *from the menu.*

- *The* **Select view** *menu will then show* **Histology**.

- *The* **GO** *button will flash green. Click on it.*

- *Click on* **TAG 1**, *and the following image will appear:*

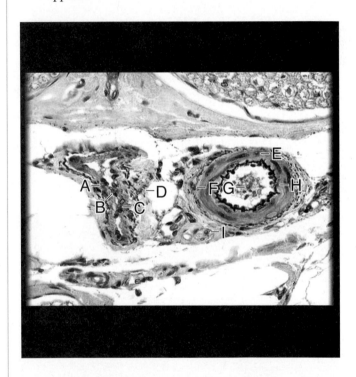

- *Mouse-over the blue pins on the screen to find the information necessary to fill in the following blanks:*

A. _____

B. _____

C. _____

D. _____

E. _____

F._____

G. _____

H. _____

I._____

IN REVIEW

What Have I Learned?

The following questions cover the material that you have just learned—the histology of arterioles and venules. Use the information in the **STRUCTURE INFORMATION** window for these vessels to answer the following questions:

1. What is the diameter of an arteriole?

2. Describe the tunica media of an arteriole.

3. What is the function of an arteriole?

4. What is the diameter of a venule?

5. What is the function of venules?

Cardiac Muscle

EXERCISE 5.36:   Cardiovascular System - Cardiac Muscle—Histology

- *Insert* Anatomy & Physiology | Revealed® **Cardiovascular, Lymphatic, and Respiratory Systems** *CD, or, if you are already in the* **Dissection** *section, click the* **CHANGE VIEW** *button at the top of the screen, and skip the next two steps.*

- *From the* **Select system** *menu, select* **Cardiovascular**.

- *In the* **Home screen**, *select the* **Dissection** *button in the left portion of the screen. You may click either on the* **Dissection** *button or on the word.*

- *In the* **SELECT A VIEW** *window that appears, click on the* **Select topic** *button.*

- *Choose* **Cardiac muscle** *from the menu.*

- *The* **Select view** *menu will then show* **Histology**.

- *The* **GO** *button will flash green. Click on it.*

- *Click on* **TAG 1**, *and the following image will appear:*

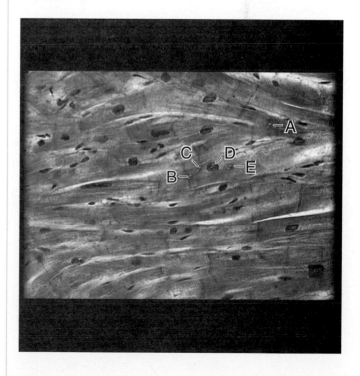

- *Mouse-over the blue pins on the screen to find the information necessary to fill in the following blanks:*

A. _____

B. _____

C. _____

D. _____

E. _____

Cardiac Muscle—Histology

1. Name three locations where cardiac muscle is found.
2. What is the function of cardiac muscle tissue?
3. This function may be modulated by _____.

IN REVIEW

What Have I Learned?

The following questions cover the material that you have just learned—the histology of cardiac muscle. Use the information in the **STRUCTURE INFORMATION** window for this tissue to answer the following questions:

1. Unlike skeletal muscle, cardiac muscle fibers

   _____ .

2. Name one way skeletal muscle is similar to cardiac muscle tissue.

3. Describe intercalated disks.

4. What is their function?

5. What is sarcoplasm?

Neurovascular Bundle

EXERCISE 5.37: Cardiovascular System - Neurovascular Bundle— Histology

- *Insert* Anatomy & Physiology | Revealed® **Cardiovascular, Lymphatic, and Respiratory Systems** *CD, or, if you are already in the* **Dissection** *section, click the* **CHANGE VIEW** *button at the top of the screen, and skip the next two steps.*

- *From the* **Select system** *menu, select* **Cardiovascular**.

- *In the* **Home screen**, *select the* **Dissection** *button in the left portion of the screen. You may click either on the* **Dissection** *button or on the word.*

- *In the* **SELECT A VIEW** *window that appears, click on the* **Select topic** *button.*

- *Choose* **Neurovascular bundle** *from the menu.*

- *The* **Select view** *menu will then show* **Histology**.

- *The* **GO** *button will flash green. Click on it.*

- *Click on* **TAG 1**, *and the following image will appear:*

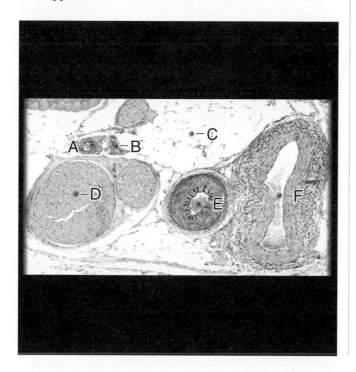

- *Mouse-over the blue pins on the screen to find the information necessary to fill in the following blanks:*

A. _____

B. _____

C. _____

D. _____

E. _____

F. _____

IN REVIEW

What Have I Learned?

The following questions cover the material that you have just learned—the histology of a neurovascular bundle. Use the information in the **STRUCTURE INFORMATION** window for these structures to answer the following questions:

1. Describe adipose connective tissue.

2. What is its function?

3. Name five structures, other than connective tissue, found in neurovascular bundles.

Hemopoiesis and Hemoglobin Breakdown

EXERCISE 5.38:   Cardiovascular System - Hemopoiesis and Hemoglobin Breakdown

Animation: Hemopoiesis

- *Insert Anatomy & Physiology | Revealed®* **Cardiovascular, Lymphatic and Respiratory Systems** *CD, or, if you are already in the* **Dissection** *section, click the* **Animation** *icon*

at the bottom of the screen, and skip the next two steps.

- *From the* **Select system** *menu, select* **Cardiovascular**.

- *In the* **Home screen**, *select the* **Animations** *button in the left portion of the screen. You may click either on the* **Animations** *button or on the word.*

- *In the* **Select topic** *menu, select* **Physiology**.

- *From the* **Select animation** *menu, select* **Hemopoiesis**.

- *Click the* **Play** *button, and the animation will run in the* **IMAGE AREA**.

- *After viewing the animation, answer the following questions:*

1. Hemopoiesis is the process of _____, which occurs primarily _____.

2. Hemopoiesis begins with undifferentiated cells called _____.

3. These _____ give rise to _____.

4. What types of hormones influence the differentiation of the blood stem cells?

5. What are the two lines of cells that differentiate from the hemocytoblast?

6. Which groups of cells arise from each of these lines?

7. Which cells are produced by the lymphoid cell line?

8. Erythropoiesis produces _____.

9. List the steps for red blood cell (erythrocyte) production.

10. Thrombopoiesis produces _____.

11. What is the name for the committed progenitor cell in thrombopoiesis?

12. In response to the hormone _____, the _____ differentiates into a _____.

13. Platelets are formed when _____.

14. Leukopoiesis is the _____.

15. The myeloid cell line gives rise to _____.

16. The lymphoid cell line produces _____.

17. Eosinophilic myelocytes differentiate into _____.

18. Basophilic myelocytes differentiate into _____.

19. Neutrophilic myelocytes differentiate into _____.

20. Monocytes are derived from _____.

21. Lymphocytes are derived from _____.

22. The two types of lymphocytes that develop are _____.

Animation: Hemoglobin Breakdown

- *Return to the animation menu at the top-left of the screen, and from the* **Select animation** *menu, select* **Hemoglobin breakdown**.

- *Click the* **Play** *button, and the animation will run in the* **IMAGE AREA**.

- *After viewing the animation, answer the following questions:*

1. What happens to the hemoglobin released by the rupture of old red blood cells?

2. The hemoglobin is broken down into two main components, which are _____ and _____.

3. The globin chains _____.

4. What is released from the heme?

5. The remaining structure of the heme goes through a two-step process, being converted to the following products sequentially _____.

6. What plasma protein transports the iron?

7. Where is it transported for storage (two locations)?

8. Where is it transported to make new hemoglobin?

9. What plasma protein transports free bilirubin?

10. Where is it transported?

11. Liver cells make _____, excreted as part of _____ into the _____.

12. _____ convert _____ into _____, which contribute to the _____.

13. Some of the _____ are absorbed into the blood and excreted from the _____ in the _____. (This product is urochrome, which gives urine its yellow color.)

IN REVIEW

What Have I Learned?

The following questions cover the material that you have just learned—hemopoiesis and hemoglobin breakdown. Use the information from the animations of these processes to answer the following questions:

1. Hemopoiesis is the process of _____, which occurs primarily _____.

2. What types of hormones influence the differentiation of the blood stem cells?

3. Erythropoiesis produces _____.

4. Describe the production of platelets.

5. Name the two types of lymphocytes. Where does each type mature?

6. What is hemopoiesis?

7. Name the oxygen-carrying molecule in red blood cells.

8. What two components of this molecule are released by hemopoiesis?

9. Describe how these components are recycled.

10. What products of hemoglobin breakdown end up in the feces? In the urine?

CHAPTER 6

The Lymphatic System

Overview: Lymphatic System

You are probably not very familiar with the lymphatic system. Without it you could not survive. It recovers fluid lost from the blood capillaries, is intricately involved with the immune system, and absorbs dietary lipids through lacteals located in the small intestine.

Let's begin our exploration of this mostly unknown organ system by taking an overview of the entire lymphatic system, and then looking at the specific structures in more detail.

Animation: Lymphatic System

- *Insert* Anatomy & Physiology | Revealed® **Cardiovascular, Lymphatic, and Respiratory Systems** *CD.*

- *From the* **Select system** *menu, select* **Lymphatic**.

- *In the* **Home screen**, *select the* **Animation** *button in the left portion of the screen. You may click either on the* **Animation** *button or on the word.*

- *In the* **Select topic** *menu, select* **Anatomy**.

- *The* **Select animation** *menu will show* **Lymphatic system overview**.

- *Click the* **Play** *button, and the animation will run in the* **IMAGE AREA**.

- *After viewing the animation, answer the following questions:*

1. What is the lymphatic system?

2. Name the fluid involved in this system.

3. As a system, what is its function?

4. Name the fluid that seeps from the blood capillaries throughout your body.

5. What percentage of this fluid becomes lymph?

6. What are lymphatic capillaries?

7. What do they form when they converge?

8. What do these vessels form when they merge?

9. Where do these vessels drain?

10. Where does the right lymphatic duct receive lymph from?

11. Where does the right lymphatic duct empty?

12. Where does the thoracic duct receive lymph from?

13. Where does it empty?

14. Which lymphatic duct is larger, the right lymphatic duct or the thoracic duct?

15. Lymphatic tissues include _____ _____.

16. What are lymphatic nodules?

17. What do they contain?

18. Where are clusters of lymphatic nodules associated?

19. Name the large groups of lymphatic nodules found in the walls of the nasal and oral cavities.

20. _____ tonsils are located in the nasopharynx. When they are inflamed, they are known as _____.

21. Where are the palatine tonsils located? The lingual tonsils?

22. Name the lymphatic organs. What is their structure?

23. What are lymph nodes? Where are prominent clusters of lymph nodes located?

24. What are the primary functions of lymph nodes?

25. What is the basic structure of a lymph node?

26. Describe the passage of lymph into, through, and out of the lymph node.

27. Where is the thymus located? What is its function?

28. When is the thymus most active?

29. What becomes of the thymus, beginning at adolescence?

30. Name the body's largest lymph organ. Where is it located?

31. What is its function?

32. How does it act like a lymph node?

The Thorax

EXERCISE 6.1:   Lymphatic System - Thorax, Anterior View

- *To begin your study of the **Dissection** view, select the **Dissection** button at the bottom of the screen.*
- *In the **SELECT A VIEW** window that appears, click on the **Select topic** button.*
- *Choose **Thorax** from the menu.*
- *The **Select view** menu will then show **Anterior**.*
- *The **GO** button will flash green. Click on it.*
- *After clicking on **TAGs 2-3** , click on **TAG 4** and the following screen will appear:*

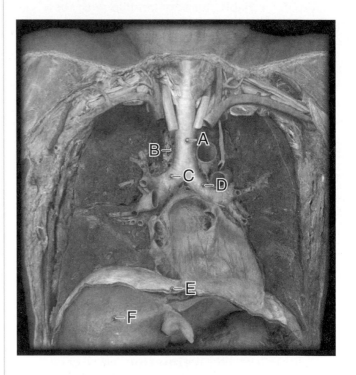

- *Mouse-over the blue pins on the screen to find the information necessary to fill in the following blanks:*

A. _____

B. _____

C. _____

D. _____

E. _____

F._____

Thorax, Anterior View

1. What are lymph nodes?
2. The mediastinal lymph nodes are clusters found along what structures?
3. Where are lymph nodes typically found?

- *Click on* **TAG 5**, *and the following image will appear:*

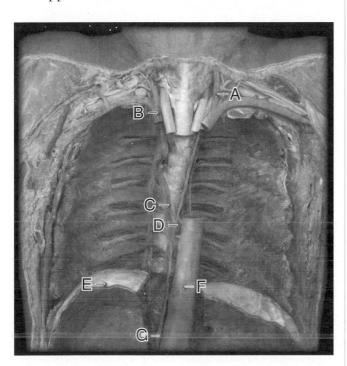

- *Mouse-over the blue pins on the screen to find the information necessary to fill in the following blanks:*

A. _____

B. _____

C. _____

D. _____

E. _____

F. _____

G. _____

Thorax, Anterior View

4. Name the small irregular-shaped lymph sac found in the abdomen.
5. From where does it receive lymph?
6. This structure in question 4 forms the origin of which lymph duct?
7. List the areas drained by the thoracic duct.

Lymph nodes

Animation: Antigen Processing

- *Insert* Anatomy & Physiology | Revealed® **Cardiovascular, Lymphatic, and Respiratory Systems** *CD.*
- *From the* **Select system** *menu, select* **Lymphatic**.
- *In the* **Home screen**, *select the* **Animation** *button in the left portion of the screen. You may click either on the* **Animation** *button or on the word.*
- *In the* **Select topic** *menu, select* **Physiology**.
- *In the* **Select animation** *menu, select* **Antigen processing**.
- *Click the* **Play** *button, and the animation will run in the* **IMAGE AREA**.
- *After viewing the animation, answer the following questions:*

1. What is an antigen?

2. After antigens are produced, where are they transported?

3. What molecules combine with the antigens there? This combination is then transported to the _____ and from there to the _____.

4. What then happens to foreign antigens? To self-antigens?

5. When an antigen originates from outside of the cell, how do the particles enter the cell?

6. What happens to the particles inside the cell?

7. The vesicle containing the foreign particles

_____

_____.

8. What then happens to the MHC class II / antigen complex?

Animation: Cytotoxic T Cells

- *Return to the animation menu at the top left of the screen, and from the* **Select animation** *menu, select* **Cytotoxic T cells**.

- *Click the* **Play** *button, and the animation will run in the* **IMAGE AREA**.

- *After viewing the animation, answer the following questions:*

1. When a virus infects a cell, what does it produce?

2. What happens to some of these proteins?

3. What molecule complexes with these fragments?

4. Where are they displayed?

5. How do the cytotoxic T cells interact with the virus-infected cells?

6. What substances are released by the cytotoxic T cells?

7. The release of these substances results in

_____

_____.

8. What is the result for self-proteins?

9. What then becomes of the cytotoxic T cells?

Animation: Helper T Cells

- *Return to the animation menu at the top left of the screen, and from the* **Select animation** *menu, select* **Helper T cells**.

- *Click the* **Play** *button, and the animation will run in the* **IMAGE AREA**.

- *After viewing the animation, answer the following questions:*

1. Proteins (antigens) require the cooperation of helper T cells for what purpose?

2. These antigens are therefore said to be

_____.

3. What does an antigen presenting cell do in the presence of the antigen?

4. The antigen is then moved _____ on a _____.

5. How does the helper T cell become activated?

6. What is the activated T cell capable of doing?

7. The antigen reacts with an _____ on the surface of the B cell and is then _____.

8. How does the B cell interact with the activated T cell?

9. The helper T cell produces _____, which stimulate the B cell to _____.

Animation: IgE Mediated Hypersensitivity

- *Return to the animation menu at the top left of the screen, and from the* **Select animation** *menu, select* **IgE mediated hypersensitivity**.

- *Click the* **Play** *button, and the animation will run in the* **IMAGE AREA**.

- *After viewing the animation, answer the following questions:*

1. Another name for an allergic reaction is

_____.

2. This is mediated by _____

_____.

3. How does sensitization occur?

4. Which tissues are rich in B cells committed to IgE production?

5. IgE producing cells are more abundant in _____ .

6. What do the helper T cells produce, and what is the effect on B cells?

7. Where and how do IgE molecules attach?

8. What are mast cells?

9. When an antigen-sensitive person is exposed a second time to the antigen, where does the antigen bind?

10. What is required to trigger a response?

11. Within seconds, what chemicals are released from the mast cells? What do they trigger?

12. What are some of those symptoms?

IN REVIEW

What Have I Learned?

The following questions cover the material that you have just learned—the lymphatic system of the thorax. Use the information in the **STRUCTURE INFORMATION** window for these structures to answer the following questions:

1. List the areas drained by the thoracic duct.

2. Lymph nodes are _____
_____ .

3. Lymph nodes are typically found in _____
_____ .

4. Where are these clusters typically found?

5. Name the structure that forms the origin of the thoracic duct.

EXERCISE 6.2: Lymphatic System - Lymph Node (low power), Histology

- *Insert* Anatomy & Physiology | Revealed® **Cardiovascular, Lymphatic, and Respiratory Systems** *CD, or, if you are already in the* **Dissection** *section, click the* **CHANGE VIEW** *button at the top of the screen, and skip the next two steps.*

- *From the* **Select system** *menu, select* **Lymphatic**.

- *In the* **Home screen**, *select the* **Dissection** *button in the left portion of the screen. You may*

click either on the **Dissection** *button or on the word.*

- *In the* **SELECT A VIEW** *window that appears, click on the* **Select topic** *button.*

- *Choose* **Lymph node (low power)** *from the menu.*

- *The* **Select view** *menu will then show* **Histology**.

- *The* **GO** *button will flash green. Click on it.*

• Click on **TAG 1**, and the following image will appear:

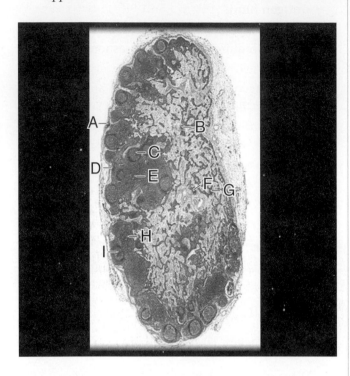

• Mouse-over the blue pins on the screen to find the information necessary to fill in the following blanks:

A. _____

B. _____

C. _____

D. _____

E. _____

F. _____

G. _____

H. _____

I. _____

CHECK POINT:

Lymph Node (low power), Histology

1. Name the dense irregular connective tissue that covers the outer surface of a lymph node.
2. Name the outer zone of the lymph node.
3. What is its function?

EXERCISE 6.3: Lymphatic System - Lymph Node (high power), Histology

• *Insert* Anatomy & Physiology | Revealed® **Cardiovascular, Lymphatic, and Respiratory Systems** *CD, or, if you are already in the* **Dissection** *section, click the* **CHANGE VIEW** *button at the top of the screen, and skip the next two steps.*

• *From the* **Select system** *menu, select* **Lymphatic**.

• *In the* **Home screen**, *select the* **Dissection** *button in the left portion of the screen. You may click either on the* **Dissection** *button or on the word.*

• *In the* **SELECT A VIEW** *window that appears, click on the* **Select topic** *button.*

• *Choose* **Lymph node (high power)** *from the menu.*

• *The* **Select view** *menu will then show* **Histology**.

• *The* **GO** *button will flash green. Click on it.*

• *Click on* **TAG 1**, *and the following image will appear:*

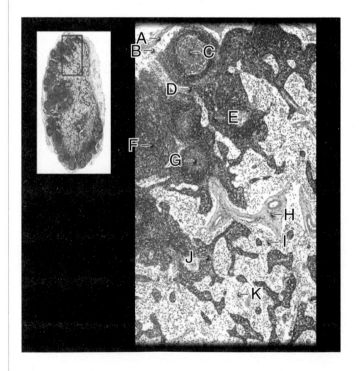

- *Mouse-over the blue pins on the screen to find the information necessary to fill in the following blanks:*

A. _____

B. _____

C. _____

D. _____

E. _____

F. _____

G. _____

H. _____

I. _____

J. _____

K. _____

IN REVIEW

What Have I Learned?

The following questions cover the material that you have just learned—the histology of a lymph node. Use the information in the **STRUCTURE INFORMATION** window for these structures to answer the following questions:

1. Name the inner zone of the lymph node.

2. What is its function?

3. What percent of the lymphocytes remain in the medulla while the rest leave the node via the lymph?

4. Name the location for memory B lymphocyte and plasma cell formation.

5. Name the site of B lymphocyte localization.

6. Name the structure of the lymph node that forms in response to antigenic challenge. What is its function?

7. Name the site of antibody production. What cells are responsible for this production?

Palatine Tonsil

EXERCISE 6.4: Lymphatic System - Palatine Tonsil (low power), Histology

- *Insert Anatomy & Physiology | Revealed® **Cardiovascular, Lymphatic, and Respiratory Systems** CD, or, if you are already in the **Dissection** section, click the **CHANGE VIEW** button at the top of the screen, and skip the next two steps.*

- *From the **Select system** menu, select **Lymphatic**.*

- *In the **Home screen**, select the **Dissection** button in the left portion of the screen. You may click either on the **Dissection** button or on the word.*

- *In the **SELECT A VIEW** window that appears, click on the **Select topic** button.*

- *Choose **Palatine tonsil (low power)** from the menu.*

- *The **Select view** menu will then show **Histology**.*

- *The **GO** button will flash green. Click on it.*

• Click on **TAG 1**, and the following image will appear:

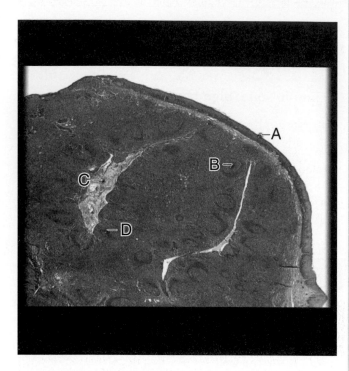

• Mouse-over the blue pins on the screen to find the information necessary to fill in the following blanks:

A. _____

B. _____

C. _____

D. _____

CHECK POINT:

Palatine Tonsil (low power), Histology

1. Name the superficial structure of the palatine tonsils.
2. What is its function?
3. What is the function of the germinal center?

EXERCISE 6.5:   Lymphatic System - Palatine Tonsil (high power), Histology

• *Insert* Anatomy & Physiology | Revealed® **Cardiovascular, Lymphatic, and Respiratory Systems** *CD, or, if you are already in the* **Dissection** *section, click the* **CHANGE**

VIEW *button at the top of the screen, and skip the next two steps.*

• *From the* **Select system** *menu, select* **Lymphatic**.

• *In the* **Home screen**, *select the* **Dissection** *button in the left portion of the screen. You may click either on the* **Dissection** *button or on the word.*

• *In the* **SELECT A VIEW** *window that appears, click on the* **Select topic** *button.*

• *Choose* **Palatine tonsil (high power)** *from the menu.*

• *The* **Select view** *menu will then show* **Histology**.

• *The* **GO** *button will flash green. Click on it.*

• *Click on* **TAG 1**, *and the following image will appear:*

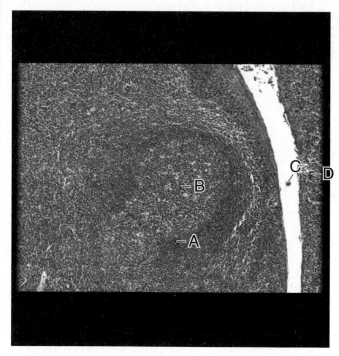

• *Mouse-over the blue pins on the screen to find the information necessary to fill in the following blanks:*

A. _____

B. _____

C. _____

D. _____

IN REVIEW

What Have I Learned?

The following questions cover the material that you have just learned - the histology of a palatine tonsil. Use the information in the **STRUCTURE INFORMATION** window for these structures to answer the following questions:

1. What is the function of the lymphatic nodules?

2. Name the structures that contain discarded epithelial cells, dead white blood cells, bacteria, and debris from the oral cavity.

3. What is the function of these structures? How many are found on each tonsil?

4. Name the structure of the palatine tonsils continuous with the epithelium of the oral cavity and oropharynx.

Peyer's Patch

EXERCISE 6.6:   Lymphatic System - Peyer's Patch (low power), Histology

- *Insert* Anatomy & Physiology | Revealed® **Cardiovascular, Lymphatic, and Respiratory Systems** *CD, or, if you are already in the* **Dissection** *section, click the* **CHANGE VIEW** *button at the top of the screen, and skip the next two steps.*

- *From the* **Select system** *menu, select* **Lymphatic**.

- *In the* **Home screen**, *select the* **Dissection** *button in the left portion of the screen. You may click either on the* **Dissection** *button or on the word.*

- *In the* **SELECT A VIEW** *window that appears, click on the* **Select topic** *button.*

- *Choose* **Peyer's patch (low power)** *from the menu.*

- *The* **Select view** *menu will then show* **Histology**.

- *The* **GO** *button will flash green. Click on it.*

- *Click on* **TAG 1**, *and the following image will appear:*

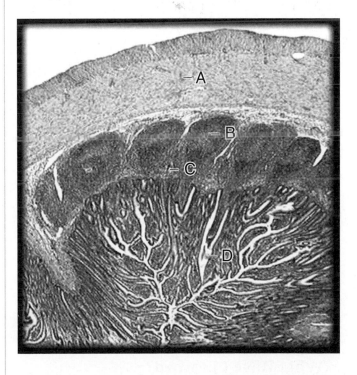

- *Mouse-over the orange pins on the screen to find the information necessary to fill in the following blanks:*

A. _____

B. _____

C. _____

D. _____

CHECK POINT:

Peyer's Patch (low power), Histology

1. Specifically, where are the Peyer's patches found?
2. What is their function?
3. How is the intestinal wall shaped at the site of a Peyer's patch?

EXERCISE 6.7: Lymphatic System - Peyer's Patch (high power), Histology

- *Insert* Anatomy & Physiology | Revealed® **Cardiovascular, Lymphatic, and Respiratory Systems** *CD, or, if you are already in the* **Dissection** *section, click the* **CHANGE VIEW** *button at the top of the screen, and skip the next two steps.*

- *From the* **Select system** *menu, select* **Lymphatic**.

- *In the* **Home screen**, *select the* **Dissection** *button in the left portion of the screen. You may click either on the* **Dissection** *button or on the word.*

- *In the* **SELECT A VIEW** *window that appears, click on the* **Select topic** *button.*

- *Choose* **Peyer's patch (high power)** *from the menu.*

- *The* **Select view** *menu will then show* **Histology**.

- *The* **GO** *button will flash green. Click on it.*

- *Click on* **TAG 1**, *and the following image will appear:*

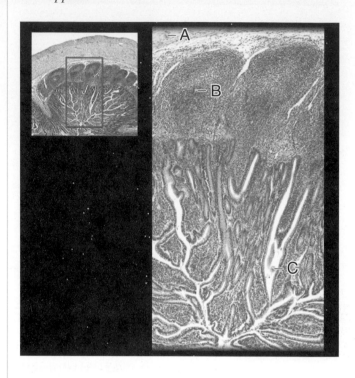

- *Mouse-over the orange pins on the screen to find the information necessary to fill in the following blanks:*

A. _____

B. _____

C. _____

IN REVIEW

What Have I Learned?

The following questions cover the material that you have just learned—the histology of a Peyer's Patch. Use the information in the **STRUCTURE INFORMATION** window for these structures to answer the following questions:

1. What does the acronym MALT stand for?

2. Where are MALT located in the body?

3. What is the function of the lymphatic nodules? Why do they form?

4. What is the function of the ileum?

The Spleen

EXERCISE 6.8:   Lymphatic System - Spleen (low power), Histology

- *Insert* Anatomy & Physiology | Revealed® **Cardiovascular, Lymphatic, and Respiratory Systems** *CD, or, if you are already in the* **Dissection** *section, click the* **CHANGE VIEW** *button at the top of the screen, and skip the next two steps.*

- *From the* **Select system** *menu, select* **Lymphatic**.

- *In the* **Home screen**, *select the* **Dissection** *button in the left portion of the screen. You may click either on the* **Dissection** *button or on the word.*

- *In the* **SELECT A VIEW** *window that appears, click on the* **Select topic** *button.*

- *Choose* **Spleen (low power)** *from the menu.*

- *The* **Select view** *menu will then show* **Histology**.

- *The* **GO** *button will flash green. Click on it.*

- *Click on* **TAG 1**, *and the following image will appear:*

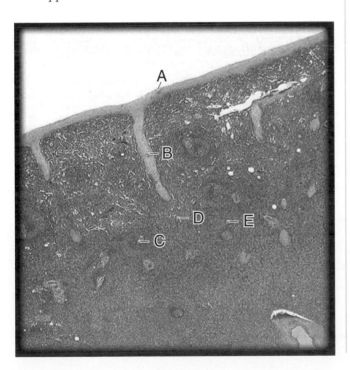

- *Mouse-over the orange pins on the screen to find the information necessary to fill in the following blanks:*

A. _____

B. _____

C. _____

D. _____

E. _____

CHECK POINT:

Spleen (low power), Histology

1. Name the outer covering of the spleen. This structure consists of what tissues?
2. What structure provides the main structural support for the spleen?
3. Name the functions of the red pulp.

EXERCISE 6.9:   Lymphatic System - Spleen (high power), Histology

- *Insert* Anatomy & Physiology | Revealed® **Cardiovascular, Lymphatic, and Respiratory Systems** *CD, or, if you are already in the* **Dissection** *section, click the* **CHANGE VIEW** *button at the top of the screen, and skip the next two steps.*

- *From the* **Select system** *menu, select* **Lymphatic**.

- *In the* **Home screen**, *select the* **Dissection** *button in the left portion of the screen. You may click either on the* **Dissection** *button or on the word.*

- *In the* **SELECT A VIEW** *window that appears, click on the* **Select topic** *button.*

- *Choose* **Spleen (high power)** *from the menu.*

- *The* **Select view** *menu will then show* **Histology**.

- *The* **GO** *button will flash green. Click on it.*

- *Click on* **TAG 1**, *and the following image will appear:*

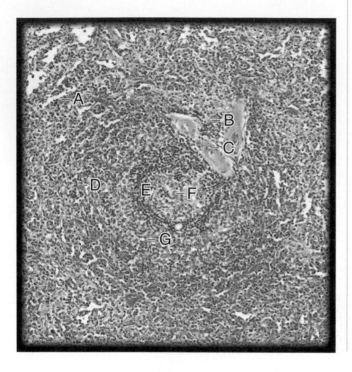

- *Mouse-over the orange pins on the screen to find the information necessary to fill in the following blanks:*

A. _____

B. _____

C. _____

D. _____

E. _____

F. _____

G. _____

CHECK POINT:

Spleen (high power), Histology

1. Name the source of arterial blood to the white pulp. These are a small branch of the _____ artery from the _____ artery.
2. What is the function of the marginal zone?
3. What is the structure of the marginal zone?

IN REVIEW

What Have I Learned?

The following questions cover the material that you have just learned—the histology of the spleen. Use the information in the **STRUCTURE INFORMATION** window for these structures to answer the following questions:

1. What is the function of the white pulp? What percent of the total mass of the spleen is white pulp?

2. Name the four structures that make up the white pulp.

3. Name the location for memory B lymphocyte and plasma cell formation. How are they affected by age?

4. What tissue/structure makes up the bulk of the spleen?

5. This tissue/structure is composed of _____ _____ _____ .

The Thymus

EXERCISE 6.10: Lymphatic System - Thymus (low power), Histology

- *Insert* Anatomy & Physiology | Revealed® **Cardiovascular, Lymphatic, and Respiratory Systems** *CD, or, if you are already in the* **Dissection** *section, click the* **CHANGE VIEW** *button at the top of the screen, and skip the next two steps.*

- *From the* **Select system** *menu, select* **Lymphatic***.*

- *In the* **Home screen***, select the* **Dissection** *button in the left portion of the screen. You may click either on the* **Dissection** *button or on the word.*

- *In the* **SELECT A VIEW** *window that appears, click on the* **Select topic** *button.*

- *Choose* **Thymus (low power)** *from the menu.*

- *The* **Select view** *menu will then show* **Histology***.*

- *The* **GO** *button will flash green. Click on it.*

- *Click on* **TAG 1***, and the following image will appear:*

- *Mouse-over the blue pins on the screen to find the information necessary to fill in the following blanks:*

A. _____

B. _____

C. _____

D. _____

E. _____

CHECK POINT:

Thymus (low power), Histology

1. Where is the cortex of the thymus located? What is its function?
2. Where is the medulla of the thymus located? What is its function?
3. What is the main subdivision of the thymus?

EXERCISE 6.11: Lymphatic System - Thymus (high power), Histology

- *Insert* Anatomy & Physiology | Revealed® **Cardiovascular, Lymphatic, and Respiratory Systems** *CD, or, if you are already in the* **Dissection** *section, click the* **CHANGE VIEW** *button at the top of the screen, and skip the next two steps.*

- *From the* **Select system** *menu, select* **Lymphatic***.*

- *In the* **Home screen***, select the* **Dissection** *button in the left portion of the screen. You may click either on the* **Dissection** *button or on the word.*

- *In the* **SELECT A VIEW** *window that appears, click on the* **Select topic** *button.*

- *Choose* **Thymus (high power)** *from the menu.*

- *The* **Select view** *menu will then show* **Histology***.*

- *The* **GO** *button will flash green. Click on it.*

- *Click on **TAG 1**, and the following image will appear:*

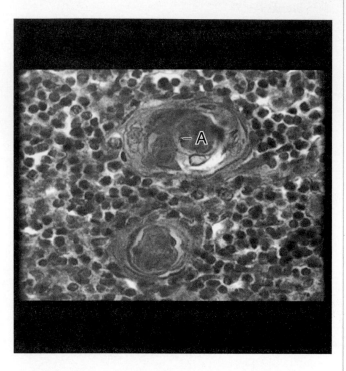

- *Mouse-over the blue pin on the screen to find the information necessary to fill in the following blank:*

A. _____

HEADS UP!

View the radiographic images for the lymphatic system.

- *Click on the **IMAGING** button at the bottom of the screen.*
- *Click on the **Select structure** menu in the **STRUCTURE LIST** box, and then on the individual structures to highlight them in the **IMAGE AREA**.*

IN REVIEW

What Have I Learned?

The following questions cover the material that you have just learned—the histology of the thymus. Use the information in the **STRUCTURE INFORMATION** window for these structures to answer the following questions:

1. What is the thymus? Where is it located? What is its function?

2. How does it change from birth to adulthood?

3. Name the subdivisions of the thymic lobe. What is its function?

4. What is the function of the septa/trabeculae?

CHAPTER 7

The Respiratory System

Overview: The Respiratory System

Take a deep breath. Now exhale. What has just occurred? You have taken in much-needed oxygen and breathed out toxic carbon dioxide. This is foundational to all of your body processes as you maintain homeostasis. Your cells require oxygen for metabolic reactions and produce carbon dioxide as a waste product of that metabolism. Without the respiratory system, metabolism, and thus homeostasis, would be impossible.

But, the function of the respiratory system doesn't stop there. It is also involved in regulating the body's acid—base balance, blood gasses, and other homeostatic controls of the circulatory system. Like so many of our organ systems, the respiratory system doesn't exist in a vacuum, but works in coordination with the circulatory and urinary systems.

We will begin with an overview of the respiratory system. Next, we will take a regional view of the respiratory structures, followed by a detailed look at these structures.

Animation: Respiratory System

- *Insert* Anatomy & Physiology | Revealed® **Cardiovascular, Lymphatic, and Respiratory Systems** *CD, or, if you are in the* **Dissection** *section, click the* **Animation** *icon at the bottom of the screen, and skip the next two steps.*

- *From the* **Select system** *menu, select* **Respiratory**.

- *In the* **Home screen,** *select the* **Animation** *button in the left portion of the screen.*

- *In the* **Select topic** *menu, select* **Anatomy.**

- *From the* **Select animation** *menu, select* **Respiratory system overview.**

- *Click the* **Play** *button, and the animation will run in the* **IMAGE AREA.**

- *After viewing the animation, answer the following questions:*

1. What are the two functions of the respiratory system?

2. Name the structures of the upper respiratory tract.

3. Name the structures of the lower respiratory tract.

4. What effect does the pharynx have on inhaled air?

5. Name the structure shared by both the respiratory and digestive systems.

6. Name the structure that contains the vestibular and vocal folds. What protective function do these folds have?

7. The vocal folds are also known as the _____. Why?

8. Name the organ that maintains an open passageway to and from the lungs. What particular structure helps to keep this passageway open?

9. This passageway divides into two _____, which upon entering the lungs, continue to divide into smaller _____ until they ultimately divide into _____.

10. Each _____ divides repetitively to form _____, _____, and _____.

11. What structure serves as the site for gas exchange? These rounded structures are surrounded by _____.

12. How do oxygen and carbon dioxide move between the blood and the alveoli?

The Upper Respiratory System

EXERCISE 7.1: Respiratory System—Upper Respiratory, Lateral View

- *Click the **DISSECTION** button that appears at the bottom of the screen to begin the dissection exercises for the respiratory system.*

- *In the **SELECT A VIEW** window that appears, click on the **Select topic** button.*

- *Choose **Upper respiratory** from the menu.*

- *The **Select view** menu will then show **Lateral**.*

- *The **GO** button will flash green. Click on it.*

- *Click on **TAG 1,** and the following image will appear:*

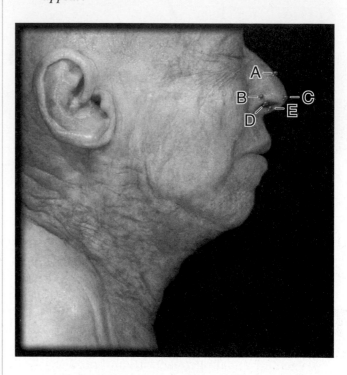

- *Mouse-over the blue pins on the screen to find the information necessary to fill in the following blanks:*

A. _____

B. _____

C. _____

D. _____

E. _____

CHECK POINT:

Upper Respitory, Lateral View

1. Name the external opening of the nose.
2. Name the structure that separates both of the openings of the nose.
3. Name the midline dorsal aspect of the nose.

• *Now click on* **TAG 2,** *and the following image will appear:*

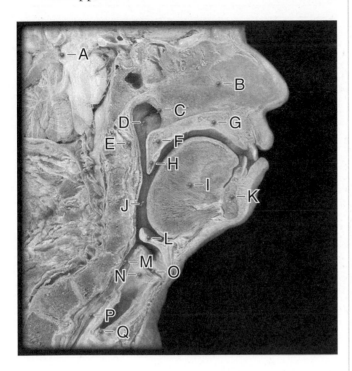

• *Mouse-over the blue pins on the screen to find the information necessary to fill in the following blanks:*

A. _____

B. _____

C. _____

D. _____

E. _____

F. _____

G. _____

H. _____

I. _____

J. _____

K. _____

L. _____

M. _____

N. _____

O. _____

P. _____

Q. _____

• *Click on* **TAG 3,** *and the following image will appear:*

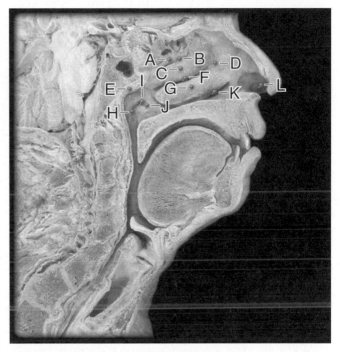

• *Mouse-over the blue pins on the screen to find the information necessary to fill in the following blanks:*

A. _____

B. _____

C. _____

D. _____

E. _____

F. _____

G. _____

H. _____

I. _____

J. _____

K. _____

L. _____

CHECK POINT:

Upper Respiratory, Lateral View

7. Name the expanded area inside of the nares. What does the Latin term mean?
8. The mucosa of what structure has both respiratory and olfactory parts?
9. Name the two shelflike projections of bone covered with mucosa and part of the ethmoid bone.

- *Click on* **TAG 4,** *and the following image will appear:*

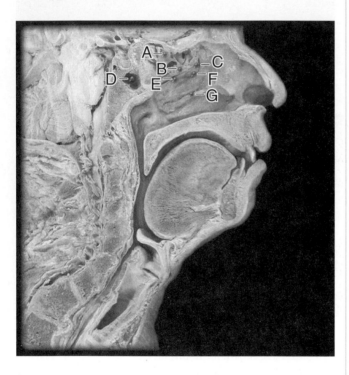

- *Mouse-over the blue pins on the screen to find the information necessary to fill in the following blanks:*

A. _____

B. _____

C. _____

D. _____

E. _____

F._____

G. _____

CHECK POINT:

Upper Respiratory, Lateral View

10. Name the structures that drain from the ethmoid sinus into the nasal cavity.
11. Name the narrow curved gap that contains the openings of the frontal and maxillary sinuses, plus the anterior ethmoid air cells.
12. Name the mucous membrane-lined cavity in the body of the sphenoid bone. Where does it drain?

⌐ **H E A D S   U P !** ────────────

View the radiographic images for the upper respiratory system.

- *Click on the* **IMAGING** *button at the bottom of the screen.*
- *Click on the* **Select region** *menu, and select* **Upper respiratory.**
- *Click on the flashing* **GO** *button.*
- *Click on the* **Select view** *menu, and select* **each option.**
- *Click on the* **Select structure** *menu in the* **STRUCTURE LIST** *box and then the individual structures to highlight them in the* **IMAGE AREA.**

IN REVIEW

What Have I Learned?

The following questions cover the material that you have just learned—the upper respiratory system. Use the information in the **STRUCTURE INFORMATION** window for these structures to answer the following questions:

1. What is the Latin term for the nose? What does this word mean?

2. Name the three structures that make up the nasal septum.

3. Name the two structures that support the nose.

4. Name the portion of the throat posterior to the oral cavity. What lymphatic organs are located there?

5. Name the structure of the soft palate that elevates during swallowing to prevent food from entering the nasopharynx.

6. Name a structure commonly deviated from the midline, impacting airflow.

7. Name the shelflike projection of mucosa-covered bone that is a separate bone.

8. Name the narrow space that contains the opening of the nasolacrimal duct.

9. What is the three-fold purpose for increasing the surface area of the nasal cavity with conchae and meatus?

10. Name the collection of lymphatic tissue in the nasopharynx.

11. Name the structure of the nasopharynx that represents the anterior end of the narrow duct between the middle ear and the pharynx. Name the structure of the nasopharynx that surrounds it.

The Lower Respiratory System

EXERCISE 7.2:   Respiratory System—Lower Respiratory, Anterior View

- *Insert* Anatomy & Physiology | Revealed® **Cardiovascular, Lymphatic, and Respiratory Systems** *CD, or, if you are already in the* **Dissection** *section, click the* **CHANGE VIEW** *button at the top of the screen, and skip the next two steps.*

- *From the* **Select system** *menu, select* **Respiratory.**

- *In the* **Home screen,** *select the* **Dissection** *button in the left portion of the screen. You may click either on the* **Dissection** *button or on the word.*

- *In the* **SELECT A VIEW** *window that appears, click on the* **Select topic** *button.*

- *Choose* **Lower respiratory** *from the menu.*

- *The* **Select view** *menu will then show* **Anterior.**

- *The* **GO** *button will flash green. Click on it.*

• *Click on* **TAG 1,** *and the following image will appear:*

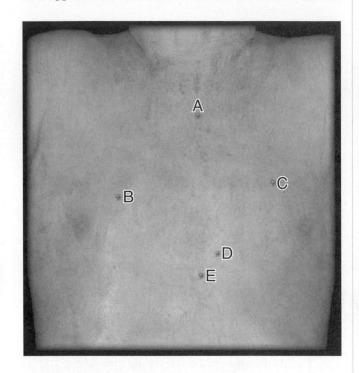

• *Mouse-over the blue pins on the screen to find the information necessary to fill in the following blanks:*

A. _____

B. _____

C. _____

D. _____

E. _____

• *Click on* **TAG 2,** *and the following image will appear:*

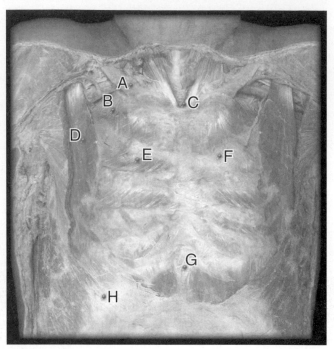

• *Mouse-over the blue pins on the screen to find the information necessary to fill in the following blanks:*

A. _____

B. _____

C. _____

D. _____

E. _____

F._____

G. _____

H._____

- *Click on* **TAG 3,** *and the following image will appear*:

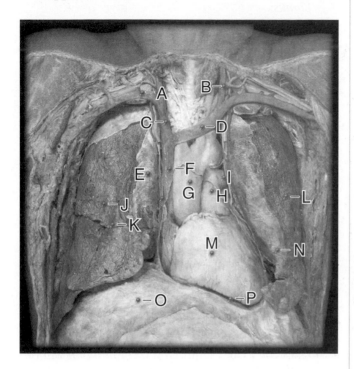

- *Mouse-over the blue pins on the screen to find the information necessary to fill in the following blanks*:

A. _____

B. _____

C. _____

D. _____

E. _____

F. _____

G. _____

H. _____

I. _____

J. _____

K. _____

L. _____

M. _____

N. _____

O. _____

P. _____

CHECK POINT:

Lower Respiratory, Anterior View

1. How many lobes are in the left lung? What are they?
2. How many lobes are in the right lung? What are they?
3. What is distinctive about the medial surface of the lungs? What structures pass through this region?

- *Click on* **TAG 4,** *and the following image will appear*:

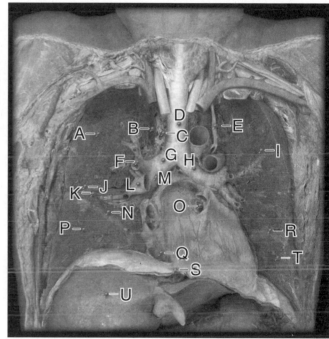

- *Mouse-over the blue pins on the screen to find the information necessary to fill in the following blanks*:

A. _____

B. _____

C. _____

D. _____

E. _____

F. _____

G. _____

H. _____

I._____

J._____

K._____

L._____

M._____

N._____

O._____

P._____

Q._____

R._____

S._____

T._____

U._____

CHECK POINT:

Lower Respiratory, Anterior View

4. Name the structures formed by the bifurcation of the trachea. Which of the two is where foreign bodies that enter the trachea tend to pass?
5. Name the branches eminating from the main bronchi. How many are found in each lung?
6. Name the branches eminating from the bronchi in question 5.

• *Click on* **TAG 5,** *and the following image will appear:*

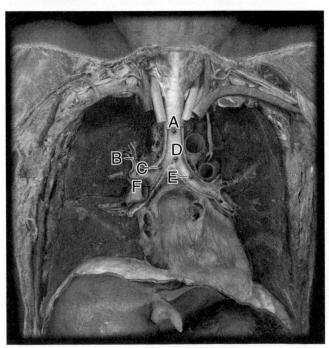

• *Mouse-over the blue pins on the screen to find the information necessary to fill in the following blanks:*

A._____

B._____

C._____

D._____

E._____

F._____

Animation: Thoracic Cavity Dimensional Changes

• *Click on the* **ANIMATIONS** *button at the bottom of the screen.*

• *In the* **Select topic** *menu, select* **Anatomy.**

• *From the* **Select animation** *menu, select* **Thoracic cavity dimensional changes.**

• *Click the* **Play** *button, and the animation will run in the* **IMAGE AREA.**

• *After viewing the animation, answer the following questions:*

1. Another term for breathing is _____.

2. Both _____ and _____ result from changes in _____ in the _____.

3. What structures drive these changes?

4. During inspiration, the thoracic cavity increases in _____. Why?

5. What regulates the length of the thoracic cavity?

6. How is the length of the thoracic cavity increased during inspiration?

7. What changes occur with the diaphragm and the thoracic cavity during expiration?

8. What regulates the depth and width of the thoracic cavity? How?

IN REVIEW

What Have I Learned?

The following questions cover the material that you have just learned—the lower respiratory system. Use the information in the **STRUCTURE INFORMATION** window for these structures to answer the following questions:

1. Name the serous membrane that covers the lungs.

2. Name the space formed between the visceral and parietal pleura.

3. What is the primary muscle of respiration?

4. What pressure changes occur when the primary muscle contracts?

5. What nerve is responsible for these contractions?

6. The oblique fissure of the left lung separates the _____ lobes, while in the right lung separates the _____ lobes.

7. The horizontal fissure of the right lung separates the _____ lobes.

9. How does the elevation of the ribs effect the thoracic cavity width? This motion is similar to _____.

10. What effect does this elevation of the ribs have on the sternum? What effect does this have on the thoracic cavity depth?

11. How does the movement of the sternum and ribs facilitate inspiration?

HEADS UP!

View the radiographic images for the lower respiratory system.

- *Click on the **IMAGING** button at the bottom of the screen.*
- *Click on the **Select region** menu, and select **Lower respiratory.***
- *Click on the flashing **GO** button.*
- *Click on the **Select image type** menu, and select each one.*
- *Click on the **Select structure** menu in the **STRUCTURE LIST** box, and then the individual structures to highlight them in the **IMAGE AREA.***

The Larynx

EXERCISE 7.3: Respiratory System, Larynx, Anterior View

- *Insert Anatomy & Physiology | Revealed® **Cardiovascular, Lymphatic, and Respiratory Systems** CD, or, if you are already in the **Dissection** section, click the **CHANGE VIEW** button at the top of the screen, and skip the next two steps.*
- *From the **Select system** menu, select **Respiratory.***
- *In the **Home screen**, select the **Dissection** button in the left portion of the screen. You may click either on the **Dissection** button or on the word.*
- *In the **SELECT A VIEW** window that appears, click on the **Select topic** button.*
- *Choose **Larynx** from the menu.*
- *Click the **Select view** menu, and choose **Anterior.***
- *The **GO** button will flash green. Click on it.*

- *Click on **TAG 1,** and the following image will appear:*

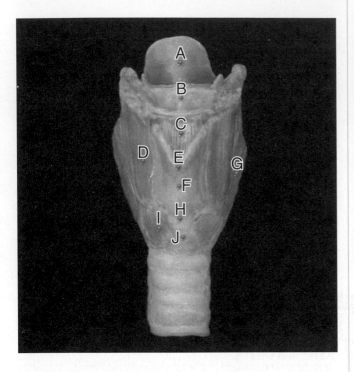

- *Click on **TAG 2,** and the following image will appear:*

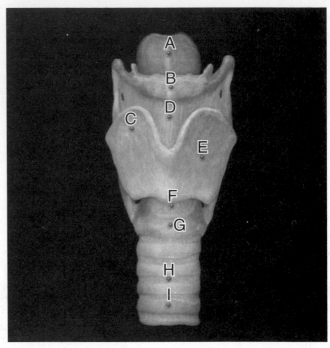

- *Mouse-over the blue pins on the screen to find the information necessary to fill in the following blanks:*

A. _____

B. _____

C. _____

D. _____

E. _____

F. _____

G. _____

H. _____

I. _____

J. _____

- *Mouse-over the blue pins on the screen to find the information necessary to fill in the following blanks:*

A. _____

B. _____

C. _____

D. _____

E. _____

F. _____

G. _____

H. _____

I. _____

CHECK POINT:

Larynx, Anterior View

1. Name the cartilaginous structure that closes over the laryngeal inlet when swallowing.
2. Name a muscle that elevates the larynx and depresses the hyoid.
3. Name a muscle that produces waves of contractions during swallowing.

CHECK POINT:

Larynx, Anterior View

4. Name the bone that does not articulate with any other bone.
5. Name the largest laryngeal cartilage.
6. Name the only laryngeal cartilage that is a complete ring.

- *Click on* **TAG 3,** *and the following image will appear:*

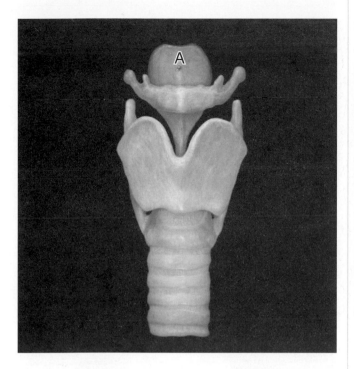

- *Mouse-over the blue pin on the screen to find the information necessary to fill in the following blank:*

A. _____

EXERCISE 7.4:   Respiratory System, Larynx, Lateral View

- *Insert* Anatomy & Physiology | Revealed® **Cardiovascular, Lymphatic, and Respiratory Systems** *CD, or, if you are already in the* **Dissection** *section, click the* **CHANGE VIEW** *button at the top of the screen, and skip the next two steps.*

- *From the* **Select system** *menu, select* **Respiratory.**

- *In the* **Home screen,** *select the* **Dissection** *button in the left portion of the screen. You may click either on the* **Dissection** *button or on the word.*

- *In the* **SELECT A VIEW** *window that appears, click on the* **Select topic** *button.*

- *Choose* **Larynx** *from the menu, if it is not already selected.*

- *Click the* **Select view** *menu, and choose* **Lateral.**

- *The* **GO** *button will flash green. Click on it.*
- *Click on* **TAG 1,** *and the following image will appear:*

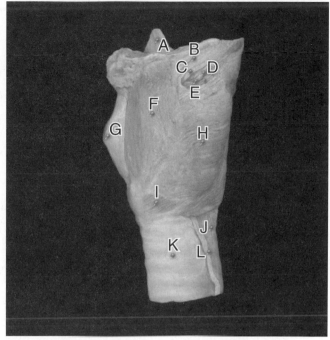

- *Mouse-over the blue pins on the screen to find the information necessary to fill in the following blanks:*

A. _____

B. _____

C. _____

D. _____

E. _____

F. _____

G. _____

H. _____

I. _____

J. _____

K. _____

L. _____

J. _____	
K. _____	
L. _____	

CHECK POINT:

Larynx, Lateral View

1. Name the sensory and motor innervations of the recurrent laryngeal nerve.
2. Name the sensory innervation of the internal laryngeal nerve.
3. Name the companion artery and vein to the internal laryngeal nerve.

- *Click on* **TAG 2,** *and the following image will appear:*

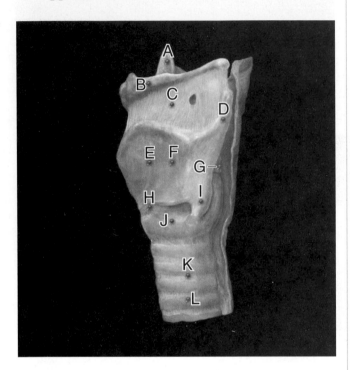

- *Mouse-over the blue pins on the screen to find the information necessary to fill in the following blanks:*

A. _____

B. _____

C. _____

D. _____

E. _____

F. _____

G. _____

H. _____

I. _____

CHECK POINT:

Larynx, Lateral View

4. Name the superior projection of the posterior thyroid cartilage.
5. Name the inferior projection of the posterior thyroid cartilage.

- *Click on* **TAG 3,** *and the following image will appear:*

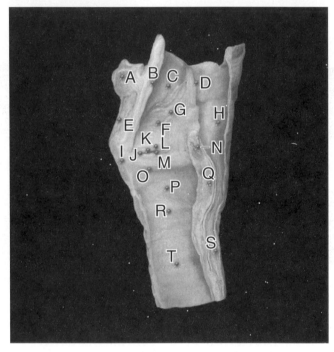

- *Mouse-over the blue pins on the screen to find the information necessary to fill in the following blanks:*

A. _____

B. _____

C. _____

D. _____

E. _____

F. _____

G. _____

H. _____

I. _____

J._____

K._____

L._____

M._____

N._____

O._____

P._____

Q._____

R._____

S._____

T._____

CHECK POINT:

Larynx, Lateral View

6. Name two structures with the major role in sound production.
7. Name the portion of the pharynx posterior to the larynx.
8. Name the muscle that abducts the vocal folds.

EXERCISE 7.5: Respiratory System, Larynx, Posterior View

- *Insert* Anatomy & Physiology | Revealed® **Cardiovascular, Lymphatic, and Respiratory Systems** *CD, or, if you are already in the* **Dissection** *section, click the* **CHANGE VIEW** *button at the top of the screen, and skip the next two steps.*

- *From the* **Select system** *menu, select* **Respiratory.**

- *In the* **Home screen,** *select the* **Dissection** *button in the left portion of the screen. You may click either on the* **Dissection** *button or on the word.*

- *In the* **SELECT A VIEW** *window that appears, click on the* **Select topic** *button.*

- *Choose* **Larynx** *from the menu, if it is not already selected.*

- *Click the* **Select view** *menu, and choose* **Posterior.**

- *The* **GO** *button will flash green. Click on it.*

- *Click on* **TAG 1,** *and the following screen will appear:*

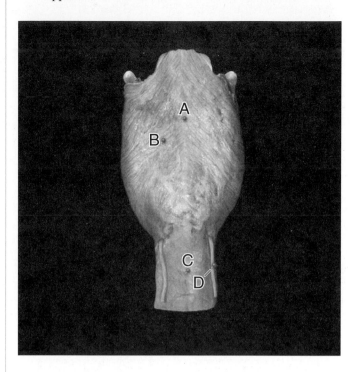

- *Mouse-over the blue pins on the screen to find the information necessary to fill in the following blanks:*

A._____

B._____

C._____

D.

CHECK POINT:

Larynx, Posterior View

1. What is a raphe?
2. Name the innervation for the inferior pharyngeal constrictor muscle.
3. Name the nerve responsible for the innervation of all but one intrinsic pharyngeal muscle.

• *Click on* **TAG 2,** *and the following image will appear:*

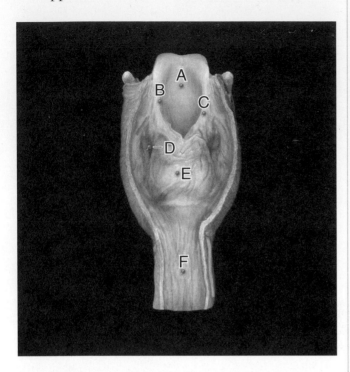

• *Mouse-over the blue pins on the screen to find the information necessary to fill in the following blanks:*

A. _____

B. _____

C. _____

D. _____

E. _____

F._____

CHECK POINT:

Larynx, Posterior View

4. Name the superior opening of the larynx.
5. When food is "caught in the throat," where is it usually lodged?
6. Name the mucous membrane fold that surrounds the structure in question 4.

• *Click on* **TAG 3,** *and the following image will appear:*

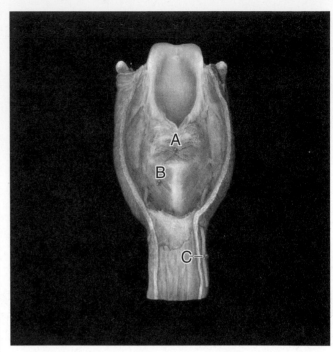

• *Mouse-over the blue pins on the screen to find the information necessary to fill in the following blanks:*

A. _____

B. _____

C. _____

CHECK POINT:

Larynx, Posterior View

7. Name two muscles that abduct the vocal folds.
8. Name a nerve of the larynx that is a branch of the vagus nerve (CN X)

- *Click on* **TAG 4,** and the following image will appear:

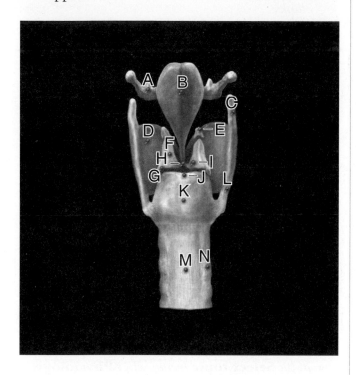

- *Mouse-over the blue pins on the screen to find the information necessary to fill in the following blanks:*

A. _____

B. _____

C. _____

D. _____

E. _____

F. _____

G. _____

H. _____

I. _____

J. _____

K. _____

L. _____

M. _____

N. _____

CHECK POINT:

Larynx, Posterior View

9. Name the process that is the posterior attachment site for the vocal ligament.
10. When is the vocal ligament called the vocal fold or vocal cord?
11. Name the cartilaginous structure that provides attachment for the lateral and posterior cricoarytenoid muscles.

IN REVIEW

What Have I Learned?

The following questions cover the material that you have just learned—the larynx. Use the information in the **STRUCTURE INFORMATION** window for these structures to answer the following questions:

1. Name three pharyngeal constrictors.

2. Name a muscle that tenses and elongates the vocal ligaments.

3. Describe the cricothyroid ligament.

4. Describe the thyrohyoid membrane.

5. Name the structure also known as the "Adam's apple."

6. Name the U-shaped structures that provide a rigid skeleton for the trachea.

7. What are the corniculate cartilages? Where are they embedded?

The Alveolus and Alveolar Duct

EXERCISE 7.6:   Respiratory System—
Alveolus, Histology

- *Insert* Anatomy & Physiology | Revealed® **Cardiovascular, Lymphatic, and Respiratory Systems** *CD, or, if you are already in the* **Dissection** *section, click the* **CHANGE VIEW** *button at the top of the screen, and skip the next two steps.*

- *From the* **Select system** *menu, select* **Respiratory.**

- *In the* **Home screen,** *select the* **Dissection** *button in the left portion of the screen. You may click either on the* **Dissection** *button or on the word.*

- *In the* **SELECT A VIEW** *window that appears, click on the* **Select topic** *button.*

- *Choose* **Alveolus** *from the menu.*

- *The* **Select view** *menu will then show* **Histology.**

- *The* **GO** *button will flash green. Click on it.*

- *Click on* **TAG 1,** *and the following image will appear:*

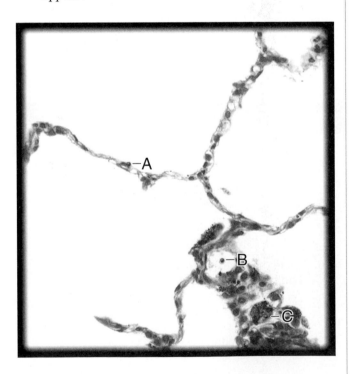

- *Mouse-over the blue pins on the screen to find the information necessary to fill in the following blanks:*

A. _____

B. _____

C. _____

Animation: Partial Pressure

- *Click on the* **ANIMATIONS** *button at the bottom of the screen.*

- *In the* **Select topic** *menu, select* **Physiology.**

- *From the* **Select animation** *menu, select* **Partial pressure.**

- *Click the* **Play** *button, and the animation will run in the* **IMAGE AREA.**

- *After viewing the animation, answer the following questions:*

1. What is the partial pressure of oxygen in fresh air entering the lungs?

2. What effect does moisture in the lungs have on this number?

3. What is the partial pressure of carbon dioxide in fresh air entering the lungs?

4. What effect does carbon dioxide delivered to the lungs from the blood have on this number?

5. Describe the direction of diffusion of oxygen and carbon dioxide in the alveoli.

6. This occurs because of _____

_____.

7. This occurs at the _____ ends of the _____ _____.

8. What occurs as a result of diffusion at the venous ends of the pulmonary capillaries?

9. With no differences in partial pressure, _____ _____ _____.

10. How do oxygen and carbon dioxide diffuse into/out of the tissue capillaries?

11. This occurs because of _____ _____.

12. What occurs at the venous ends of the tissue capillaries?

13. The blood now carries the _____ to the _____.

14. In the body, all of these exchanges occur _____ _____ —.

Animation: Alveolar Pressure Changes

- *Return to the animation menu at the top left of the screen, and from the **Select animation** menu, select **Alveolar pressure changes.***

- *Click the **Play** button, and the animation will run in the **IMAGE AREA.***

- *After viewing the animation, answer the following questions:*

1. At the end of expiration, _____ _____.

2. Therefore, _____ _____.

3. Inspiration begins with _____ of the _____ to _____.

4. This results in _____ _____.

5. The increased alveolar pressure causes a _____ in _____ below _____ and _____ flows _____.

6. At the end of inspiration, _____ _____.

7. Air flow into the lungs causes _____ _____.

8. Because the pressure becomes equal, _____ _____.

9. During expiration, the _____ of the _____ _____ as the _____ _____, and the _____ and the _____ _____.

10. This results in a _____ in _____ and an _____ in _____.

11. The _____ is now _____ than _____, so air flows _____ of the lungs.

12. Air continues to flow out of the lungs until _____.

Animation: Diffusion Across Respiratory Membrane

- *Return to the animation menu at the top left of the screen, and from the **Select animation** menu, select **Diffusion across respiratory membrane.***

- *Click the **Play** button, and the animation will run in the **IMAGE AREA.***

- *After viewing the animation, answer the following questions:*

1. In the lungs, gas exchange takes place _____ _____.

2. What are alveoli? Where are they located?

3. What is the diameter of an alveolus? How many are in each lung?

4. What are the two types of specialized cells in the wall of the alveoli?

5. The cells that form 90 percent of the alveolar wall are _____ _____.

6. Describe these cells.

7. Name and describe the second type of cells.

8. What do these cells secrete? What is the function of this secretion?

9. _____ form a network around each alveolus.

10. Name the thin structure that separates the capillary blood from the air in the alveolus.

11. What is this thin structure comprised of?

12. This structure has a thickness of only _____, which facilitates _____ _____.

13. What causes the diffusion of gases across this membrane? Explain it for both oxygen and carbon dioxide.

Animation: Capillary Exchange

- *Return to the animation menu at the top left of the screen, and from the* **Select animation** *menu, select* **Capillary exchange.**

- *Click the* **Play** *button, and the animation will run in the* **IMAGE AREA.**

- *After viewing the animation, answer the following questions:*

1. What are capillaries?

2. What makes them optimal for capillary exchange? What substances are exchanged?

3. In general, where do fluids move out of capillaries?

4. What happens to that fluid?

5. Name and describe the two pressures involved with capillary exchange.

6. The difference between these two pressures is called _____.

7. What does this regulate?

8. What is the net filtration pressure at the arterial end of the capillary. What does this favor?

9. What is caused by a positive net filtration pressure?

10. What causes the shift from filtration to reabsorption at the venous end of the capillary?

11. What is caused by a negative net filtration pressure?

12. What is the destination of the 10 percent of the fluid that leaves the blood capillaries? Where is its ultimate destination?

13. Duplicate the following Fluid Exchange table to help you learn the concepts involved.

FLUID EXCHANGE SUMMARY		
ARTERIAL END		**VENOUS END**
___ mmHg (out)	**Net Hydrostatic Pressure**	___ mmHg (out)
___ mmHg (in)	**Oncotic Pressure**	___ mmHg (out)
___ mmHg (out)	**Net Filtration Pressure**	___ mmHg (in)

EXERCISE 7.7:   Respiratory System— Alveolar Duct, Histology

- *Insert Anatomy & Physiology | Revealed*® **Cardiovascular, Lymphatic, and Respiratory Systems** *CD, or, if you are already in the* **Dissection** *section, click the* **CHANGE VIEW** *button at the top of the screen, and skip the next two steps.*

- *From the* **Select system** *menu, select* **Respiratory.**

- *In the* **Home screen,** *select the* **Dissection** *button in the left portion of the screen. You may click either on the* **Dissection** *button or on the word.*

- *In the* **SELECT A VIEW** *window that appears, click on the* **Select topic** *button.*

- *Choose* **Alveolar duct** *from the menu.*

- The **Select view** menu will then show **Histology.**
- The **GO** button will flash green. Click on it.
- Click on **TAG 1,** and the following image will appear:

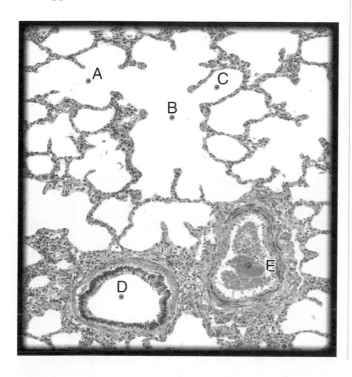

- Mouse-over the blue pins on the screen to find the information necessary to fill in the following blanks:

A. _____

B. _____

C. _____

D. _____

E. _____

CHECK POINT:

Alveolar Duct, Histology

1. Name the structure that provides the respiratory surface for gas exchange.
2. Name the structure that is a common space for the openings of several alveoli.
3. Name the structure that conducts air to the alveoli.

IN REVIEW

What Have I Learned?

The following questions cover the material that you have just learned—the alveolus and alveolar duct. Use the information in the **STRUCTURE INFORMATION** window for these structures to answer the following questions:

1. What are alveoli? Where are they located?

2. What is the diameter of an alveolus? How many are in each lung?

3. What are the two types of specialized cells in the wall of the alveoli?

4. What are capillaries?

5. What makes capillaries optimal for capillary exchange? What substances are exchanged?

6. What is the partial pressure of oxygen in fresh air entering the lungs?

7. What is the partial pressure of carbon dioxide in fresh air entering the lungs?

8. Describe the direction of diffusion of oxygen and carbon dioxide in the alveoli.

9. How do oxygen and carbon dioxide diffuse into/out of the tissue capillaries?

IN REVIEW

10. Name the vessel that transports deoxygenated blood to the alveolar capillary plexus.

11. Name the structure that conducts air to the terminal bronchiole.

12. Name the vessel that transports deoxygenated blood to the alveolar capillary plexus.

The Trachea

EXERCISE 7.8: Respiratory System—Trachea (low power), Histology

- *Insert* Anatomy & Physiology | Revealed® **Cardiovascular, Lymphatic, and Respiratory Systems** *CD, or, if you are already in the* **Dissection** *section, click the* **CHANGE VIEW** *button at the top of the screen, and skip the next two steps.*

- *From the* **Select system** *menu, select* **Respiratory.**

- *In the* **Home screen,** *select the* **Dissection** *button in the left portion of the screen. You may click either on the* **Dissection** *button or on the word.*

- *In the* **SELECT A VIEW** *window that appears, click on the* **Select topic** *button.*

- *Choose* **Trachea (low power)** *from the menu.*

- *The* **Select view** *menu will then show* **Histology.**

- *The* **GO** *button will flash green. Click on it.*

- *Click on* **TAG 1,** *and the following image will appear:*

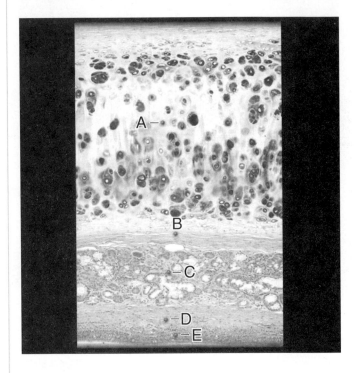

- *Mouse-over the blue pins on the screen to find the information necessary to fill in the following blanks:*

A. _____

B. _____

C. _____

D. _____

E. _____

Trachea (low power), Histology

1. What type of cartilage provides support for the majority of the respiratory tract?
2. Name the fibrocartilage layer that covers the hyaline cartilage.
3. Name the layer of the trachea that contains numerous seromucous glands and is rich in blood and lymph vessels.
4. Name the layer of the trachea that provides support for the epithelial cells. Name the lymphoid elements it may contain.
5. What tissue type makes up the epithelial layer? What is its function?

EXERCISE 7.9: Respiratory System—Trachea (high power), Histology

- *Insert* Anatomy & Physiology | Revealed® **Cardiovascular, Lymphatic, and Respiratory Systems** *CD, or, if you are already in the* **Dissection** *section, click the* **CHANGE VIEW** *button at the top of the screen, and skip the next two steps.*

- *From the* **Select system** *menu, select* **Respiratory.**

- *In the* **Home screen,** *select the* **Dissection** *button in the left portion of the screen. You may click either on the* **Dissection** *button or on the word.*

- *In the* **SELECT A VIEW** *window that appears, click on the* **Select topic** *button.*

- *Choose* **Trachea (high power)** *from the menu.*

- *The* **Select view** *menu will then show* **Histology.**

- *The* **GO** *button will flash green. Click on it.*

- *Click on* **TAG 1,** *and the following image will appear:*

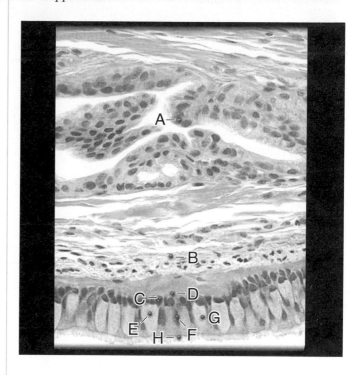

- *Mouse-over the blue pins on the screen to find the information necessary to fill in the following blanks:*

A. _____

B. _____

C. _____

D. _____

E. _____

F. _____

G. _____

H. _____

Trachea (high power), Histology

1. Name the tissue layer that provides support for most epithelia in the body.
2. Name the stem cell that can produce new epithelial cells.
3. Name the mucous-producing epithelial cells. What is the purpose of the mucus?
4. What is the function of the cilia on the ciliated epithelial cells?

IN REVIEW

What Have I Learned?

The following questions cover the material that you have just learned—the trachea. Use the information in the **STRUCTURE INFORMATION** window for this structure to answer the following questions:

1. What type of cartilage provides support for the majority of the respiratory tract?

2. What tissue type makes up the epithelial layer of the trachea? What is its function?

3. Name the tissue layer that provides support for most epithelia in the body.

4. With your knowledge of the function of the cilia in the respiratory tract, explain why cigarette smoking, which paralyzes these cilia, causes the smoker's hacking cough.

CHAPTER 8

The Digestive System

Overview: The Digestive System

Think of everything that you have eaten today, or this week for that matter. After you swallowed it, where did it go? Out of sight, out of mind. We don't even think about food again until it's meal time or our stomach growls during class (or worse yet, during a test). What happens to the food that you eat, between the time you swallow it and the time that the waste is eliminated? What structures are involved with this transformation?

The nutrients in your food are mostly unavailable to your body when they are swallowed. The primary nutrients—proteins, carbohydrates, and fats must be broken down into their basic building blocks before they can be absorbed into the bloodstream.

Only then can your body build your own proteins, carbohydrates, and fats from these available building blocks. It's like moving a large desk into a small office. The desk won't fit through the door, so you need to disassemble it first into smaller parts that *will* fit through the door. Once the parts have passed through the door and inside the office, they can be reassembled into a desk. Likewise, proteins for example, are too large to pass through the epithelium of the small intestine and into the capillaries located there. Digestive enzymes must first break these proteins down into their smallest building blocks—amino acids. These amino acids can then pass through the epithelium of the small intestine and into the bloodstream. Then, when your body needs to make a protein molecule—say for example your growing hair, it puts together the required amino acids and constructs hair. This is where the office desk anal-

ogy breaks down. When you reassembled the desk in the small office, it was more or less the same desk that you had taken apart. The only difference would be any missing pieces lost in the move. In your body, the proteins that you create to build structures are unique and completely unlike the original protein in your food. But the point is the same, they must be broken down into smaller pieces to make it through the "door" of your digestive system.

To learn the structures of the digestive system, we will begin with an animated overview of the entire system. Then, we will look at the anatomical structures of the alimentary canal sequentially, from proximal to distal. Next we will look at the accessory digestive organs, and we will conclude by learning how we get energy out of the food we eat.

Animation: Digestive System Overview

Before beginning your study of the digestive system, view the correlated *Anatomy & Physiology | Revealed*® animation.

- *Insert the* Anatomy & Physiology | Revealed® **Digestive, Urinary, Reproductive, and Endocrine Systems** *CD.*

- *From the* **Select system** *menu, select* **Digestive.**

- *In the* **Home screen,** *select the* **Animation** *button in the left portion of the screen. You may click either on the* **Animation** *button or on the word.*

- *In the* **Select topic** *menu, select* **Anatomy.**

- *From the* **Select animation** *menu, select* **Digestive system overview.**

- *Click the* **Play** *button, and the animation will run in the* **IMAGE AREA.**

- *After viewing the animation, answer the following questions:*

1. What are the four main functions of the digestive system?

2. Name the two types of digestion.

3. Where does digestion begin? How does this occur?

4. Another name for chewing is _____.

5. What is the function of the salivary glands? Of saliva?

6. What prevents food from entering the nasal cavity while swallowing?

7. What muscles push food particles into the pharynx?

8. Name the structure that prevents food from entering the respiratory system.

9. Name the structure that connects the pharynx to the stomach.

10. Once it has been swallowed, the food mass is called a _____.

11. What is the term for the involuntary wavelike contractions that propel the bolus to the stomach?

12. What are rugae? What are their functions?

13. The stomach cells secrete _____.

14. What effect do these secretions have on the bolus?

15. The bolus, mixed with stomach secretions, is now called _____.

16. _____ exits the stomach through the _____ and enters the _____.

17. Name the major site of nutrient absorption.

18. Name the three parts of the small intestine, from proximal to distal.

19. What digestive aids enter the duodenum? Where do they originate?

20. How are nutrients absorbed?

21. What is the destination of the chyme not absorbed in the small intestine?

22. List the sequence of structures the chyme passes through as it becomes feces. What has been absorbed from the chyme as it passes through the colon?

23. Where are the feces stored? What causes fecal elimination?

The Oral Cavity

EXERCISE 8.1:   Digestive System—Oral Cavity—Pharynx, Lateral View

- *Insert* Anatomy & Physiology | Revealed® **Digestive, Urinary, Reproductive, and Endocrine Systems** *CD, or, if you are in the* **Animation** *section, click the* **DISSECTION** *button at the bottom of the screen, and skip the next two steps.*

- *From the* **Select system** *menu, select* **Digestive.**

- *In the* **Home screen,** *select the* **Dissection** *button in the left portion of the screen. You may click either on the* **Dissection** *button or on the word.*

- *In the* **SELECT A VIEW** *window that appears, click on the* **Select topic** *button.*

- *Choose* **Oral cavity—pharynx** *from the menu.*
- **Lateral** *will appear in the* **Select view** *menu.*
- *The* **GO** *button will flash green. Click on it.*
- *Click on* **TAG 1,** *and the following screen will appear:*

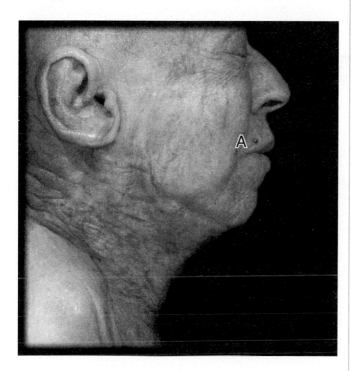

- *Mouse-over the blue pin on the screen to find the information necessary to fill in the following blank:*

A. _____

CHECK POINT

Oral Cavity—Pharynx, Lateral View

1. Name the fleshy folds surrounding the mouth.
2. Name the midline vertical groove of the upper lip.
3. Name the muscle contained in the lips.

- *Click on* **TAG 2,** *and the following image will appear:*

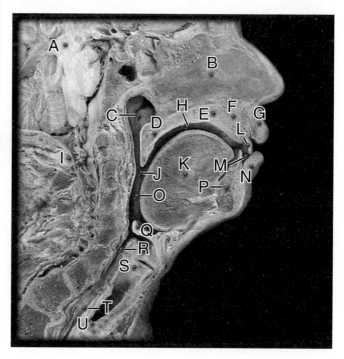

- *Mouse-over the blue pins on the screen to find the information necessary to fill in the following blanks:*

A. _____

B. _____

C. _____

D. _____

E. _____

F. _____

G. _____

H. _____

I. _____

J. _____

K. _____

L. _____

M. _____

N. _____

O. _____

P. _____

Q. _____

R. _____

S. _____

T. _____

U. _____

IN REVIEW

What Have I Learned?

The following questions cover the material that you have just learned—the oral cavity. Use the information in the **STRUCTURE INFORMATION** window for these structures to answer the following questions:

1. Name the three functions of the lips.

2. What is the function of the oral cavity?

3. What is the pharynx? What are the three subdivisions?

4. Name the divisions that are part of the respiratory system, the digestive system, or both.

5. Describe the structure of the tongue.

6. What is the tongue's function?

7. Name the structure involved in oral breathing.

8. The Latin term pharynx means _____.

9. Name the structure that separates the oropharynx from the nasopharynx.

10. Name the structure that separates the oral cavity from the nasal cavity. Which bones form this structure?

11. Name the tube-shaped structure located between the oropharynx and the esophagus.

The Salivary Glands

EXERCISE 8.2:   Digestive System—Salivary Glands, Lateral View

- *Insert the* Anatomy & Physiology | Revealed® **Digestive, Urinary, Reproductive, and Endocrine Systems** *CD, or, if you are already in the* **Dissection** *section, click the* **CHANGE VIEW** *button at the top of the screen, and skip the next two steps.*

- *From the* **Select system** *menu, select* **Digestive.**

- *In the* **Home screen,** *select the* **Dissection** *button in the left portion of the screen. You may click either on the* **Dissection** *button or on the word.*

- *In the* **SELECT A VIEW** *window that appears, click on the* **Select topic** *button.*

- *Choose* **Salivary glands** *from the menu.*

- **Lateral** *will appear in the* **Select view** *menu.*

- *The* **GO** *button will flash green. Click on it.*

- *Click on* **TAG 1,** *and the following image will appear:*

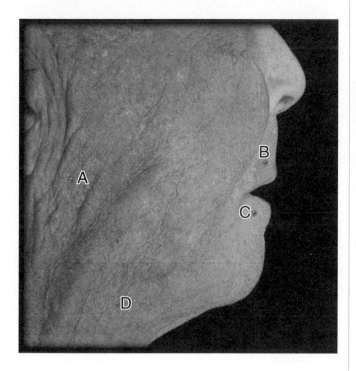

- *Mouse-over the blue pins on the screen to find the information necessary to fill in the following blanks:*

A. _____

B. _____

C. _____

D. _____

CHECK POINT:

Salivary Glands, Lateral View

1. The superficial part of which salivary gland is located anterior to the auricle?
2. Name the salivary gland located inferior and medial to the body of the mandible.
3. Name the three major paired salivary glands.

- *Click on* **TAG 2,** *and the following image will appear:*

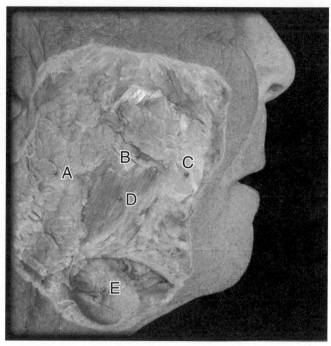

- *Mouse-over the blue pins on the screen to find the information necessary to fill in the following blanks:*

A. _____

B. _____

C. _____

D. _____

E. _____

CHECK POINT

Salivary Glands, Lateral View

4. Which salivary gland produces 25–30 percent of your saliva? Where is it located?
5. Which salivary gland produces 60–70 percent of your saliva? Where is it located?
6. Name the salivary duct that ends in the vestibule of the oral cavity opposite the second maxillary molar.

- *Click on* **TAG 3,** and the following image will appear:

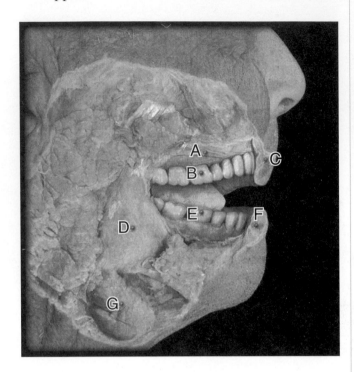

- *Mouse-over the blue pins on the screen to find the information necessary to fill in the following blanks:*

A. _____

B. _____

C. _____

D. _____

E. _____

F. _____

G. _____

C H E C K   P O I N T :

Salivary Glands, Lateral View

7. Describe the gingivae. What is another name for them? What is gingivitis?
8. Describe the permanent mandibular teeth. What is their function?
9. Describe the permanent maxillary teeth. Are they identical in number and function as the mandibular teeth? What do you suppose is the reason for this?

- *Click on* **TAG 4,** and the following image will appear:

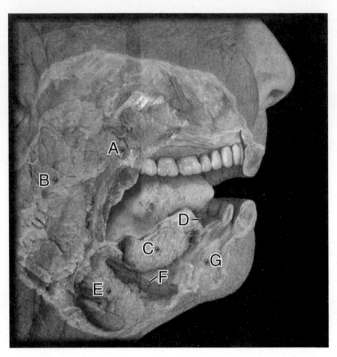

- *Mouse-over the blue pins on the screen to find the information necessary to fill in the following blanks:*

A. _____

B. _____

C. _____

D. _____

E. _____

F. _____

G. _____

C H E C K   P O I N T :

Salivary Glands, Lateral View

10. Name the salivary gland inferior to the tongue. What percentage of your saliva does it secrete?
11. What is the sublingual papilla?
12. Name a muscle responsible for the elevation of the floor of the mouth.

- *Click on **TAG 5,** and the following image will appear:*

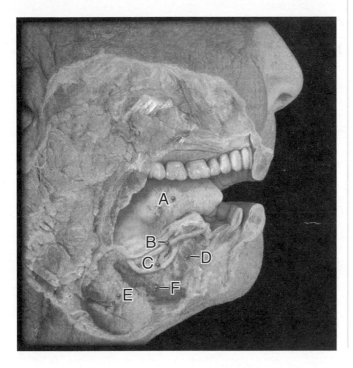

- *Mouse-over the blue pins on the screen to find the information necessary to fill in the following blanks:*

A. _____

B. _____

C. _____

D. _____

E. _____

F. _____

IN REVIEW

What Have I Learned?

The following questions cover the material that you have just learned—the salivary glands. Use the information in the **STRUCTURE INFORMATION** window for these structures to answer the following questions:

1. Name the globular, encapsulated fat body prominent in the cheeks of infants.

2. What is its function?

3. Name a muscle involved with elevation of the mandible.

4. Name the secretions of each extrinsic salivary gland.

5. What is mastication?

6. What is deglutition?

7. What is the name for the "sockets" of the teeth?

8. Describe the submandibular duct. What is its function?

9. What is the sensory innervation of the lingual nerve?

10. The lingual frenulum has been mentioned several times in this topic. What is the lingual frenulum?

The Esophagus

EXERCISE 8.3:   Digestive System—
Esophagus, Anterior View

- *Insert the* Anatomy & Physiology | Revealed® **Digestive, Urinary, Reproductive, and Endocrine Systems** *CD, or, if you are already in the* **Dissection** *section, click the* **CHANGE VIEW** *button at the top of the screen, and skip the next two steps.*

- *From the* **Select system** *menu, select* **Digestive.**

- *In the* **Home screen,** *select the* **Dissection** *button in the left portion of the screen. You may click either on the* **Dissection** *button or on the word.*

- *In the* **SELECT A VIEW** *window that appears, click on the* **Select topic** *button.*

- *Choose* **Esophagus** *from the menu.*

- **Anterior** *will appear in the* **Select view** *menu.*

- *The* **GO** *button will flash green. Click on it.*

- *Click on* **TAG 1,** *and the following image will appear:*

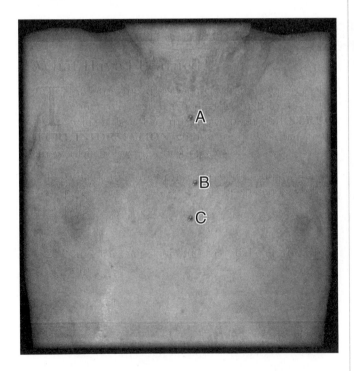

- *Mouse-over the blue pins on the screen to find the information necessary to fill in the following blanks:*

A. _____

B. _____

C. _____

CHECK POINT:

Esophagus, Anterior View

1. Where is the esophagus located?
2. Describe the esophagus.
3. What is its function?

- *Click on* **TAG 2,** *and the following image will appear:*

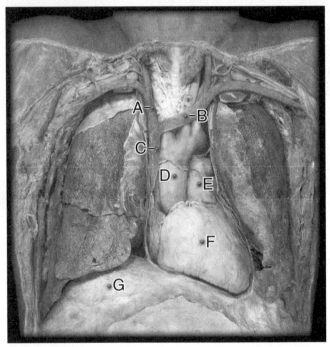

- *Mouse-over the blue pins on the screen to find the information necessary to fill in the following blanks:*

A. _____

B. _____

C. _____

D. _____

E. _____

F. _____

G. _____

- *Click on* **TAG 3,** *and the following image will appear:*

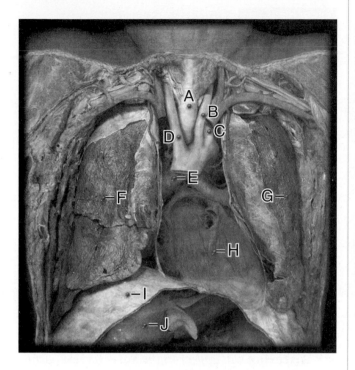

- *Mouse-over the blue pins on the screen to find the information necessary to fill in the following blanks:*

A. _____

B. _____

C. _____

D. _____

E. _____

F._____

G. _____

H._____

I._____

J._____

- *Click on* **TAG 4,** *and the following image will appear:*

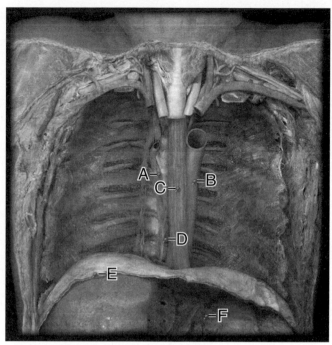

- *Mouse-over the blue pins on the screen to find the information necessary to fill in the following blanks:*

A. _____

B. _____

C. _____

D. _____

E. _____

F._____

• *Click on* **TAG 5,** *and the following image will appear:*

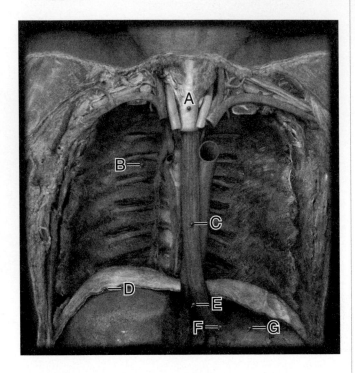

• *Mouse-over the blue pins on the screen to find the information necessary to fill in the following blanks:*

A. _____

B. _____

C. _____

D. _____

E. _____

F._____

G. _____

CHECK POINT:

Esophagus, Anterior View

10. The esophagus conveys food to what digestive organ?
11. What is reflux esophagitis?
12. Name the "hole" in the diaphragm for the passage of the esophagus.

The Abdominal Cavity

EXERCISE 8.4: Digestive System—Abdominal Cavity, Anterior View

• *Insert the* Anatomy & Physiology | Revealed® **Digestive, Urinary, Reproductive, and Endocrine Systems** *CD, or, if you are already in the* **Dissection** *section, click the* **CHANGE VIEW** *button at the top of the screen, and skip the next two steps.*

• *From the* **Select system** *menu, select* **Digestive.**

• *In the* **Home screen,** *select the* **Dissection** *button in the left portion of the screen. You may click either on the* **Dissection** *button or on the word.*

• *In the* **SELECT A VIEW** *window that appears, click on the* **Select topic** *button.*

• *Choose* **Abdominal cavity** *from the menu.*

• **Anterior** *will appear in the* **Select view** *menu.*

• *The* **GO** *button will flash green. Click on it.*

• *Click on* **TAG 1,** *and the following image will appear:*

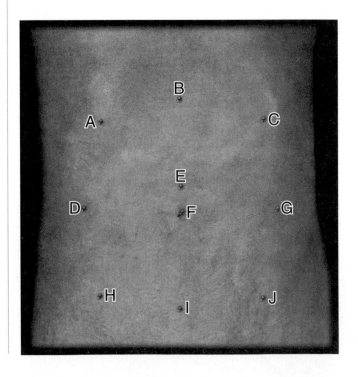

• *Mouse-over the blue pins on the screen to find the information necessary to fill in the following blanks:*

A. _____

B. _____

C. _____

D. _____

E. _____

F. _____

G. _____

H. _____

I. _____

J. _____

CHECK POINT:

Abdominal Cavity, Anterior View

1. What is the umbilicus? Which abdominal region contains the umbilicus?
2. Name the abdominal region located superior to that region.
3. Name the abdominal region located inferior to the region in question 1.

• *Click on* **TAG 2,** and the following image will appear:

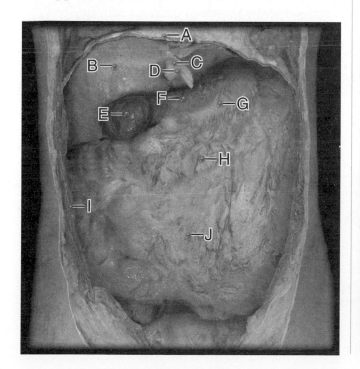

• *Mouse-over the blue pins on the screen to find the information necessary to fill in the following blanks:*

A. _____

B. _____

C. _____

D. _____

E. _____

F. _____

G. _____

H. _____

I. _____

J. _____

CHECK POINT:

Abdominal Cavity, Anterior View

4. What is the greater omentum? What is its function?
5. What is the lesser omentum? Where is it located?
6. Describe the structure of the gallbladder. Where is it located?

• *Click on* **TAG 3,** and the following image will appear:

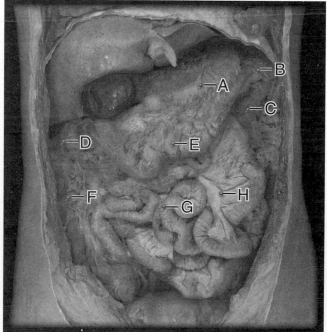

- *Mouse-over the blue pins on the screen to find the information necessary to fill in the following blanks:*

A. _____

B. _____

C. _____

D. _____

E. _____

F. _____

G. _____

H. _____

CHECK POINT:

Abdominal Cavity, Anterior View

7. Name the four parts of the stomach from proximal to distal.
8. Name the location where the ascending colon joins the transverse colon. What is another name for this location? To what does this name refer?
9. Name the location where the transverse colon joins the descending colon. What is another name for this location? What does this name refer to?

- *Click on* **TAG 4,** *and the following image will appear:*

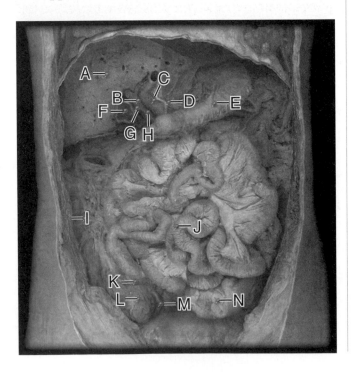

- *Mouse-over the blue pins on the screen to find the information necessary to fill in the following blanks:*

A. _____

B. _____

C. _____

D. _____

E. _____

F. _____

G. _____

H. _____

I. _____

J. _____

K. _____

L. _____

M. _____

N. _____

CHECK POINT:

Abdominal Cavity, Anterior View

10. Name the structure at the junction of the small intestine and large intestine. What part of each intestine is joined here?
11. Name the pouch of the large intestine into which the small intestine empties. What muscle regulates the flow from one to the other?
12. Name the slender hollow appendage attached to this pouch. What is its function? What structures does it contain? What is it called when the lumen of this structure becomes obstructed?

- *Click on* **TAG 5,** and the following image will appear:

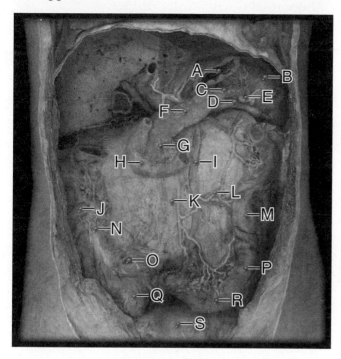

- *Mouse-over the blue pins on the screen to find the information necessary to fill in the following blanks:*

A. _____

B. _____

C. _____

D. _____

E. _____

F. _____

G. _____

H. _____

I. _____

J. _____

K. _____

L. _____

M. _____

N. _____

O. _____

P. _____

Q. _____

R. _____

S. _____

CHECK POINT:

Abdominal Cavity, Anterior View

13. What are the taeniae coli? What is their function?
14. What are haustra?
15. Name the peritoneal appendages filled with fat. What is their function?

- *Click on* **TAG 6,** and the following image will appear:

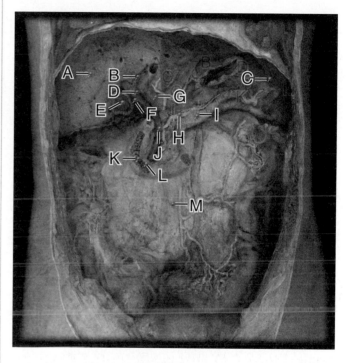

- *Mouse-over the blue pins on the screen to find the information necessary to fill in the following blanks:*

A. _____

B. _____

C. _____

D. _____

E. _____

F. _____

G. _____

H. _____

I._____

J._____

K._____

L._____

M._____

Abdominal Cavity, Anterior View

16. Name and describe the largest lymphatic organ.
17. What is its function?
18. What protects this delicate structure?

IN REVIEW

What Have I Learned?

The following questions cover the material that you have just learned—the abdominal cavity. Use the information in the **STRUCTURE INFORMATION** window for these structures to answer the following questions:

1. Using the following figure, label the nine abdominal regions.

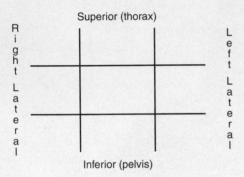

2. What is the function of the gallbladder?

3. Name the muscles of the abdominal wall from superficial to deep.

4. Name the structure that separates the thoracic cavity from the abdominal cavity.

5. Name the four parts of the colon from proximal to distal.

6. Name the three parts of the small intestine, from proximal to distal.

7. Name the most mobile part of the large intestine.

8. Which parts of the colon absorb water? Which parts of the colon absorb electrolytes? Which parts of the colon store feces?

The Stomach

Animation: Stomach

- *Insert* Anatomy & Physiology | Revealed® **Digestive, Urinary, Reproductive, and Endocrine Systems** *CD, or, if you are in the* **Dissection** *section, click the* **ANIMATIONS** *button at the bottom of the screen, and skip the next two steps.*

- *From the* **Select system** *menu, select* **Digestive.**

- *In the* **Home screen,** *select the* **Animation** *button in the left portion of the screen. You may click either on the* **Animation** *button or on the word.*

- *In the* **Select topic** *menu, select* **Anatomy.**

- *From the* **Select animation** *menu, select* **Stomach.**

- *Click the* **Play** *button, and the animation will run in the* **IMAGE AREA.**

- *After viewing the animation, answer the following questions:*

1. Where is the stomach located? Between which two organs?

2. What is the function of the stomach? What two processes contribute to this function?

3. What structures of the stomach form the superior and inferior borders?

4. Describe the four stomach regions.

5. How is the distal part of the stomach subdivided?

6. What is the function of the pyloric sphincter?

7. What are gastric rugae? What is their function?

8. Describe the four layers of the stomach.

9. Name the layers of the stomach muscularis. How does it compare to the rest of the digestive tract?

10. What substance is secreted by the mucous cells of the epithelium?

11. List three functions of gastric mucus.

12. Describe the gastric pits. What are the four different cells found there? What are their functions?

EXERCISE 8.5:   Digestive System—Stomach, Histology

- *Insert* Anatomy & Physiology | Revealed® **Digestive, Urinary, Reproductive, and Endocrine Systems** CD, *or, if you are in the* **Animation** *section, click the* **DISSECTION** *button at the bottom of the screen, and skip the next two steps.*

- *From the* **Select system** *menu, select* **Digestive.**

- *In the* **Home screen,** *select the* **Dissection** *button in the left portion of the screen. You may click either on the* **Dissection** *button or on the word.*

- *In the* **SELECT A VIEW** *window that appears, click on the* **Select topic** *button.*

- *Choose* **Stomach** *from the menu.*

- **Histology** *will appear in the* **Select view** *menu.*

- *The* **GO** *button will flash green. Click on it.*

- *Click on* **TAG 1,** *and the following image will appear:*

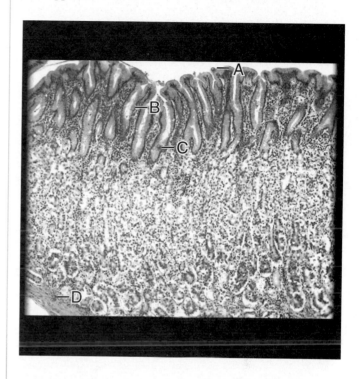

- *Mouse-over the blue pins on the screen to find the information necessary to fill in the following blanks:*

A. _____

B. _____

C. _____

D. _____

C H E C K   P O I N T :

Stomach, Histology

1. Describe the muscularis mucosae. What is its function?
2. Describe the gastric pits. What is their function?
3. Describe the gastric glands. What is their function?

EXERCISE 8.6: Digestive System—Stomach and Duodenum, Anterior View

- *Insert* Anatomy & Physiology | Revealed® **Digestive, Urinary, Reproductive, and Endocrine Systems** *CD, or, if you are already in the* **Dissection** *section, click the* **CHANGE VIEW** *button at the top of the screen, and skip the next two steps.*

- *From the* **Select system** *menu, select* **Digestive.**

- *In the* **Home screen,** *select the* **Dissection** *button in the left portion of the screen. You may click either on the* **Dissection** *button or on the word.*

- *In the* **SELECT A VIEW** *window that appears, click on the* **Select topic** *button.*

- *Choose* **Stomach and duodenum** *from the menu.*

- **Anterior** *will appear in the* **Select view** *menu.*

- *The* **GO** *button will flash green. Click on it.*

- *Click on* **TAG 1,** *and the following image will appear:*

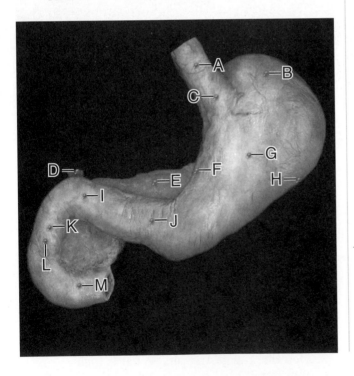

- *Mouse-over the blue pins on the screen to find the information necessary to fill in the following blanks.*

A. _____

B. _____

C. _____

D. _____

E. _____

F. _____

G. _____

H. _____

I. _____

J. _____

K. _____

L. _____

M. _____

CHECK POINT

Stomach and Duodenum, Anterior View

4. Name the part of the stomach located at the junction with the esophagus. What structure does it contain?
5. Name the dome-shaped superior part of the stomach. What is usually retained there?
6. Name the terminal part of the stomach. Name the muscular ring of muscle located there. What is the function of this part of the stomach *and* the muscular ring?

- *Click on* **TAG 2,** and the following image will appear:

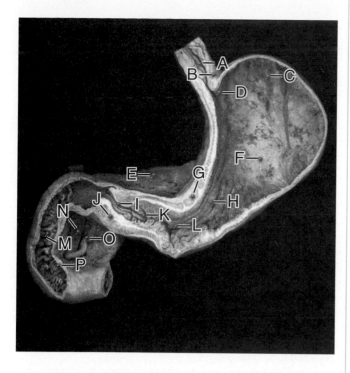

- *Mouse-over the blue pins on the screen to find the information necessary to fill in the following blanks:*

A. _____

B. _____

C. _____

D. _____

E. _____

F. _____

G. _____

H. _____

I. _____

J. _____

K. _____

L. _____

M. _____

N. _____

O. _____

P. _____

7. Name the muscular structure that prevents reflux of stomach contents. Name two times that this muscle relaxes.
8. Name the structures that allow the stomach to expand as it fills.
9. What is the major duodenal papilla? What is its function?

Animation: Gastric Secretion

- *Insert* Anatomy & Physiology | Revealed® **Digestive, Urinary, Reproductive, and Endocrine Systems** *CD, or, if you are in the* **Dissection** *section, click the* **ANIMATIONS** *button at the bottom of the screen, and skip the next two steps.*

- *From the* **Select system** *menu, select* **Digestive.**

- *In the* **Home screen,** *select the* **Animation** *button in the left portion of the screen. You may click either on the* **Animation** *button or on the word.*

- *In the* **Select topic** *menu, select* **Physiology.**

- *From the* **Select animation** *menu, select* **Gastric secretion.**

- *Click the* **Play** *button, and the animation will run in the* **IMAGE AREA.**

- *After viewing the animation, answer the following questions:*

1. List the three phases of gastric secretion.

2. What initiates the cephalic phase?

3. Describe the parasympathetic response involved in gastric secretion.

4. What role does gastrin play?

5. What initiates the gastric phase?

6. Describe the parasympathetic reflex and the direct stimulation that results in this phase.

7. What is the net result of these nervous stimuli?

8. What initiates the intestinal phase?

9. What two events inhibit gastric secretion?

10. Describe the three steps involved in this inhibition.

Animation: HCl Production

- *Return to the animation menu at the top left of the screen, and from the* **Select animation** *menu, select* **HCl production.**

- *Click the* **Play** *button, and the animation will run in the* **IMAGE AREA.**

- *After viewing the animation, answer the following questions:*

1. Which cells produce hydrochloric acid in the stomach? Where are these cells located?

2. What enzyme is involved in the formation of carbonic acid?

3. What two substrates are involved in this reaction?

4. Carbonic acid dissociates into _____ _____.

5. Which ion is transported back to the bloodstream?

6. What exchange occurs at the ion exchange molecule in the plasma membrane?

7. How are the hydrogen ions transported into the duct of the gastric gland?

8. What movement occurs with the chloride ions?

9. What ions are transported into the parietal cell in exchange for the hydrogen ions?

10. What is the result of this movement?

> **HEADS UP!**
> View the radiographic images for the digestive system.

- *Click on the* **IMAGING** *button at the bottom of the screen.*

- *Click on the* **Select topic** *menu, and select* **Stomach and small intestine.**

- *Click the flashing* **GO** *button.*

- *Click on the* **Select structure** *menu in the* **STRUCTURE LIST** *box, and then the individual structures to highlight them in the* **IMAGE AREA.**

IN REVIEW

What Have I Learned?

The following questions cover the material that you have just learned—the stomach. Use the information in the **STRUCTURE INFORMATION** window for these structures to answer the following questions:

1. Name each different type of gastric gland and the products that they produce.

2. Name the structure that conducts mucus and secretions from the gastric glands to the lumen of the stomach.

3. What is a lumen?

4. Name the elongated, nodular gland posterior to the stomach. Describe it.

IN REVIEW

5. What are the functions of this elongated, nodular gland?

6. Name the part of the small intestine to receive ingested material from the stomach. What is that ingested material called?

7. List the functions of the structure in question 6.

8. Name the four parts of this structure.

9. What is the function of the bile duct? Where is it located?

10. What is the function of the pancreatic duct? Where is it located?

The Small Intestine

EXERCISE 8.7: Digestive System—Small Intestine (Duodenum), Histology

- *Insert* Anatomy & Physiology | Revealed® **Digestive, Urinary, Reproductive, and Endocrine Systems** *CD, or, if you are already in the* **Dissection** *section, click the* **CHANGE VIEW** *button at the top of the screen, and skip the next two steps.*

- *From the* **Select system** *menu, select* **Digestive.**

- *In the* **Home screen,** *select the* **Dissection** *button in the left portion of the screen. You may click either on the* **Dissection** *button or on the word.*

- *In the* **SELECT A VIEW** *window that appears, click on the* **Select topic** *button.*

- *Choose* **Small intestine (duodenum)** *from the menu.*

- **Histology** *will appear in the* **Select view** *menu.*

- *The* **GO** *button will flash green. Click on it.*

- *Click on* **TAG 1,** *and the following image will appear:*

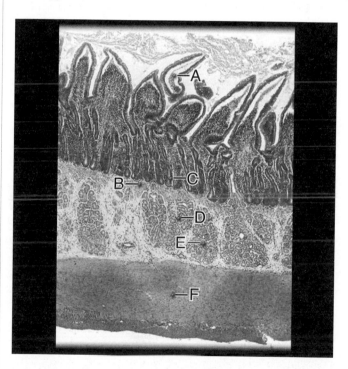

- *Mouse-over the blue pins on the screen to find the information necessary to fill in the following blanks:*

A. _____

B. _____

C. _____

D. _____

E. _____

F. _____

CHECK POINT:

Small Intestine (Duodenum), Histology

1. Name and describe the structures that increase the surface area of the small intestine. What is their function?
2. Describe the structures also known as the crypts of Lieberkühn. What are their functions?
3. What are duodenal glands? Where are they located? What is their function?

EXERCISE 8.8:   Digestive System—Small Intestine (Jejunum/Ileum), Histology

- *Insert* Anatomy & Physiology | Revealed® **Digestive, Urinary, Reproductive, and Endocrine Systems** *CD, or, if you are already in the* **Dissection** *section, click the* **CHANGE VIEW** *button at the top of the screen, and skip the next two steps.*

- *From the* **Select system** *menu, select* **Digestive.**

- *In the* **Home screen,** *select the* **Dissection** *button in the left portion of the screen. You may click either on the* **Dissection** *button or on the word.*

- *In the* **SELECT A VIEW** *window that appears, click on the* **Select topic** *button.*

- *Choose* **Small intestine (jejunum/ileum)** *from the menu.*

- **Histology** *will appear in the* **Select view** *menu.*

- *The* **GO** *button will flash green. Click on it.*

- *Click on* **TAG 1,** *and the following image will appear:*

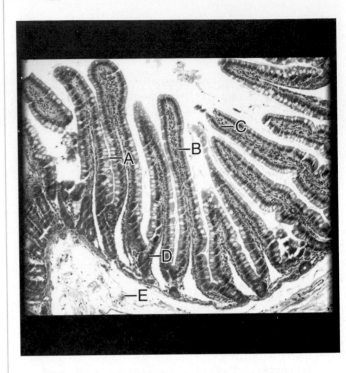

- *Mouse-over the orange pins on the screen to find the information necessary to fill in the following blanks:*

A. _____

B. _____

C. _____

D. _____

E. _____

CHECK POINT:

Small Intestine (jejunum/ileum), Histology

4. Name the cells that produce and release mucin into the intestinal lumen.
5. Where are these cells located?
6. What tissue-type covers the villi of the small intestine?

HEADS UP!

View the radiographic images for the digestive system.

- Click on the **IMAGING** button at the bottom of the screen.
- Click on the **Select region** menu, and select **Stomach and small intestine.**

- Click the flashing **GO** button.
- Click on the **Select structure** menu in the **STRUCTURE LIST** box, and then the individual structures to highlight them in the **IMAGE AREA.**

IN REVIEW

What Have I Learned?

The following questions cover the material that you have just learned—the small intestine. Use the information in the **STRUCTURE INFORMATION** window for these structures to answer the following questions:

1. List from superficial to deep the four layers of the small intestine.

2. Which layer regulates the length of the villi?

3. The contraction of which layer compresses the glands of the mucosa to expel their contents?

4. Name and describe the layer responsible for peristalsis.

5. Which layer contains blood and lymph vessels, lymphoid nodules, and nerve plexus?

6. What is the function of the submucosal glands?

7. What cell types are found on the epithelium of the small intestine?

8. What are the functions of these cells?

9. Name two phenomena that change the shape of the villi.

10. What layer of the small intestine contains Peyer's patches? In what part of the small intestine are they found? What is their function?

The Colon

EXERCISE 8.9: Digestive System— Colon, Histology

- *Insert* Anatomy & Physiology | Revealed® **Digestive, Urinary, Reproductive, and Endocrine Systems** *CD, or, if you are already in the* **Dissection** *section, click the* **CHANGE VIEW** *button at the top of the screen, and skip the next two steps.*

- *From the* **Select system** *menu, select* **Digestive.**

- *In the* **Home screen,** *select the* **Dissection** *button in the left portion of the screen. You may click either on the* **Dissection** *button or on the word.*

- *In the* **SELECT A VIEW** *window that appears, click on the* **Select topic** *button.*

- *Choose* **Colon** *from the menu.*

- **Histology** *will appear in the* **Select view** *menu.*

- *The* **GO** *button will flash green. Click on it.*

- *Click on* **TAG 1,** *and the following image will appear:*

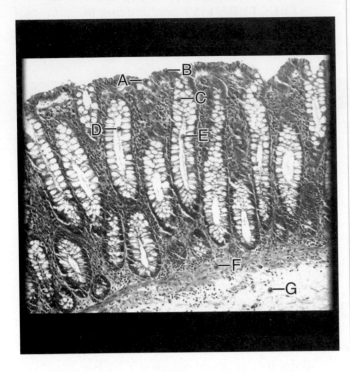

- *Mouse-over the blue pins on the screen to find the information necessary to fill in the following blanks:*

A. _____

B. _____

C. _____

D. _____

E. _____

F. _____

G. _____

CHECK POINT:

Colon, Histology

1. Name and describe the tubular glands located in the colon.
2. What is their function?
3. What is the dominant cell-type in these glands?
4. Name the tissue-type that lines the colon. What is its function?
5. The contraction of which muscles compresses the glands of the mucosa and expels their contents?
6. Name the layer of the colon containing blood and lymph vessels.

- *Click on the* **IMAGING** *button at the bottom of the screen.*
- *Click on the* **Select region** *menu, and select* **Colon.**
- *Click the flashing* **GO** *button.*
- *Click on the* **Select structure** *menu in the* **STRUCTURE LIST** *box, and then the individual structures to highlight them in the* **IMAGE AREA.**

The Liver

Animation: Liver

- *Insert the* Anatomy & Physiology | Revealed® **Digestive, Urinary, Reproductive, and Endocrine Systems** *CD, or, if you are in the* **Dissection** *section, click the* **ANIMATIONS** *button at the bottom of the screen, and skip the next two steps.*
- *From the* **Select system** *menu, select* **Digestive.**
- *In the* **Home screen,** *select the* **Animation** *button in the left portion of the screen. You may click either on the* **Animation** *button or on the word.*
- *In the* **Select topic** *menu, select* **Anatomy.**
- *From the* **Select animation** *menu, select* **Liver.**
- *Click the* **Play** *button, and the animation will run in the* **IMAGE AREA.**
- *After viewing the animation, answer the following questions:*

1. The liver is the _____ internal organ of the body.

2. Where is it located?

3. Name the four lobes of the liver. Which two are part of another lobe?

4. Name and describe the structure that separates the two anterior lobes.

5. What are the two fetal remnants found on the liver?

6. What structures are located in the porta hepatis?

7. Histologically, the liver is composed of functional units called _____.

8. The hub of these functional units is _____.

9. Describe the portal triad.

10. Describe the sinusoids. How are they arranged?

11. How are the hepatocytes arranged?

12. What are the two basic functions of the liver?

13. Name the three categories of this task.

14. Describe these three categories of tasks.

EXERCISE 8.10: Digestive System—Liver, Anterior and Posteroinferior Views

- *Insert the* Anatomy & Physiology | Revealed® **Digestive, Urinary, Reproductive, and Endocrine Systems** *CD, or, if you are in the* **Animation** *section, click the* **DISSECTION** *button at the bottom of the screen, and skip the next two steps.*

- *From the* **Select system** *menu, select* **Digestive.**

- *In the* **Home screen,** *select the* **Dissection** *button in the left portion of the screen. You may click either on the* **Dissection** *button or on the word.*

- *In the* **SELECT A VIEW** *window that appears, click on the* **Select topic** *button.*

- *Choose* **Liver** *from the menu.*

- *Click the* **Select view** *menu, and choose* **Anterior and posteroinferior.**

- *The* **GO** *button will flash green. Click on it.*

- *Click on* **TAG 1,** *and the following image will appear:*

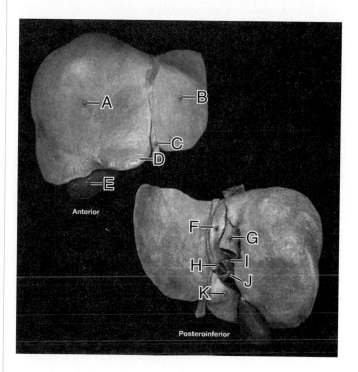

- *Mouse-over the blue pins on the screen to find the information necessary to fill in the following blanks:*

A. _____

B. _____

C. _____

D. _____

E. _____

F. _____

G. _____

H. _____

I. _____

J. _____

K. _____

Liver, Anterior and Posteroinferior Views

1. Name the largest lobe of the liver. What two lobes make up this lobe?
2. Name the structures that separate the left and right lobes of the liver.
3. Name the functions of the liver.

EXERCISE 8.11:   Digestive System—Liver, Histology

- *Insert the* Anatomy & Physiology | Revealed® **Digestive, Urinary, Reproductive, and Endocrine Systems** *CD, or, if you are already in the* **Dissection** *section, click the* **CHANGE VIEW** *button at the top of the screen, and skip the next two steps.*

- *From the* **Select system** *menu, select* **Digestive.**

- *In the* **Home screen,** *select the* **Dissection** *button in the left portion of the screen. You may click either on the* **Dissection** *button or on the word.*

- *In the* **SELECT A VIEW** *window that appears, click on the* **Select topic** *button.*

- *Choose* **Liver** *from the menu, if it is not already selected.*

- *Click the* **Select view** *menu, and choose* **Histology.**

- *The* **GO** *button will flash green. Click on it.*

- *Click on* **TAG 1,** *and the following image will appear:*

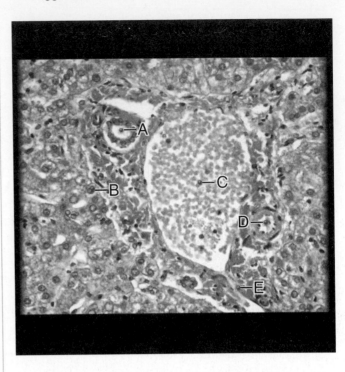

- *Mouse-over the blue pins on the screen to find the information necessary to fill in the following blanks:*

A. _____

B. _____

C. _____

D. _____

E. _____

Liver, Histology

1. Name the liver cells arranged in cords within a liver lobule. What are their functions?
2. What are hepatic sinusoids? What is their function?
3. Name the structure that brings venous blood to the liver sinusoids.

EXERCISE 8.12:   Digestive System—Biliary Ducts, Anterior View

- *Insert the* Anatomy & Physiology | Revealed® **Digestive, Urinary, Reproductive, and Endocrine Systems** *CD, or, if you are already in the* **Dissection** *section, click the* **CHANGE**

VIEW *button at the top of the screen, and skip the next two steps.*

- *From the* **Select system** *menu, select* **Digestive.**

- *In the* **Home screen,** *select the* **Dissection** *button in the left portion of the screen. You may click either on the* **Dissection** *button or on the word.*

- *In the* **SELECT A VIEW** *window that appears, click on the* **Select topic** *button.*

- *Choose* **Biliary ducts** *from the menu.*

- **Anterior** *will appear in the* **Select view** *menu.*

- *The* **GO** *button will flash green. Click on it.*

- *Click on* **TAG 1,** *and the following image will appear:*

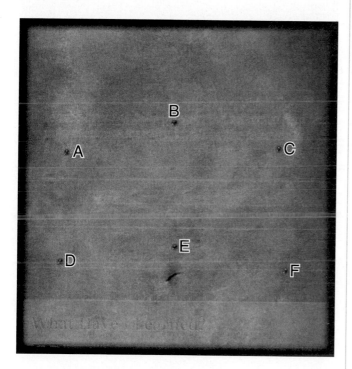

- *Mouse-over the blue pins on the screen to find the information necessary to fill in the following blanks:*

A. _____

B. _____

C. _____

D. _____

E. _____

F. _____

- *Click on* **TAG 2,** *and the following image will appear:*

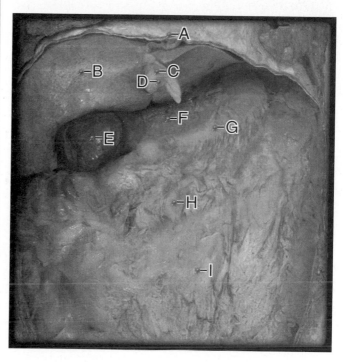

- *Mouse-over the blue pins on the screen to find the information necessary to fill in the following blanks:*

A. _____

B. _____

C. _____

D. _____

E. _____

F. _____

G. _____

H. _____

I. _____

• *Click on* **TAG 3,** and the following image will appear:

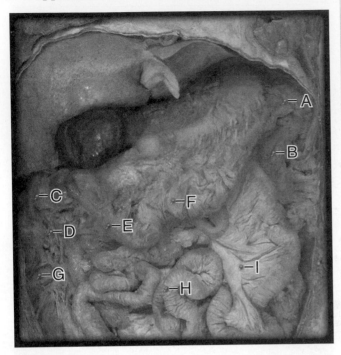

• *Click on* **TAG 4,** and the following image will appear:

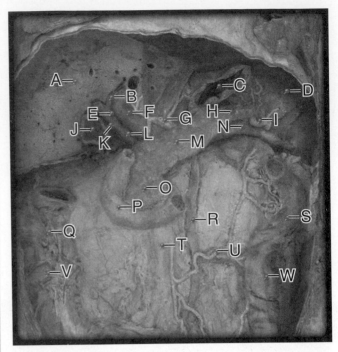

• *Mouse-over the blue pins on the screen to find the information necessary to fill in the following blanks:*

A. _____

B. _____

C. _____

D. _____

E. _____

F. _____

G. _____

H. _____

I. _____

• *Mouse-over the blue pins on the screen to find the information necessary to fill in the following blanks:*

A. _____

B. _____

C. _____

D. _____

E. _____

F. _____

G. _____

H. _____

I. _____

J. _____

K. _____

L. _____

M. _____

N. _____

O. _____

P. _____

Q. _____

R. _____

S. _____

T. _____

U._____

V. _____

W. _____

CHECK POINT:

Biliary Ducts, Anterior View

1. Name the two structures that receive bile from the liver. What structure do they form when they merge?
2. Bile passes from the _____ to the _____ for storage and from the _____ to the _____ for _____ of _____.
3. What two ducts unite to form the bile duct?

- *Click on* **TAG 5,** and the following image will appear:

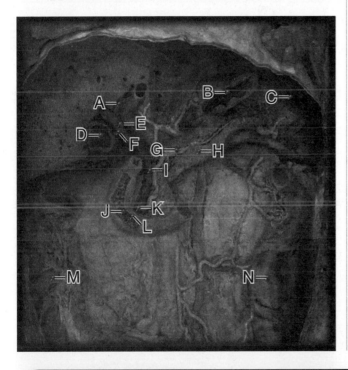

- *Mouse-over the blue pins on the screen to find the information necessary to fill in the following blanks:*

A. _____

B. _____

C. _____

D. _____

E. _____

F. _____

G. _____

H. _____

I. _____

J. _____

K. _____

L. _____

M. _____

N. _____

CHECK POINT:

Biliary Ducts, Anterior View

4. Name the structure that transmits pancreatic secretions.
5. Name the structure formed by the joining of this structure and the bile duct.
6. Into which section of the small intestine does the structure in number 5 empty?

IN REVIEW

What Have I Learned?

The following questions cover the material that you have just learned—the liver. Use the information in the **STRUCTURE INFORMATION** window for this organ to answer the following questions:

1. Name the structure that is the remnant of the umbilical vein of the fetus.

2. Name the blood vessel that carries absorbed products of digestion to the liver.

3. Name the organ that stores, concentrates, and releases bile.

IN REVIEW

4. Name the structure that brings arterial blood to the liver sinusoids.

5. Name the structure in the liver lobule that collects and transports bile toward the bile duct.

6. What is the hepatic portal triad?

7. Name the elevation on the wall of the duodenum that contains the hepatopancreatic ampulla.

8. What is the exocrine function of the pancreas?

9. What is the endocrine function of the pancreas?

10. What hormones are produced by the pancreas?

The Pancreas

EXERCISE 8.13: Digestive System—Pancreas (Exocrine), Histology

- *Insert the* Anatomy & Physiology | Revealed® **Digestive, Urinary, Reproductive, and Endocrine Systems** *CD, or, if you are already in the* **Dissection** *section, click the* **CHANGE VIEW** *button at the top of the screen, and skip the next two steps.*

- *From the* **Select system** *menu, select* **Digestive.**

- *In the* **Home screen,** *select the* **Dissection** *button in the left portion of the screen. You may click either on the* **Dissection** *button or on the word.*

- *In the* **SELECT A VIEW** *window that appears, click on the* **Select topic** *button.*

- *Choose* **Pancreas (exocrine)** *from the menu.*

- **Histology** *will appear in the* **Select view** *menu.*

- *The* **GO** *button will flash green. Click on it.*

- *Click on* **TAG 1,** *and the following image will appear:*

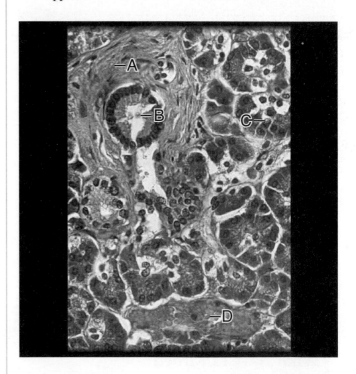

- *Mouse-over the orange pins on the screen to find the information necessary to fill in the following blanks:*

A. _____

B. _____

C. _____

D. _____

Pancreas, (Exocrine), Histology

1. Name the oval collection of secretory (acinar) cells in the exocrine pancreas.
2. What do these cells secrete? How are these secretions regulated?
3. Name the series of ducts that transport digestive enzymes through the exocrine pancreas. Where is the destination of these enzymes?
4. Name the structure that drains blood from, and contains hormones from, the endocrine pancreas.
5. Name the structure that separates the pancreatic lobules.

ATP Synthesis
Animation: Hydrolysis of Sucrose

We have been discovering the structures involved in digestion—the process of breaking down our food into smaller, simple molecules that can be absorbed into the bloodstream. But that is not the end of the story. Amino acids will be recombined by your body to build proteins, fatty acids, and glycerol will be recombined into energy storing fat molecules, and glucose will either be stored as glycogen molecules in your liver and muscle tissues or used to fuel the production of energy in the form of ATP. Any of these molecules can serve to supply the fuel to feed the ATP production process, but they must be converted into glucose first. The following animations will walk you through the process of harvesting energy from the foods you eat. As you view these animations, fill in the blanks in the chart to keep track of the products produced in this process. Then, answer the questions that follow each animation to check your comprehension of the information presented.

- *Insert* Anatomy & Physiology | Revealed® **Digestive, Urinary, Reproductive, and Endocrine Systems** CD, *or, if you are in the* **Dissection** *section, click the* **ANIMATIONS** *button at the bottom of the screen, and skip the next two steps.*

- *From the* **Select system** *menu, select* **Digestive.**

- *In the* **Home screen,** *select the* **Animation** *button in the left portion of the screen. You may click either on the* **Animation** *button or on the word.*

- *In the* **Select topic** *menu, select* **Physiology.**

- *From the* **Select animation** *menu, select* **Hydrolysis of sucrose.**

- *Click the* **Play** *button, and the animation will run in the* **IMAGE AREA.**

- *After viewing the animation, answer the following questions:*

1. The enzyme sucrase breaks the disaccharide _____ into two monosaccharides: _____, or _____ sugar, and _____, or _____ sugar.

2. Where does this reaction occur?

3. For hydrolysis to occur, the sucrose must bind to what part of the sucrase enzyme?

4. What happens to the enzyme when this occurs? What happens to the sucrose?

5. What molecule breaks the bond? Do you see why the process is called hydrolysis?

Produced	Glycolysis	Krebs Cycle	Electron Transport Chain	Total
ATP				
NADH				
FADH$_2$				

hydro = water and lysis = to break. The bond is broken by water.

6. After the bond is broken and the two monosaccharides are released, what happens to the enzyme?

7. How many times can this process be repeated?

8. What three events can occur to end this process?

Animation: NADH/Oxidation-Reduction Reaction

- *Return to the animation menu at the top left of the screen, and from the **Select animation** menu, select **NADH/oxidation-reduction reactions.***

- *Click the **Play** button, and the animation will run in the **IMAGE AREA**.*

- *After viewing the animation, answer the following questions:*

1. Cells obtain energy through the process of _____.

2. How is this energy obtained? What is the energy molecule that results from this process?

3. What is oxidation?

4. A hydrogen atom consists of _____ _____.

5. Whenever a molecule is oxidized, another molecule must be _____, which means _____.

6. What is the enzyme's role in the oxidation-reduction reaction in the cell?

7. What structures does the enzyme have that facilitate this reaction?

8. In this reaction, the substrate is _____ (loses a hydrogen) and the _____ is reduced to _____.

9. What two things occur after the reaction is complete?

10. What is NADH? What is this molecule available to do now?

Animation: Glycolysis

- *Return to the animation menu at the top left of the screen, and from the **Select animation** menu, select **Glycolysis**.*

- *Click the **Play** button, and the animation will run in the **IMAGE AREA**.*

- *After viewing the animation, answer the following questions:*

1. Cells derive energy from the _____ of nutrients, such as _____.

2. The oxidation of _____ to _____ occurs through a series of steps called _____.

3. How many carbons are in a molecule of glucose?

4. The energy released during these _____ reactions is used to form _____ (_____), the _____ of the cell.

5. Name the two initial steps in glycolysis.

6. What are the three molecules that result?

7. What then occurs to the 6-carbon molecule?

8. These 3-carbon molecules are converted to _____.

9. What happens to the electrons in this reaction? What two molecules are formed?

10. What happens to the pyruvate under aerobic conditions?

11. What happens to the pyruvate under anaerobic conditions?

Animation: Krebs Cycle

- *Return to the animation menu at the top left of the screen, and from the **Select animation** menu, select **Krebs cycle.***

- *Click the **Play** button, and the animation will run in the **IMAGE AREA.***

- *After viewing the animation, answer the following questions:*

1. Name the product of glycolysis that enters the Krebs cycle. How many of these molecules are produced per glucose molecule?

2. Name the two-carbon fragment that enters the Krebs cycle.

3. What is another name for the Krebs cycle? Where in the cell does it occur? Where in the cell did glycolysis occur?

4. What two products are formed during the conversion of pyruvate to acetyl-CoA?

5. What reaction occurs to release the CoA carrier molecule?

6. What two products are formed as the 6-carbon molecule is converted to a 5-carbon molecule?

7. What three products are produced during the second oxidation and decarboxylation? This process forms a _____-carbon molecule.

8. Finally, the _____-carbon molecule is further oxidized, and the _____ removed are used to form _____ and _____. What is regenerated by these reactions?

9. Each glucose molecule is broken down into _____ _____ molecules during glycolysis. Then each _____ molecule is converted to _____ and enters the Krebs cycle.

10. For each glucose molecule, how many circuits of the Krebs cycle are completed to completely break down the two pyruvate molecules?

11. For each circuit of the Krebs cycle, how many ATP are produced? Therefore, how many ATP are produced in the Krebs cycle for every glucose molecule?

Animation: Electron Transport and ATP Synthesis

- *Return to the animation menu at the top left of the screen, and from the **Select animation** menu, select **Electron transport and ATP synthesis.***

- *Click the **Play** button, and the animation will run in the **IMAGE AREA.***

- *After viewing the animation, answer the following questions:*

1. In the mitochondrion, the energy stored in NADH is _____.

2. This energy of the _____ is used to form _____.

3. What changes occur to NAD^+ and FAD during glycolysis and the Krebs cycle?

4. Where are the electrons from NADH transferred? How? What happens to the protons?

5. Where are the electrons carried? By what molecule? What else occurs as this is happening?

6. Where are the electrons from $FADH_2$ transferred? And the protons?

7. Where do the electrons travel next? And the protons?

8. What is the terminal electron acceptor? What molecule forms when the electrons are transferred (accepted)?

9. What force is generated by the transfer of protons to the intermembrane space? Where is this force generated?

10. Name the special proton channel proteins that allow the protons to reenter the matrix of the mitochondrion.

11. What energy is used to synthesize the ATP molecules? What products are used to make the ATP?

12. What name is given to this method of ATP production?

CHAPTER **9**

The Urinary System

Overview: The Urinary System

Metabolic wastes, toxins, drugs, hormones, salts, hydrogen ions, and water are all excreted through the urinary system of your body. To do this, it must filter your blood. All of your blood plasma is filtered 60 times per day. That's once every 24 minutes. The average plasma filtration rate is 125 mL per minute for adults with two kidneys. This produces 180 liters or 45 gallons of filtrate every day. Aren't you glad that you don't have to urinate all of that fluid? Fortunately, your body reabsorbs 99 percent of this fluid, so that you only excrete 1.8 liters or 0.45 gallons of urine daily.

The urinary system also regulates blood volume and pressure; regulates the amounts of dissolved solutes in body fluids—called osmolarity; helps to control blood pressure, red blood cell count and the blood's ability to carry oxygen, the pH of body fluids, calcium homeostasis, and, free radical detoxification; and has the ability to convert amino acids to glucose during times of starvation. Talk about multi-tasking!

We will begin our study with an overview of the structures of the urinary system. We will then take a sequential look at the organs in great detail, from the kidneys through the urethra. Finally, we will look at how molecules move through the body and pass through the cell membrane.

Animation: Urinary System Overview

Before beginning your study of the urinary system, view the correlated *Anatomy & Physiology | Revealed®* animation.

- *Insert the* Anatomy & Physiology | Revealed® **Digestive, Urinary, Reproductive, and Endocrine Systems** *CD.*

- *From the* **Select system** *menu, select* **Urinary.**

- *In the* **Home screen,** *select the* **Animation** *button in the left portion of the screen. You may click either on the* **Animation** *button or on the word.*

- *In the* **Select topic** *menu, select* **Anatomy.**

- *From the* **Select animation** *menu, select* **Urinary system overview.**

- *Click the* **Play** *button, and the animation will run in the* **IMAGE AREA.**

- *After viewing the animation, answer the following questions:*

1. What are the major functions of the urinary system?

2. What are the organs of the urinary system?

3. Where are the kidneys located?

4. What is the renal hilum?

5. Blood to be filtered is transported to the kidney by the _____.

6. Filtered blood leaves the kidney via the _____.

7. Name the structure that covers the outer surface of the kidney.

8. Describe the structure of the interior of a kidney.

9. Fluids from the _____ ultimately are funneled into the _____ of the _____.

10. What is the function of the ureters?

11. How is urine propelled through the ureters?

12. What is the urinary bladder? Where is it located?

13. Where do the ureters drain into the urinary bladder?

14. Name the muscle of the urinary bladder wall.

15. Another name for urination is _____.

16. How is urine expelled from the urinary bladder?

17. Compare the functions of the male and female urethras.

18. What is the function of the internal urethral sphincter muscle? Is it under voluntary or involuntary control?

19. What is the function of the external urethral sphincter muscle? Is it under voluntary or involuntary control?

20. How are both of these sphincters involved with urination?

The Upper Urinary System

EXERCISE 9.1:   Urinary System—Upper Urinary, Anterior View

- *Insert the* Anatomy & Physiology | Revealed® **Digestive, Urinary, Reproductive, and**

Endocrine Systems *CD, or, if you are in the* **Animation** *section, click the* **DISSECTION** *button at the bottom of the screen, and skip the next two steps.*

- *From the* **Select system** *menu, select* **Urinary.**

- *In the* **Home screen,** *select the* **Dissection** *button in the left portion of the screen. You may click either on the* **Dissection** *button or on the word.*

- *In the* **SELECT A VIEW** *window that appears, click on the* **Select topic** *button.*

- *Choose* **Upper urinary** *from the menu.*

- **Anterior** *will appear in the* **Select view** *menu.*

- *The* **GO** *button will flash green. Click on it.*

- *After clicking on the first three tags to orient yourself as to location, click on* **TAG 4** *and the following image will appear:*

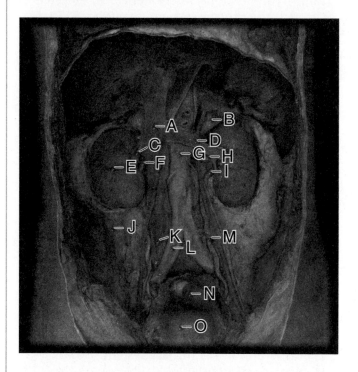

- *Mouse-over the blue pins on the screen to find the information necessary to fill in the following blanks:*

A. _____

B. _____

C. _____

D. _____

E. _____

F. _____

G._____

H._____

I._____

J._____

K._____

L._____

M._____

N._____

O._____

CHECK POINT:

Upper Urinary, Anterior View

1. Describe the structure of the kidneys.
2. What is their function?
3. Name the structure that conveys urine from the renal pelvis.

• *Click on* **TAG 5,** *and the following image will appear:*

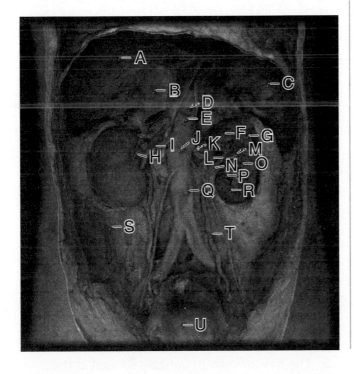

• *Mouse-over the blue pins on the screen to find the information necessary to fill in the following blanks:*

A._____

B._____

C._____

D._____

E._____

F._____

G._____

H._____

I._____

J._____

K._____

L._____

M._____

N._____

O._____

P._____

Q._____

R._____

S._____

T._____

U._____

CHECK POINT:

Upper Urinary, Anterior View

4. What is the function of the renal pyramids? How many are located in each kidney?
5. Name the tip of the renal pyramids. Where does it project?
6. What is a minor calyx? What is its function?

IN · REVIEW

What Have I Learned?

The following questions cover the material that you have just learned—the upper urinary system. Use the information in the **STRUCTURE INFORMATION** window for these structures to answer the following questions:

1. Name the arteries that supply blood to the kidneys.

2. Name the veins that drain blood from the kidneys.

3. Which is anterior to the other?

4. Name the funnel-shaped structure that drains urine from the kidneys to the ureter.

5. Name the structure that conducts urine from the minor calyx to the renal pelvis.

6. How many of each calyx are located in each kidney?

7. Name the outer layer of the kidney. What are its extensions into the middle part of the kidney called? What structures are contained in this outer layer?

8. Name the inner layer of the kidney.

9. Name the part of the kidney where blood vessels and the renal pelvis enter and exit the kidney.

10. Name the part of the kidney between the structure in question 9 and the renal papillae.

The Kidney

Animation: Kidney—Gross Anatomy

- *Insert the* Anatomy & Physiology | Revealed® **Digestive, Urinary, Reproductive, and Endocrine Systems** *CD, or, if you are in the* **Dissection** *section, click the* **ANIMATIONS** *button at the bottom of the screen, and skip the next two steps.*

- *From the* **Select system** *menu, select* **Urinary.**

- *In the* **Home screen,** *select the* **Animation** *button in the left portion of the screen. You may click either on the* **Animation** *button or on the word.*

- *In the* **Select topic** *menu, select* **Anatomy.**

- *From the* **Select animation** *menu, select* **Kidney-gross anatomy.**

- *Click the* **Play** *button, and the animation will run in the* **IMAGE AREA.**

- *After viewing the animation, answer the following questions:*

1. Where are the kidneys located?

2. What is the renal hilum? Name three structures that pass through the hilum.

3. The hilum is continuous with the _____, which is _____.

4. How is the kidney tissue divided? Name the structures of the one tissue division that project into the other.

5. The base of each pyramid is located at the junction of these two tissues, the _____.

6. Name the structure at the apex of the renal pyramids. Where does it project?

7. Several _____ _____ merge to form the larger _____ _____, which merge to form a single, funnel-shaped _____.

8. How many lobes are in each kidney _____? Each lobe consists of _____.

9. How is the blood to be filtered carried to the kidneys?

10. List the series of branches that form from this artery up to and including the glomerulus. Give the locations where each artery is found.

11. Name the initial filtering component of the kidney. Name the functional filtration unit of the kidney. What are its components?

12. Name the arteriole that leaves each glomerulus. _____ Upon leaving the glomerulus, this vessel enters _____.

13. For nephrons in the renal cortex, a _____ forms around the _____.

14. List the series of vessels that carries blood as it drains from these capillary networks.

15. What is the destination of the efferent arterioles associated with nephrons at the corticomedullary junction?

16. These capillaries are known as the _____.

17. Name the sequence of vessels for the blood draining these capillaries.

EXERCISE 9.2: Urinary System—Kidney, Anterior View

- *Insert* Anatomy & Physiology | Revealed® **Digestive, Urinary, Reproductive, and Endocrine Systems** *CD, or, if you are in the*

Animation *section, click the* **DISSECTION** *button at the bottom of the screen, and skip the next two steps.*

- *From the* **Select system** *menu, select* **Urinary.**
- *In the* **Home screen,** *select the* **Dissection** *button in the left portion of the screen. You may click either on the* **Dissection** *button or on the word.*
- *In the* **SELECT A VIEW** *window that appears, click on the* **Select topic** *button.*
- *Choose* **Kidney** *from the menu.*
- **Anterior** *will appear in the* **Select view** *menu.*
- *The* **GO** *button will flash green. Click on it.*
- *Click on* **TAG 1,** *and the following image will appear:*

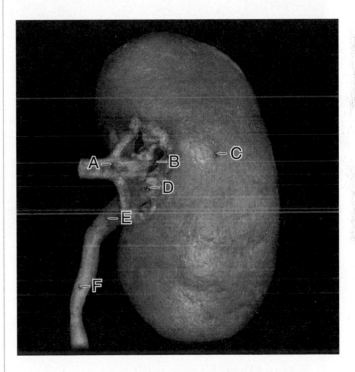

- *Mouse-over the blue pins on the screen to find the information necessary to fill in the following blanks:*

A. _____

B. _____

C. _____

D. _____

E. _____

F. _____

CHECK POINT:

Kidney, Anterior View

1. Name the connective tissue coat of the kidney. What is its function?
2. Describe the structure of the renal hilum. What is its function?
3. What is the function of the renal pelvis?

- *Click on* **TAG 2,** and the following image will appear:

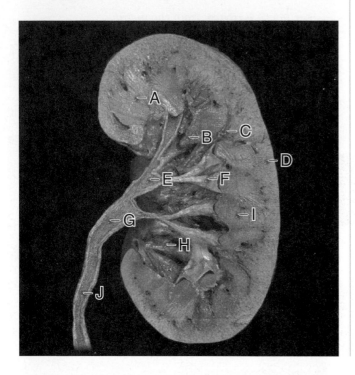

- *Mouse-over the blue pins on the screen to find the information necessary to fill in the following blanks:*

A. _____

B. _____

C. _____

D. _____

E. _____

F. _____

G. _____

H. _____

I. _____

J. _____

CHECK POINT:

Kidney, Anterior View

4. Describe the structure of the renal pyramids.
5. Name the structures that separate the renal pyramids. These structures are an extension of what part of the kidney?
6. Which nephron parts are located in the renal cortex? Which are located in the renal medulla?

IN REVIEW

What Have I Learned?

The following questions cover the material that you have just learned—the kidney. Use the information in the **STRUCTURE INFORMATION** window for these structures to answer the following questions:

1. What is the distribution of the renal artery?

2. Describe its branching before it enters the kidney.

3. List the structures drained by the renal vein.

4. Describe the ureter. What is its function?

5. Trace the pathway of urine as it passes through and then out of the kidney.

The Renal Corpuscle

EXERCISE 9.3:   Urinary System—Renal Corpuscle, Histology

- *Insert the* Anatomy & Physiology | Revealed® **Digestive, Urinary, Reproductive, and Endocrine Systems** *CD, or, if you are already in the* **Dissection** *section, click the* **CHANGE VIEW** *button at the top of the screen, and skip the next two steps.*

- *From the* **Select system** *menu, select* **Urinary.**

- *In the* **Home screen,** *select the* **Dissection** *button in the left portion of the screen. You may click either on the* **Dissection** *button or on the word.*

- *In the* **SELECT A VIEW** *window that appears, click on the* **Select topic** *button.*

- *Choose* **Renal corpuscle** *from the menu.*

- **Histology** *will appear in the* **Select view** *menu.*

- *The* **GO** *button will flash green. Click on it.*

- *Click on* **TAG 1,** *and the following image will appear:*

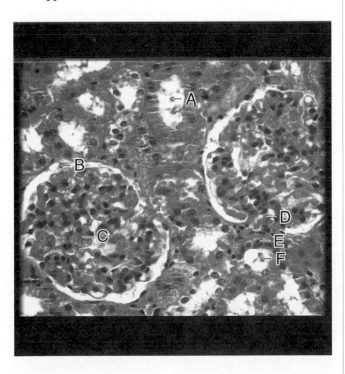

- *Mouse-over the blue pins on the screen to find the information necessary to fill in the following blanks:*

A. _____

B. _____

C. _____

D. _____

E. _____

F._____

CHECK POINT:

Renal Corpuscle, Histology

1. Name the two structures that make up the renal corpuscle.
2. What is the function of the renal corpuscle?
3. Name the structures that constitute a nephron.

Animation: Kidney—Microscopic Anatomy

- *Insert* Anatomy & Physiology | Revealed® **Digestive, Urinary, Reproductive, and Endocrine Systems** *CD, or, if you are in the* **Dissection** *section, click the* **ANIMATIONS** *button at the bottom of the screen, and skip the next two steps.*

- *From the* **Select system** *menu, select* **Urinary.**

- *In the* **Home screen,** *select the* **Animation** *button in the left portion of the screen. You may click either on the* **Animation** *button or on the word.*

- *In the* **Select topic** *menu, select* **Anatomy.**

- *From the* **Select animation** *menu, select* **Kidney—microscopic anatomy.**

- *Click the* **Play** *button, and the animation will run in the* **IMAGE AREA.**

- *After viewing the animation, answer the following questions:*

1. Name the functional filtration unit of the kidney. Approximately how many of these are located in each kidney?

2. Name the two parts to a nephron.

3. Describe the structure of the renal corpuscle. What are its two poles? What structures are located at each pole?

4. Describe the visceral layer of the glomerular capsule. What are podocytes and pedicels?

5. Describe the parietal layer of the glomerular capsule. What is the capsular space?

6. Name and describe the three components of the filtration membrane of the glomerulus.

7. Name and describe the three parts of the renal tubule.

8. What are the structural differences between the thick and thin segments of the nephron loop?

9. Several distal convoluted tubules drain into _____ that pass through the _____. These merge to form _____ that drain into a _____.

10. What events occur as the fluid passes through the renal tubule?

11. Describe the two types of nephrons. What percent of the total number of nephrons consist of each type?

12. Describe the blood pathway to the vascular pole of the glomerulus. Name the apparatus located at this point.

13. What are the three parts of this apparatus? What is its function?

14. The efferent arteriole from each glomerulus enters a _____.

15. For cortical nephrons, where is the peritubular capillary network located? Where does it drain?

16. The efferent arterioles associated with juxtaglomerular nephrons are known as _____ that surround _____. Where does this blood drain?

Animation: Urine Formation

- *Return to the animation menu at the top left of the screen.*
- *In the **Select topic** menu, select **Physiology**.*
- *From the **Select animation** menu, select **Urine formation**.*
- *Click the **Play** button, and the animation will run in the **IMAGE AREA**.*
- *After viewing the animation, answer the following questions:*

1. What are the primary functions of the kidneys?

2. In the process of these functions, they regulate _____. How?

3. Where is urine formed? What are the three processes in its formation?

4. Where does glomerular filtration occur? How does this occur?

5. What is tubular fluid? What occurs to it as it flows through the renal tubule?

6. What occurs to the fluid during tubular reabsorption?

7. How do the solutes move across the tubule wall into the interstitial fluid? Where do they go from there?

8. How does water move through the tubule wall? What percent of water is reabsorbed from each portion of the renal tubule?

9. What percent of the water in the glomerular filtrate returns to the bloodstream? What happens to the remaining water?

10. What event occurs during the process of tubular secretion?

11. Give some examples of waste products involved in this process.

12. What happens to the waste products to prepare them for excretion?

13. Tubular fluid that enters the collecting ducts is called _____.

14. What determines whether the urine produced is dilute or concentrated?

15. Describe the urine produced when water intake is high? When water intake is limited?

16. What are the two key factors that determine the kidneys' ability to concentrate urine?

IN REVIEW

What Have I Learned?

The following questions cover the material that you have just learned—the renal corpuscle. Use the information in the **STRUCTURE INFORMATION** window for these structures to answer the following questions:

1. Name the three parts of the renal tubule.

2. Name the specialized cells of the distal convoluted tubule that are part of the juxtaglomerular device.

3. What is the function of the juxtaglomerular device?

4. Name the structure that regulates the final volume and electrolyte content of the urine. What hormone influences this regulation?

5. What is the function of the capsular space?

The Ureter

EXERCISE 9.4:   Urinary System—Ureter, Histology

- *Insert the* Anatomy & Physiology | Revealed® **Digestive, Urinary, Reproductive, and Endocrine Systems** *CD, or, if you are already in the* **Dissection** *section, click the* **CHANGE VIEW** *button at the top of the screen, and skip the next two steps.*

- *From the* **Select system** *menu, select* **Urinary.**

- *In the* **Home screen,** *select the* **Dissection** *button in the left portion of the screen. You may click either on the* **Dissection** *button or on the word.*

- *In the* **SELECT A VIEW** *window that appears, click on the* **Select topic** *button.*

- *Choose* **Ureter** *from the menu.*

- **Histology** *will appear in the* **Select view** *menu.*

- *The* **GO** *button will flash green. Click on it.*

- *Click on* **TAG 1,** *and the following image will appear:*

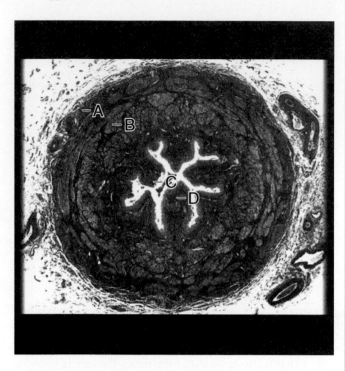

- *Mouse-over the blue pins on the screen to find the information necessary to fill in the following blanks:*

A. _____

B. _____

C. _____

D. _____

CHECK POINT:

Ureter, Histology

1. Name the tissue that loosely attaches the ureter to other structures.
2. What structures are found in this tissue?
3. How does the muscularis change in composition along the length of the ureter? What is its function?
4. What is the mucosa? What is its function?
5. Describe the transitional epithelium. What is unique about its function?

The Female Lower Urinary System

EXERCISE 9.5:   Urinary System—Lower Urinary—Female, Sagittal View

- *Insert the* Anatomy & Physiology | Revealed® **Digestive, Urinary, Reproductive, and Endocrine Systems** *CD, or, if you are already in the* **Dissection** *section, click the* **CHANGE VIEW** *button at the top of the screen, and skip the next two steps.*

- *From the* **Select system** *menu, select* **Urinary.**

- *In the* **Home screen,** *select the* **Dissection** *button in the left portion of the screen. You may click either on the* **Dissection** *button or on the word.*

- *In the* **SELECT A VIEW** *window that appears, click on the* **Select topic** *button.*

- *Choose* **Lower urinary—female** *from the menu.*

- **Sagittal** *will appear in the* **Select view** *menu.*

- *The* **GO** *button will flash green. Click on it.*

- *Click on* **TAG 1,** *and the following image will appear:*

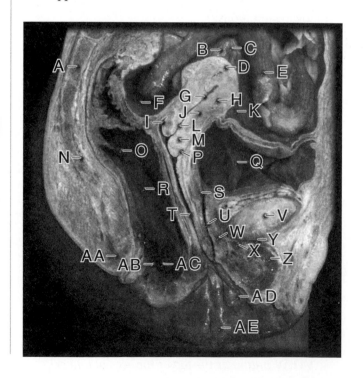

- *Mouse-over the blue pins on the screen to find the information necessary to fill in the following blanks:*

A. _____

B. _____

C. _____

D. _____

E. _____

F. _____

G. _____

H. _____

I. _____

J. _____

K. _____

L. _____

M. _____

N. _____

O. _____

P. _____

Q. _____

R. _____

S. _____

T. _____

U. _____

V. _____

W. _____

X. _____

Y. _____

Z. _____

AA. _____

AB. _____

AC. _____

AD. _____

AE. _____

CHECK POINT:

Lower Urinary—Female, Sagittal View

1. Describe the structure of the urinary bladder.
2. What are the two functions of this organ?
3. What is the function of the female urethra?
4. What effect does volume of urine have on the urinary bladder? What effect does it have on other organs?
5. Give two reasons why urinary tract infections are more common in females.

The Male Lower Urinary System

EXERCISE 9.6: Urinary System—Lower Urinary—Male, Sagittal View

- *Insert the* Anatomy & Physiology | Revealed® **Digestive, Urinary, Reproductive, and Endocrine Systems** *CD, or, if you are already in the* **Dissection** *section, click the* **CHANGE VIEW** *button at the top of the screen, and skip the next two steps.*

- *From the* **Select system** *menu, select* **Urinary.**

- *In the* **Home screen,** *select the* **Dissection** *button in the left portion of the screen. You may click either on the* **Dissection** *button or on the word.*

- *In the* **SELECT A VIEW** *window that appears, click on the* **Select topic** *button.*

- *Choose* **Lower urinary—male** *from the menu.*

- **Sagittal** *will appear in the* **Select view** *menu.*

- *The* **GO** *button will flash green. Click on it.*

• *Click on* **TAG 1,** *and the following image will appear:*

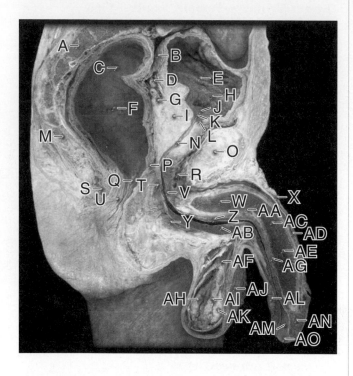

• *Mouse-over the blue pins on the screen to find the information necessary to fill in the following blanks:*

A. _____

B. _____

C. _____

D. _____

E. _____

F. _____

G. _____

H. _____

I. _____

J. _____

K. _____

L. _____

M. _____

N. _____

O. _____

P. _____

Q. _____

R. _____

S. _____

T. _____

U. _____

V. _____

W. _____

X. _____

Y. _____

Z. _____

AA. _____

AB. _____

AC. _____

AD. _____

AE. _____

AF. _____

AG. _____

AH. _____

AI. _____

AJ. _____

AK. _____

AL. _____

AM. _____

AN. _____

AO. _____

CHECK POINT:

Lower Urinary—Male, Sagittal View

1. Name the structures that form the trigone of the urinary bladder.
2. Name the first part of the male urethra.
3. Where is semen formed? What combines to form semen?
4. Name the three different parts of the male urethra. What is the function of the male urethra?
5. Name the external orifice of the urinary tract.

The Urinary Bladder

EXERCISE 9.7: Urinary System—Urinary Bladder (Low Power), Histology

- *Insert the* Anatomy & Physiology | Revealed® **Digestive, Urinary, Reproductive, and Endocrine Systems** *CD, or, if you are already in the* **Dissection** *section, click the* **CHANGE VIEW** *button at the top of the screen, and skip the next two steps.*

- *From the* **Select system** *menu, select* **Urinary.**

- *In the* **Home screen,** *select the* **Dissection** *button in the left portion of the screen. You may click either on the* **Dissection** *button or on the word.*

- *In the* **SELECT A VIEW** *window that appears, click on the* **Select topic** *button.*

- *Choose* **Urinary bladder—low power** *from the menu.*

- **Histology** *will appear in the* **Select view** *menu.*

- *The* **GO** *button will flash green. Click on it.*

- *Click on* **TAG 1,** *and the following image will appear:*

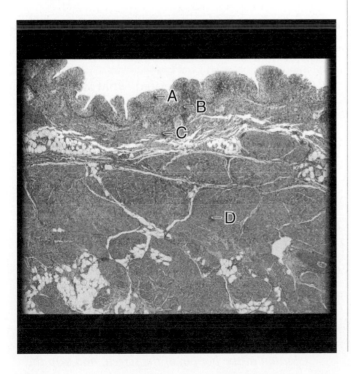

- *Mouse-over the blue pins on the screen to find the information necessary to fill in the following blanks:*

A. _____

B. _____

C. _____

D. _____

CHECK POINT:

Urinary Bladder (Low Power), Histology

1. Name the fibroblastic layer of the urinary bladder that contains blood, lymph vessels, and nerves.
2. Name the three-layered muscle of the urinary bladder.
3. What is this muscle's function?

EXERCISE 9.8: Urinary System—Urinary Bladder (High Power), Histology

- *Insert the* Anatomy & Physiology | Revealed® **Digestive, Urinary, Reproductive, and Endocrine Systems** *CD, or, if you are already in the* **Dissection** *section, click the* **CHANGE VIEW** *button at the top of the screen, and skip the next two steps.*

- *From the* **Select system** *menu, select* **Urinary.**

- *In the* **Home screen,** *select the* **Dissection** *button in the left portion of the screen. You may click either on the* **Dissection** *button or on the word.*

- *In the* **SELECT A VIEW** *window that appears, click on the* **Select topic** *button.*

- *Choose* **Urinary bladder—high power** *from the menu.*

- **Histology** *will appear in the* **Select view** *menu.*

- *The* **GO** *button will flash green. Click on it.*

- *Click on* **TAG 1,** *and the following image will appear:*

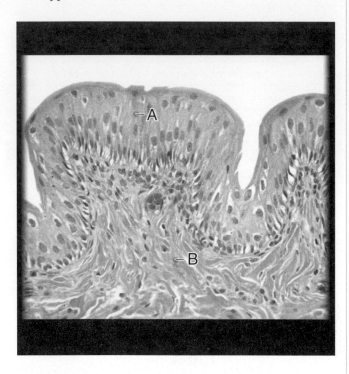

- *Mouse-over the blue pins on the screen to find the information necessary to fill in the following blanks:*

A. _____

B. _____

Animation: Micturition Reflex

- *Insert* Anatomy & Physiology | Revealed® **Digestive, Urinary, Reproductive, and Endocrine Systems** *CD, or, if you are in the* **Dissection** *section, click the* **ANIMATIONS** *button at the bottom of the screen, and skip the next two steps.*

- *From the* **Select system** *menu, select* **Urinary.**

- *In the* **Home screen,** *select the* **Animation** *button in the left portion of the screen. You may*

click either on the **Animation icon** *or on the word itself.*

- *In the* **Select topic** *menu, select* **Physiology.**

- *From the* **Select animation** *menu, select* **Micturition reflex.**

- *Click the* **Play** *button, and the animation will run in the* **IMAGE AREA.**

- *After viewing the animation, answer the following questions:*

1. What kind of reflex is the micturition reflex? What pathway do the impulses take?

2. How is the reflex coordinated? What signals can influence this reflex?

3. What causes the increase in the frequency of action potentials along this pathway?

4. What is the parasympathetic response to these action potentials? How does the urinary bladder respond to these stimuli?

5. What causes a conscious desire to urinate?

6. What happens if urination is not appropriate?

7. What causes the external urethral sphincter to remain contracted? What does this prevent?

8. When urination is desired, what stimulates the micturition reflex?

9. How does the brain then initiate urination?

Movement Through the Cell Membrane

Animation: Diffusion

- *Insert the* Anatomy & Physiology | Revealed® **Digestive, Urinary, Reproductive, and Endocrine Systems** *CD, or, if you are in the* **Dissection** *section, click the* **ANIMATIONS** *button at the bottom of the screen, and skip the next two steps.*

- *From the **Select system** menu, select **Urinary**.*

- *In the **Home screen**, select the **Animation** button in the left portion of the screen. You may click either on the **Animation** button or on the word.*

- *In the **Select topic** menu, select **Physiology**.*

- *From the **Select animation** menu, select **Diffusion**.*

- *Click the **Play** button, and the animation will run in the **IMAGE AREA**.*

- *After viewing the animation, answer the following questions:*

1. What causes the constant random motion of dissolved molecules?

2. What is one result of this random motion?

3. This tendency of molecules to spread out is an example of _____.

4. Even in a solid state, molecules are _____ _____.

5. What two motions characterize the movement of the dissolving sugar molecules?

6. As they dissolve, the sugar molecules move from the area where they are _____ to the area where they are _____.

7. This type of motion, from an area of higher concentration to an area of lower concentration, is called _____.

8. Diffusion continues until _____ _____.

9. What three things affect the rate of diffusion?

Animation Osmosis

- *Return to the animation menu at the top left of the screen, and from the **Select animation** menu, select **Osmosis**.*

- *Click the **Play** button, and the animation will run in the **IMAGE AREA**.*

- *After viewing the animation, answer the following questions:*

1. What is diffusion?

2. What does this process allow?

3. Do most polar molecules freely cross the lipid cell membrane? Name two groups of polar molecules.

4. What is the name for the special case of diffusion that involves the movement of water molecules across a membrane?

5. Why is a molecule of urea unable to diffuse across the membrane?

6. How does a urea molecule interact with water molecules? Why?

7. Why is there now a net movement of water molecules? Which direction do they move?

8. What happens to the water level on the side of the beaker where the water molecules are moving into?

9. Define isotonic, hypertonic, and hypotonic.

Animation: Facilitated Diffusion

- *Return to the animation menu at the top left of the screen, and from the **Select animation** menu, select **Facilitated diffusion**.*

- *Click the **Play** button, and the animation will run in the **IMAGE AREA**.*

- *After viewing the animation, answer the following questions:*

1. What occurs in the process of facilitated diffusion?

2. What is unique about the carrier molecules and the molecules to which they bind?

3. Once the molecule binds to the carrier protein, the protein will facilitate the diffusion process by _____
_____ .

4. Facilitated diffusion and simple diffusion are similar in that both _____
_____ .

5. How is facilitated diffusion different from simple diffusion?

6. What determines which direction facilitated diffusion occurs?

- *Return to the animation menu at the top left of the screen, and from the* **Select animation** *menu, select* **Cotransport.**
- *Click the* **Play** *button, and the animation will run in the* **IMAGE AREA.**
- *After viewing the animation, answer the following questions:*

1. Which direction can small molecules, such as sugars and amino acids, be transported?

2. How does the sugar move? How does the concentration of sugar compare inside and outside the cell?

3. How is this transport of sugar driven through a coupled transport protein? Are these counterions moving from a higher to a lower concentration or from a lower to a higher concentration?

4. What is a symport? What occurs there?

5. How is a low concentration of sodium maintained inside the cell? How is it powered?

6. What is counter-transport?

7. What is an antiport? What occurs there? How is this different than what occurs at a symport?

8. How does the sodium-potassium pump come into play in this process?

CHAPTER 10

The Reproductive System

Overview: The Reproductive System

Without the reproductive system, you wouldn't be here! This is the one organ system that doesn't demand much attention until those wonderful years of puberty—and then it takes over like a raging fire. Hormone levels fluctuate and structures change as the body transitions from childhood into adulthood. This is a time when life takes on a whole different perspective.

Our perspective with *Anatomy & Physiology | Revealed®* will be to look at the reproductive system from a macroscopic as well as a microscopic level. We will begin with the female reproductive system, focusing on the structures unique to the human female. We will then explore how the body orchestrates her monthly reproductive cycles. We will conclude with the male reproductive system, which, although not as complex as the female's, is every bit the product of the same precision engineering.

Animation: Female Reproductive System Overview

Before beginning your study of the reproductive system, view the correlated *Anatomy & Physiology | Revealed®* animation:

- *Insert the* Anatomy & Physiology | Revealed® **Digestive, Urinary, Reproductive, and Endocrine Systems** CD.

- *From the* **Select system** *menu, select* **Reproductive.**

- *In the* **Home screen,** *select the* **Animation** *button in the left portion of the screen. You may click either on the* **Animation** *button or on the word.*

- *In the* **Select topic** *menu, select* **Anatomy.**

- *From the* **Select animation** *menu, select* **Female reproductive system overview.**

- *Click the* **Play** *button, and the animation will run in the* **IMAGE AREA.**

- *After viewing the animation, answer the following questions:*

1. Name the internal organs of the female reproductive system.

2. Name the other organs of the female reproductive system.

3. Describe the structure of the ovaries.

4. What are the ligaments of the ovaries?

5. Name the three structures housed in each suspensory ligament.

6. What is the mesovarium? What is its function?

7. Name the capsule that surrounds each ovary. What tissue type is it?

8. Name the two regions of the ovary. Which region contains the ovarian follicles?

9. What is located within each follicle?

10. What event triggers development of some follicles?

11. What events occur during ovulation?

12. The remnant of the ruptured follicle is the _____, which later degenerates into a _____.

13. Where are the uterine tubes located? What is the mesosalpinx?

14. Into where do the uterine tubes directly open? Why?

15. Name the four segments of each uterine tube.

16. What are the fimbriae? What is their function?

17. Where does fertilization normally occur?

18. What is the shape and function of the uterus?

19. Name the three regions of the uterus.

20. The uterus' hollow lumen connects to the vagina via _____.

21. Describe the three layers of the uterus.

22. Describe the two layers of the lamina propria. Which layer is shed during menstruation? How is it regenerated?

23. Name the three functions of the uterus.

24. Describe the three layers of the vaginal wall.

25. How do bacteria help inhibit the growth of pathogens in the vagina?

The Female Breast

EXERCISE 10.1:   Reproductive System— Breast, Female, Anterior View

- *Insert the* Anatomy & Physiology | Revealed® **Digestive, Urinary, Reproductive, and Endocrine Systems** *CD, or, if you are in the* **Animation** *section, click the* **Dissection** *button at the bottom of the screen, and skip the next two steps.*

- *From the* **Select system** *menu, select* **Reproductive.**

- *In the* **Home screen,** *select the* **Dissection** *button in the left portion of the screen. You may click either on the* **Dissection** *button or on the word.*

- *In the* **SELECT A VIEW** *window that appears, click on the* **Select topic** *button.*

- *Choose* **Breast—female** *from the menu.*

- **Anterior** *will appear in the* **Select view** *menu.*

- *The* **GO** *button will flash green. Click on it.*

- *Click on* **TAG 1,** *and the following image will appear:*

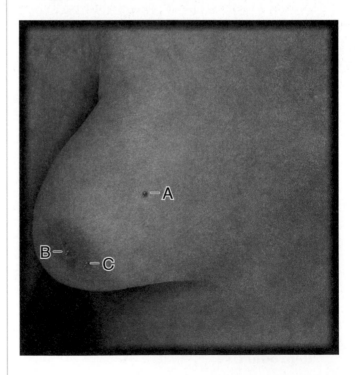

- *Mouse-over the blue pins on the screen to find the information necessary to fill in the following blanks:*

A. _____

B. _____

C. _____

Breast, Female, Anterior View

1. Name the elevation of skin containing lactiferous ducts for transferring milk during nursing.
2. Name the structure that maintains a permanent pigment increase during the first pregnancy.
3. What is the function of this structure?

- *Click on* **TAG 2,** *and the following image will appear:*

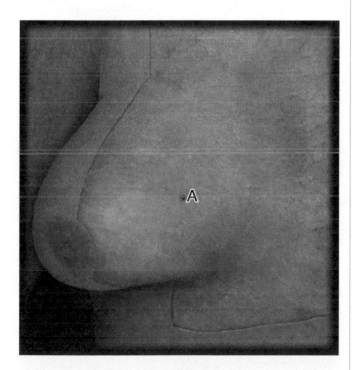

- *Mouse-over the blue pin on the screen to find the information necessary to fill in the following blank:*

A. _____

Breast, Female, Anterior View

4. What cells form most of the nonlactating breast tissue?
5. What are the functions of this tissue?
6. What determines the amount of this tissue in the breast?

- *Click on* **TAG 3,** *and the following image will appear:*

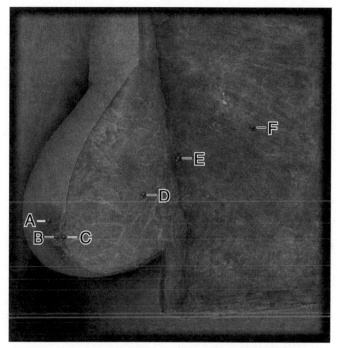

- *Mouse-over the blue pins on the screen to find the information necessary to fill in the following blanks:*

A. _____

B. _____

C. _____

D. _____

E. _____

F. _____

Breast, Female, Anterior View

7. Describe the structure of the lactiferous ducts.
8. What is their function?
9. What is the lactiferous sinus?

331

IN REVIEW

What Have I Learned?

The following questions cover the material that you have just learned—the breast. Use the information in the **STRUCTURE INFORMATION** window for these structures to answer the following questions:

1. What is the composition of the breast?

2. What muscle is deep to the base of the breast?

3. The "tail" of the breast extends into the _____.

4. Into what structures do the suspensory ligaments divide the breast?

5. What tissue partially replaces the breast fat tissue during pregnancy and lactation?

6. The _____ allows movement of the breast independent of the thoracic wall muscles.

7. What are the functions of the suspensory ligaments?

The Female Pelvis

EXERCISE 10.2:   Reproductive System— Pelvis, Female, Sagittal View

- *Insert the* Anatomy & Physiology | Revealed® **Digestive, Urinary, Reproductive, and Endocrine Systems** *CD, or, if you are already in the* **Dissection** *section, click the* **CHANGE VIEW** *button at the top of the screen, and skip the next two steps.*

- *From the* **Select system** *menu, select* **Reproductive.**

- *In the* **Home screen,** *select the* **Dissection** *button in the left portion of the screen. You may click either on the* **Dissection** *button or on the word.*

- *In the* **SELECT A VIEW** *window that appears, click on the* **Select topic** *button.*

- *Choose* **Pelvis—female** *from the menu.*

- *Select* **Sagittal** *from the* **Select view** *menu.*

- *The* **GO** *button will flash green. Click on it.*

- *Click on* **TAG 1,** *and the following image will appear:*

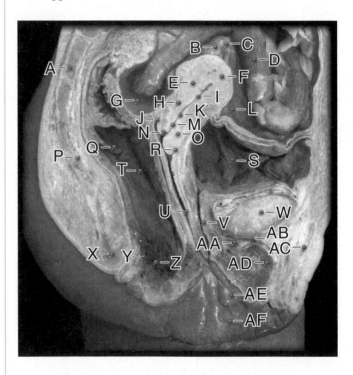

- *Mouse-over the blue pins on the screen to find the information necessary to fill in the following blanks:*

A. _____

B. _____

C. _____

D._____

E._____

F._____

G._____

H._____

I._____

J._____

K._____

L._____

M._____

N._____

O._____

P._____

Q._____

R._____

S._____

T._____

U._____

V._____

W._____

X._____

Y._____

Z._____

AA._____

AB._____

AC._____

AD._____

AE._____

AF._____

CHECK POINT:

Pelvis, Female, Sagittal View

1. Name the primary organ of female sexual response. Describe its structures.
2. Describe the structure of the crus of the clitoris. What is its function?
3. Describe the structure of the labium minus. What is its function?

EXERCISE 10.3:   Reproductive System—Pelvis, Female, Superior View

- *Insert Anatomy & Physiology | Revealed® **Digestive, Urinary, Reproductive, and Endocrine Systems** CD, or, if you are already in the **Dissection** section, click the **CHANGE VIEW** button at the top of the screen, and skip the next two steps.*

- *From the **Select system** menu, select **Reproductive.***

- *In the **Home screen**, select the **Dissection** button in the left portion of the screen. You may click either on the **Dissection** button or on the word.*

- *In the **SELECT A VIEW** window that appears, click on the **Select topic** button.*

- *Choose **Pelvis—female** from the menu, if it is not already selected.*

- *Select **Superior** from the **Select view** menu.*

- *The **GO** button will flash green. Click on it.*

- *Click on **TAG 1,** and the following image will appear:*

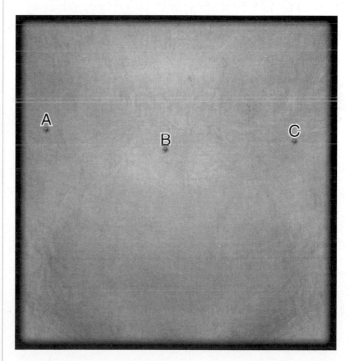

- *Mouse-over the blue pins on the screen to find the information necessary to fill in the following blanks:*

A._____

B._____

C._____

CHECK POINT:

Pelvis, Female, Superior View

1. Name the lower medial abdominal region.
2. What is another name for this region?
3. Name the two regions that flank this region on either side.

- *Click on* **TAG 2,** *and the following image will appear:*

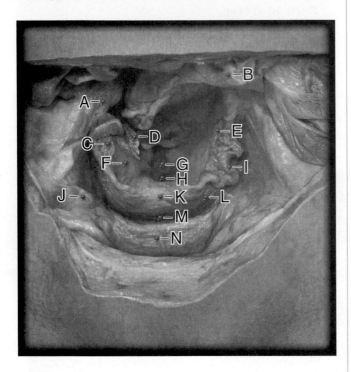

- *Mouse-over the blue pins on the screen to find the information necessary to fill in the following blanks:*

A. _____

B. _____

C. _____

D. _____

E. _____

F. _____

G. _____

H. _____

I. _____

J. _____

K. _____

L. _____

M. _____

N. _____

CHECK POINT:

Pelvis, Female, Superior View

4. What is a hysterectomy?
5. Name the blind recess that lies between the urinary bladder and the vagina.
6. Name the structure that is the embryonic remnant of the gubernaculum of the ovary. What was its embryonic function?

HEADS UP!

View the radiographic images for the reproductive system.

- *Click on the* **Imaging** *button at the bottom of the screen.*
- **Uterus and vagina** *appear in the* **Select region** *menu.*
- *Click the flashing* **GO** *button.*
- *Click on the* **Select structure** *menu in the* **STRUCTURE LIST** *box and then the individual structures to highlight them in the* **IMAGE AREA.**

IN REVIEW

What Have I Learned?

The following questions cover the material that you have just learned—the female pelvis. Use the information in the **STRUCTURE INFORMATION** window for these structures to answer the following questions:

1. Name the region between the labia minora.

2. What is the urogenital triangle?

3. Describe the structure of the labium minus. What is its function?

4. Describe the structure of the labium majus. What is its function?

5. Describe the structure of the vagina. What is its function?

6. Define parturition. What is the standard measure for timing parturition?

7. Name the structure that varies its position with the fullness of the urinary bladder and rectum. What is the function of that structure?

8. The junction of the cervical canal and what structure forms the internal os? What is the function of that structure?

9. Name the female gonads. What is their function?

10. Name the location for a tubal ligation. What is this procedure?

11. Name the location where the cartilage softens in late pregnancy to allow a slight separation of the pubic bones.

12. Name the structure that conducts the ovarian vessels and nerves.

The Uterus

EXERCISE 10.4: Reproductive System— Uterus (Proliferative Phase), Histology

- *Insert the* Anatomy & Physiology | Revealed® **Digestive, Urinary, Reproductive, and Endocrine Systems** *CD, or, if you are already in the* **Dissection** *section, click the* **CHANGE VIEW** *button at the top of the screen, and skip the next two steps.*

- *From the* **Select system** *menu, select* **Reproductive.**

- *In the* **Home screen,** *select the* **Dissection** *button in the left portion of the screen. You may click either on the* **Dissection** *button or on the word.*

- *In the* **SELECT A VIEW** *window that appears, click on the* **Select topic** *button.*

- *Choose* **Uterus (proliferative phase)** *from the menu.*

- *The* **Select view** *menu will then show* **Histology.**

- *The* **GO** *button will flash green. Click on it.*

• *Click on* **TAG 1,** *and the following image will appear:*

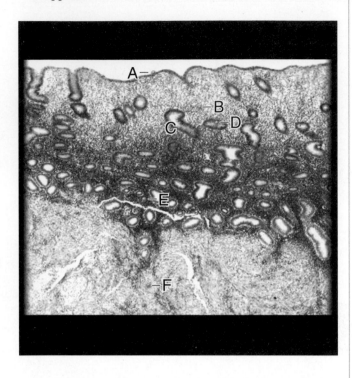

• *Mouse-over the orange pins on the screen to find the information necessary to fill in the following blanks:*

A. _____

B. _____

C. _____

D. _____

E. _____

F. _____

C H E C K   P O I N T :

Uterus (Proliferative Phase), Histology

1. What is unique for the cervical endometrium compared to the endometrium of the body and fundus of the uterus?
2. Describe the endometrial glands. What is their function?
3. The endometrium is formed from what two layers?
4. Name the thick, smooth muscle layer of the uterus.
5. Name the superficial part of the endometrium sloughed during menstruation.

The Overian Follicle

Animation: Comparison of Meiosis and Mitosis

• *Insert the* Anatomy & Physiology | Revealed® **Digestive, Urinary, Reproductive, and Endocrine Systems** *CD, or, if you are in the* **Dissection** *section, click the* **ANIMATIONS** *button at the bottom of the screen, and skip the next two steps.*

• *From the* **Select system** *menu, select* **Reproduction.**

• *In the* **Home screen,** *select the* **Animation** *button in the left portion of the screen. You may click either on the* **Animation** *button or on the word.*

• *In the* **Select topic** *menu, select* **Physiology.**

• *From the* **Select animation** *menu, select* **Comparison of meiosis and mitosis.**

• *Click the* **Play** *button, and the animation will run in the* **IMAGE AREA.**

• *After viewing the animation, answer the following questions:*

1. What is the normal human chromosome number? This number is referred to as _____.

2. There are _____ pairs of chromosomes, known as _____ pairs.

3. What process occurs before both meiosis and mitosis? What happens to the chromosomes afterward?

4. How many successive divisions occur in meiosis?

5. List the events of the first division.

6. List the events of the second division.

7. Define haploid.

8. How many cell divisions occur during mitosis? What is the result of mitosis?

Animation: Meiosis

- *Return to the animation menu at the top left of the screen, and from the* **Select animation** *menu, select* **Meiosis.**

- *Click the* **Play** *button, and the animation will run in the* **IMAGE AREA.**

- *After viewing the animation, answer the following questions:*

1. Sperm and egg cells are formed through the process of _____. Where are these cells found?

2. Are the germ-cell lines diploid or haploid? Through the process of meiosis, the cells produced will be _____, having _____ set(s) of chromosomes.

3. During fertilization, these _____ cells _____ to form a _____ offspring.

4. Name the phase of meiosis when the DNA replicates and each chromosome becomes doubled.

5. These doubled chromosomes consist of _____.

6. The first division of meiosis, meiosis I, _____.

7. In the second division, meiosis II, _____ _____.

8. What is the result of meiosis?

9. What changes have occurred to the chromosomes during prophase I?

10. Name the structure that joins identical sister chromatids.

11. What is crossing-over? What events occur just prior to crossing-over?

12. As a result of crossing-over, how do the two sister chromatids of one chromosome compare?

13. List the events of metaphase I.

14. What is independent assortment?

15. List the events of anaphase I. Describe the chromosomes during this phase.

16. How many members of the homologous chromosome pair will be in each new cell at the end of anaphase I? Are these new cells haploid or diploid?

17. List the events in telophase I.

18. Define cytokinesis.

19. What are the results of meiosis I?

20. How does meiosis II compare to mitosis?

21. List the events of prophase II.

22. What action occurs among the chromosomes during metaphase II? How does metaphase II compare with metaphase I?

23. What events occur during anaphase II? During telophase II?

24. How many cells result from meiosis? How do they compare genetically?

Animation: Unique Features of Meiosis

- *Return to the animation menu at the top left of the screen, and from the* **Select animation** *menu, select* **Unique features of meiosis.**

- *Click the* **Play** *button, and the animation will run in the* **IMAGE AREA.**

- *After viewing the animation, answer the following questions:*

1. Name the three unique features of meiosis.

2. Describe the events of synapsis.

3. Describe the events of crossing-over.

4. What is another name for crossing-over?

5. The name for the first nuclear division is
_____.

6. What is the result of reduction division?

7. What effect does crossing-over have on the sister chromatids of each daughter cell?

EXERCISE 10.5: Reproductive System— Primordial Follicle, Histology

- *Insert the* Anatomy & Physiology | Revealed® **Digestive, Urinary, Reproductive, and Endocrine Systems** *CD, or, if you are in the* **Animation** *section, click the* **DISSECTION** *button at the bottom of the screen, and skip the next two steps.*

- *From the* **Select system** *menu, select* **Reproductive.**

- *In the* **Home screen,** *select the* **Dissection** *button in the left portion of the screen. You may click either on the* **Dissection** *button or on the word.*

- *In the* **SELECT A VIEW** *window that appears, click on the* **Select topic** *button.*

- *Choose* **Primordial follicle** *from the menu.*

- **Histology** *will appear in the* **Select view** *menu.*

- *The* **GO** *button will flash green. Click on it.*

- *Click on* **TAG 1,** *and the following image will appear:*

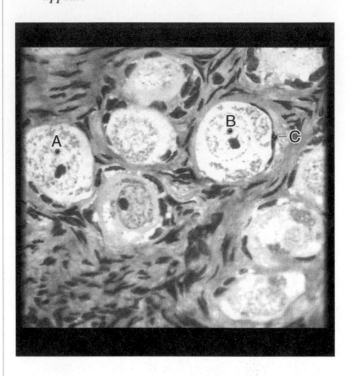

- *Mouse-over the blue pins on the screen to find the information necessary to fill in the following blanks:*

A. _____

B. _____

C. _____

CHECK POINT:

Primordial Follicle, Histology

1. A primordial follicle is one that has not responded to _____ to become a _____.
2. Where are these follicles located?
3. What structure surrounds the oocyte?

Animation: Female Reproductive Cycles

- *Insert the* Anatomy & Physiology | Revealed® **Digestive, Urinary, Reproductive, and Endocrine Systems** *CD, or, if you are in the* **Dissection** *section, click the* **ANIMATIONS** *button at the bottom of the screen, and skip the next two steps.*

- *From the* **Select system** *menu, select* **Reproduction.**

- *In the **Home screen**, select the **Animation** button in the left portion of the screen. You may click either on the **Animation** button or on the word.*
- *In the **Select topic** menu, select **Physiology**.*
- *From the **Select animation** menu, select **Female reproductive cycles**.*
- *Click the **Play** button, and the animation will run in the **IMAGE AREA**.*
- *After viewing the animation, answer the following questions:*

1. Female reproductive cycles are initiated usually between the ages of _____ and _____, a period known as _____.

2. How long is the average female reproductive cycle? Where do the changes occur during the cycle?

3. During each cycle, the hypothalamus releases _____, which stimulates the _____ to release two _____ hormones, _____ and _____.

4. Name the target organ for these hormones.

5. Name the two components of the female reproductive cycle.

6. Ovulation occurs on day _____ of the _____-day ovarian cycle.

7. What are the 14 days prior to ovulation called in the ovarian cycle? The 14 days following ovulation?

8. What regulates the ovarian cycle?

9. Ovulation occurs on day _____ of the _____-day uterine cycle.

10. Describe the two subdivisions of the uterine cycle that occur before ovulation. And the 14 days following ovulation?

11. What controls the uterine cycle?

12. Describe the first five days of the follicular stage.

13. What occurs during this time in the uterine cycle? This is the _____ phase, commonly referred to as _____.

14. Describe the ovarian events during days 6–13 of the follicular phase.

15. Describe the ovarian events two days before ovulation. What hormones influences these events?

16. What causes the final maturation of the follicle? How soon before ovulation does this occur? What is another name for the mature follicle?

17. What event occurs in the primary oocyte just prior to ovulation? What does this form?

18. What event occurs in the uterus during the proliferative stage? Name the hormone responsible for this event and where that hormone is produced.

19. Describe the events of ovulation. What hormone peaks at this time?

20. During the luteal stage, what structure is formed from the remaining luteal cells? What is the function of this structure and the hormones that it produces?

21. After the ovum is fertilized and implants in the uterine wall, what hormone is produced by the cells of the implantation site? What is the function of this hormone?

22. How long does the corpus luteum continue hormone production? What transition takes place after this time? What structures produce hormones in the place of the corpus luteum?

23. If fertilization does not occur, when does the corpus luteum become the corpus albicans? What effect does this have on hormone levels? What events do these changes initiate?

EXERCISE 10.6: Reproductive System—Primary Follicle, Histology

• Insert the Anatomy & Physiology | Revealed® **Digestive, Urinary, Reproductive, and Endocrine Systems** *CD, or, if you are already in the* **Dissection** *section, click the* **CHANGE VIEW** *button at the top of the screen, and skip the next two steps.*

• *From the* **Select system** *menu, select* **Reproductive.**

• *In the* **Home screen,** *select the* **Dissection** *button in the left portion of the screen. You may click either on the* **Dissection** *button or on the word.*

• *In the* **SELECT A VIEW** *window that appears, click on the* **Select topic** *button.*

• *Choose* **Primary follicle** *from the menu.*

• **Histology** *will appear in the* **Select view** *menu.*

• *The* **GO** *button will flash green. Click on it.*

• *Click on* **TAG 1,** *and the following image will appear:*

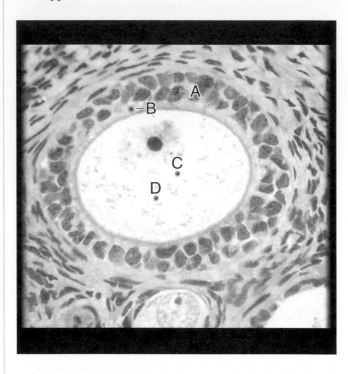

• *Mouse-over the blue pins on the screen to find the information necessary to fill in the following blanks:*

A. _____

B. _____

C. _____

D. _____

CHECK POINT:

Primary Follicle, Histology

1. Name the cells responsible for estrogen secretion. Where are they located?
2. Name the glycoprotein produced by the oocyte.
3. Is the primary oocyte diploid or haploid? What do these terms mean?
4. Name and describe the structure that surrounds the primary oocyte.

EXERCISE 10.7: Reproductive System—Secondary Follicle, Histology

• *Insert the* Anatomy & Physiology | Revealed® **Digestive, Urinary, Reproductive, and Endocrine Systems** *CD, or, if you are already* *in the* **Dissection** *section, click the* **CHANGE VIEW** *button at the top of the screen, and skip the next two steps.*

- *From the* **Select system** *menu, select* **Reproductive.**

- *In the* **Home screen,** *select the* **Dissection** *button in the left portion of the screen. You may click either on the* **Dissection** *button or on the word.*

- *In the* **SELECT A VIEW** *window that appears, click on the* **Select topic** *button.*

- *Choose* **Secondary follicle** *from the menu.*

- **Histology** *will appear in the* **Select view** *menu.*

- *The* **GO** *button will flash green. Click on it.*

- *Click on* **TAG 1,** *and the following image will appear:*

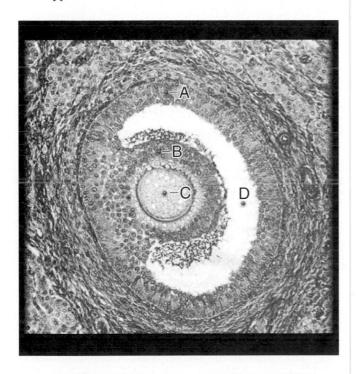

- *Mouse-over the blue pins on the screen to find the information necessary to fill in the following blanks:*

A. _____

B. _____

C. _____

D. _____

CHECK POINT:

Secondary Follicle, Histology

1. Is the secondary follicle haploid or diploid?
2. Name the fluid-filled cavity rich in estrogen. Where is it located?
3. Name the innermost layer of granulosa cells. These cells are in contact with which layer?

Animation: Ovulation Through Implantation

- *Insert the* Anatomy & Physiology | Revealed® **Digestive, Urinary, Reproductive, and Endocrine Systems** *CD, or, if you are in the* **Dissection** *section, click the* **ANIMATIONS** *button at the bottom of the screen, and skip the next two steps.*

- *From the* **Select system** *menu, select* **Reproductive.**

- *In the* **Home screen,** *select the* **Animation** *button in the left portion of the screen. You may click either on the* **Animation** *button or on the word.*

- *In the* **Select topic** *menu, select* **Physiology.**

- *From the* **Select animation** *menu, select* **Ovulation through implantation.**

- *Click the* **Play** *button, and the animation will run in the* **IMAGE AREA.**

- *After viewing the animation, answer the following questions:*

1. What are oocytes? Where are they produced? What is the process of their production?

2. How many primary oocytes are produced before birth? Are they diploid or haploid?

3. How many primary oocytes survive until puberty? Of these, how many will become secondary oocytes?

4. What meiotic stage do the primary oocytes remain in until just before ovulation?

5. Describe the structure of the follicle.

6. What events occur in the primary oocyte just before ovulation? What do these events form?

7. How do the ovaries alternate between ovulations?

8. What events occur during ovulation? Name the structures that receive the ovulated secondary oocyte.

9. What processes move the secondary oocyte toward the uterus?

10. Where does fertilization occur?

11. How many sperm are deposited in the vagina during ejaculation? How many reach the ampulla? How soon do they arrive?

12. How many sperm penetrate the secondary oocyte? What does this event trigger?

13. What events occur on a cellular level during fertilization? What cell do they form? How many chromosomes does this cell contain? Is it diploid or haploid?

14. What is the term for a fertilized oocyte (ovum)? What term describes the mitotic divisions of this fertilized secondary oocyte? Where is the fertilized ovum located while these mitotic divisions occur?

15. What is a morula? What is the term for the hollow sphere of cells into which the morula transforms? Where is the morula located when this transition occurs?

16. What are a trophoblast and embryoblast? What structures are formed from each?

17. What role do the hormones estrogen and progesterone play in the implantation of the blastocyst?

18. What is implantation? When does it occur?

19. What changes occur in the trophoblast as implantation occurs? What are the functions of these new tissues?

20. When does the blastocyst become embedded superficially in the endometrium?

EXERCISE 10.8: Reproductive System—Corpus Albicans, Histology

- *Insert the* Anatomy & Physiology | Revealed® **Digestive, Urinary, Reproductive, and Endocrine Systems** *CD, or, if you are already in the* **Dissection** *section, click the* **CHANGE VIEW** *button at the top of the screen, and skip the next two steps.*

- *From the* **Select system** *menu, select* **Reproductive.**

- *In the* **Home screen,** *select the* **Dissection** *button in the left portion of the screen. You may click either on the* **Dissection** *button or on the word.*

- *In the* **SELECT A VIEW** *window that appears, click on the* **Select topic** *button.*

- *Choose* **Corpus albicans** *from the menu.*

- **Histology** *will appear in the* **Select view** *menu.*

- *The* **GO** *button will flash green. Click on it.*

- *Click on* **TAG 1,** *and the following image will appear:*

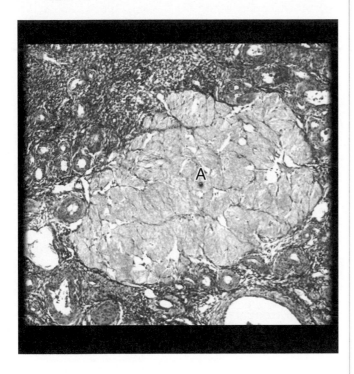

- *Mouse-over the blue pin on the screen to find the information necessary to fill in the following blank:*

A. _____

CHECK POINT:

Corpus Albicans, Histology

1. The corpus albicans is a connective tissue _____ at the surface of the _____.
2. The corpus albicans is a remnant of _____.
3. What does it identify?

Did you know that the name corpus albicans means "white body" and the name corpus luteum means "yellow body"? These are descriptive terms for their appearance.

The Uterine Tube

EXERCISE 10.9: Reproductive System—
Uterine Tube (Low Power),
Histology

- *Insert the* Anatomy & Physiology | Revealed®
Digestive, Urinary, Reproductive, and

Endocrine Systems *CD, or, if you are already in the* **Dissection** *section, click the* **CHANGE VIEW** *button at the top of the screen, and skip the next two steps.*

- *From the* **Select system** *menu, select* **Reproductive.**

- *In the* **Home screen,** *select the* **Dissection** *button in the left portion of the screen. You may click either on the* **Dissection** *button or on the word.*

- *In the* **SELECT A VIEW** *window that appears, click on the* **Select topic** *button.*

- *Choose* **Uterine tube—low power** *from the menu.*

- **Histology** *will appear in the* **Select view** *menu.*

- *The* **GO** *button will flash green. Click on it.*

- *Click on* **TAG 1,** *and the following image will appear:*

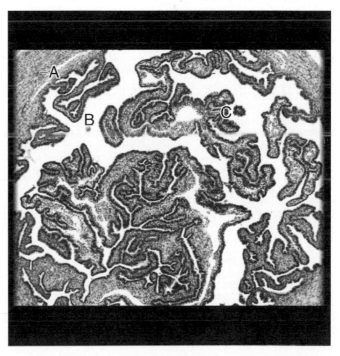

343

- *Mouse-over the orange pins on the screen to find the information necessary to fill in the following blanks:*

A. _____

B. _____

C. _____

CHECK POINT:

Uterine Tube (Low Power), Histology

1. What is the central tubular cavity of the uterine tube? What is its function?
2. What is the lining of the uterine tube? It consists of which tissue type?
3. Name the smooth muscle layers in the wall of the uterine tube. What is its function?

EXERCISE 10.10: Reproductive System— Uterine Tube (High Power), Histology

- *Insert the* Anatomy & Physiology | Revealed® **Digestive, Urinary, Reproductive, and Endocrine Systems** *CD, or, if you are already in the* **Dissection** *section, click the* **CHANGE VIEW** *button at the top of the screen, and skip the next two steps.*

- *From the* **Select system** *menu, select* **Reproductive.**

- *In the* **Home screen,** *select the* **Dissection** *button in the left portion of the screen. You may click either on the* **Dissection** *button or on the word.*

- *In the* **SELECT A VIEW** *window that appears, click on the* **Select topic** *button.*

- *Choose* **Uterine tube—high power** *from the menu.*

- **Histology** *will appear in the* **Select view** *menu.*

- *The* **GO** *button will flash green. Click on it.*

- *Click on* **TAG 1,** *and the following image will appear:*

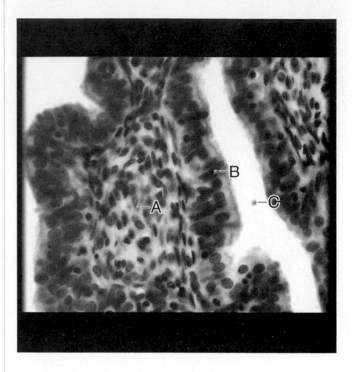

- *Mouse-over the orange pins on the screen to find the information necessary to fill in the following blanks:*

A. _____

B. _____

C. _____

CHECK POINT:

Uterine Tube (High Power), Histology

1. Name the tissue type of the cells lining the uterine tubes.
2. Name the two types of cells located within this tissue. What is the function of each cell type?
3. Name the layer that attaches the epithelium to the muscularis of the uterine tube. Name two structural layers not present in the uterine tube.

The Female Perineum

EXERCISE 10.11: Reproductive System—
Perineum—Female, Inferior
View

- *Insert the* Anatomy & Physiology | Revealed® **Digestive, Urinary, Reproductive, and Endocrine Systems** *CD, or, if you are already in the* **Dissection** *section, click the* **CHANGE VIEW** *button at the top of the screen, and skip the next two steps.*

- *From the* **Select system** *menu, select* **Reproductive.**

- *In the* **Home screen,** *select the* **Dissection** *button in the left portion of the screen. You may click either on the* **Dissection** *button or on the word.*

- *In the* **SELECT A VIEW** *window that appears, click on the* **Select topic** *button.*

- *Choose* **Perineum—female** *from the menu.*

- **Inferior** *will appear in the* **Select view** *menu.*

- *The* **GO** *button will flash green. Click on it.*

- *Click on* **TAG 1,** *and the following image will appear:*

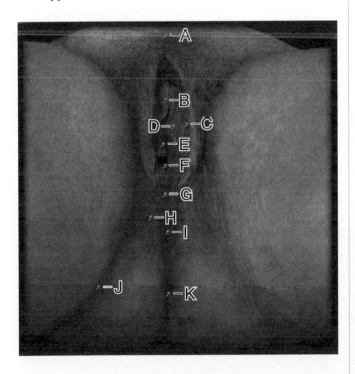

- *Mouse-over the blue pins on the screen to find the information necessary to fill in the following blanks:*

A. _____

B. _____

C. _____

D. _____

E. _____

F. _____

G. _____

H. _____

I. _____

J. _____

K. _____

CHECK POINT:

Perineum—Female, Inferior View

1. Name the subcutaneous fat pad anterior to the pubic symphysis and pubic bones. When does the fat layer increase and decrease?
2. Name the limiting structures of the urogenital triangle. What is contained in this area?
3. What is the hymen? What is its function?

- *Click on* **TAG 2,** *and the following image will appear:*

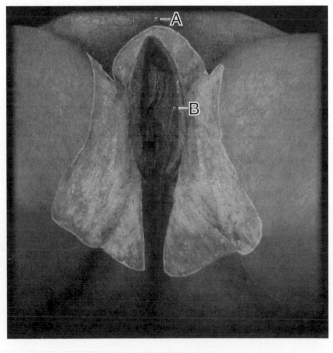

- *Mouse-over the blue pins on the screen to find the information necessary to fill in the following blanks:*

A. _____

B. _____

- *Click on **TAG 3**, and the following image will appear:*

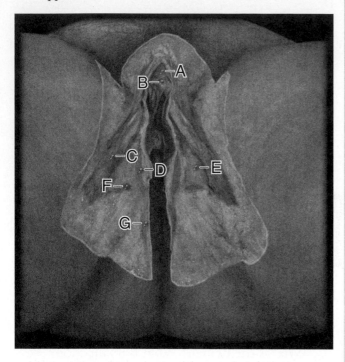

- *Mouse-over the blue pins on the screen to find the information necessary to fill in the following blanks:*

A. _____

B. _____

C. _____

D. _____

E. _____

F._____

G. _____

CHECK POINT:

Perineum—Female, Inferior View

4. Name the hood of thin skin over the glans clitoris. What are its functions?
5. Describe the structure of the glans clitoris. What is its function?
6. What is the perineal membrane? What is its function?

- *Click on **TAG 4**, and the following image will appear:*

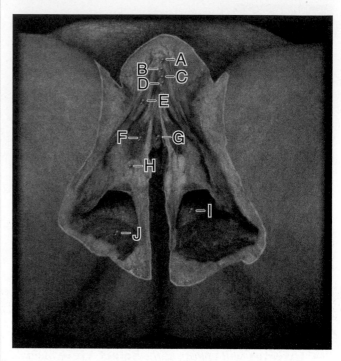

- *Mouse-over the blue pins on the screen to find the information necessary to fill in the following blanks:*

A. _____

B. _____

C. _____

D. _____

E. _____

F._____

G._____

H._____

I._____

J._____

CHECK POINT:

Perineum—Female, Inferior View

7. What is the function of the clitoris? What occurs within the clitoris during sexual arousal?
8. Name the external orifice of the urinary tract. Compare the function in the male and the female.
9. What is the function of the bulb of the vestibule?

IN REVIEW

What Have I Learned?

The following questions cover the material that you have just learned—the female perineum. Use the information in the **STRUCTURE INFORMATION** window for these structures to answer the following questions:

1. What is an episiotomy? Why is this procedure done?

2. What is the anal triangle?

3. What is the intergluteal cleft?

4. What is the gluteal fold?

5. Name the potential space between the labia minora. Name the contents of this space from anterior to posterior.

6. Name the inferior opening of the vagina.

7. Name the structure that protects the external genitalia.

8. What are the functions of the labia minus?

9. What are the actions of the ischiocavernosus muscles?

10. What are the actions of the bulbospongiosus muscles?

11. Name the paired mucus-producing glands lateral to the vaginal opening. What is their function?

12. Describe the structures of the clitoris. What is the function of each?

The Male Pelvis

Animation: Male Reproductive System Overview

- *Insert the* Anatomy & Physiology | Revealed® **Digestive, Urinary, Reproductive, and Endocrine Systems** *CD, or, if you are in the* **Dissection** *section, click the* **ANIMATIONS** *button at the bottom of the screen, and skip the next two steps.*

- *From the* **Select system** *menu, select* **Reproductive.**

- *In the* **Home screen,** *select the* **Animation** *button in the left portion of the screen. You may click either on the* **Animation** *button or on the word.*

- *In the* **Select topic** *menu, select* **Anatomy.**

- *From the* **Select animation** *menu, select* **Male reproductive system overview.**

- *Click the* **Play** *button, and the animation will run in the* **IMAGE AREA.**

- *After viewing the animation, answer the following questions:*

1. The primary male sex organs, or _____, are the _____.

2. List the pathway of sperm from the primary sex organs through the penis.

3. Name the accessory sex glands. What is their function?

4. Describe the location of the testes. How does the temperature there compare to core body temperature? Why?

5. How is the temperature of the testes regulated? What muscles are involved? What are the actions of these muscles when the temperature is too warm *and* too cold?

6. Describe the counter-current heat exchanger mechanism. What effect does this mechanism have on the arterial blood temperature as it reaches the testes?

7. The testes are the site of what events?

8. Name the two connective tissue coats of the testes?

9. Internally, the testes are divided into _____ wedge-shaped _____. Each _____ contains up to _____ ducts called _____. What occurs in these ducts?

10. What is equivalent to the combined length of these ducts in both testes?

11. How many sperm are in the average ejaculate?

12. Describe the epithelium that lines the seminiferous tubule.

13. Name the cells responsible for testosterone production. Where are they located?

14. Describe the pathway of the sperm as they leave the seminiferous tubule, mature, and are stored.

15. How many sperm form each day?

16. What structure is continuous with the tail of the epididymis? Where is this structure located?

17. The _____ joins the duct of the _____ to form the _____, which passes through the _____ to empty into the _____ part of the _____.

18. Name the accessory gland that produces 60 percent of the semen volume. Describe the secretion produced.

19. Name the accessory gland that produces 30 percent of the semen volume. Describe the secretion produced.

20. Name the accessory gland that produces mucus that neutralizes the acidic urethra. What percent of the total semen volume consists of this fluid?

21. What are the dual functions of the penis?

22. What are the divisions of the penis?

23. Describe the erectile bodies of the penis.

24. Which structure encloses the spongy urethra? What is the terminal end of this structure?

25. How is the size and shape of the penis determined?

26. What events occur in the process of an erection?

EXERCISE 10.12: Reproductive System— Pelvis-Male, Sagittal View

- *Insert the* Anatomy & Physiology | Revealed® **Digestive, Urinary, Reproductive, and Endocrine Systems** *CD, or, if you are already in the* **Animation** *section, click the* **Dissection icon** *at the bottom of the screen, and skip the next two steps.*

- *From the* **Select system** *menu, select* **Reproductive.**

- *In the* **Home screen,** *select the* **Dissection** *button in the left portion of the screen. You may click either on the* **Dissection** *button or on the word.*

- *In the **SELECT A VIEW** window that appears, click on the **Select topic** button.*

- *Choose **Pelvis-male** from the menu.*

- ***Sagittal** will appear in the **Select view** menu.*

- *The **GO** button will flash green. Click on it.*

- *Click on **TAG 1**, and the following image will appear:*

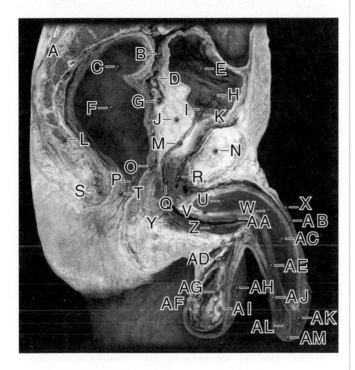

- *Mouse-over the blue pins on the screen to find the information necessary to fill in the following blanks:*

A. _____

B. _____

C. _____

D. _____

E. _____

F. _____

G. _____

H. _____

I. _____

J. _____

K. _____

L. _____

M. _____

N. _____

O. _____

P. _____

Q. _____

R. _____

S. _____

T. _____

U. _____

V. _____

W. _____

X. _____

Y. _____

Z. _____

AA. _____

AB. _____

AC. _____

AD. _____

AE. _____

AF. _____

AG. _____

AH. _____

AI. _____

AJ. _____

AK. _____

AL. _____

AM. _____

CHECK POINT:

Pelvis—Male, Sagittal View

1. Name the paired and unpaired erectile tissues of the penis. What is their function?
2. What structure is responsible for most of the increased size and rigidity of the penis during an erection?
3. What is circumcision? How does the structure of the prepuce compare between circumcised and uncircumcised males?

IN REVIEW

What Have I Learned?

The following questions cover the material that you have just learned—the male pelvis. Use the information in the **STRUCTURE INFORMATION** window for these structures to answer the following questions:

1. Describe the structure of the crus of the penis. What is its function?

2. Describe the structure of the tunica albuginea of the penis. What is its function?

3. Describe the structure of the bulb of the penis. What is its function?

4. What are the actions of the bulbospongiosus muscle in the male? What are the results of these actions?

5. Describe the parts of the male urethra from proximal to distal. What is the function of the male urethra?

6. Name the accessory reproductive organ that contributes 30 percent of the semen volume. What is BPH?

7. Name the accessory reproductive organ that contributes 60 percent of the semen volume. What is the pH of this contribution?

8. Name the network of veins that drain the erectile tissues of the penis. What other structures are drained by these veins?

The Prostate

EXERCISE 10.13:   Reproductive System—
Prostate, Histology

- *Insert the* Anatomy & Physiology | Revealed® **Digestive, Urinary, Reproductive, and Endocrine Systems** *CD, or, if you are already in the* **Dissection** *section, click the* **CHANGE VIEW** *button at the top of the screen, and skip the next two steps.*

- *From the* **Select system** *menu, select* **Reproductive.**

- *In the* **Home screen,** *select the* **Dissection** *button in the left portion of the screen. You may click either on the* **Dissection** *button or on the word.*

- *In the* **SELECT A VIEW** *window that appears, click on the* **Select topic** *button.*

- *Choose* **Prostate** *gland from the menu.*

- **Histology** *will appear in the* **Select view** *menu.*

- *The* **GO** *button will flash green. Click on it.*

- *Click on* **TAG 1,** *and the following image will appear:*

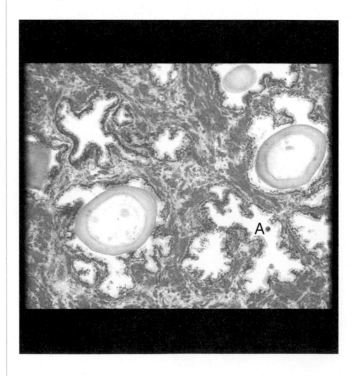

- Mouse-over the blue pin on the screen to find the information necessary to fill in the following blank:

A. _____

The Male Perineum

EXERCISE 10.14: Reproductive System—Perineum—Male, Inferior View

- Insert the Anatomy & Physiology | Revealed® **Digestive, Urinary, Reproductive, and Endocrine Systems** CD, or, if you are already in the **Dissection** section, click the **CHANGE VIEW** button at the top of the screen, and skip the next two steps.

- From the **Select system** menu, select **Reproductive.**

- In the **Home screen,** select the **Dissection** button in the left portion of the screen. You may click either on the **Dissection** button or on the word.

- In the **SELECT A VIEW** window that appears, click on the **Select topic** button.

- Choose **Perineum—male** from the menu.

- **Inferior** will appear in the **Select view** menu.

- The **GO** button will flash green. Click on it.

- Click on **TAG 1,** and the following image will appear:

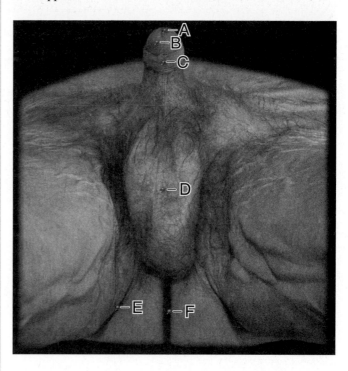

- Mouse-over the blue pins on the screen to find the information necessary to fill in the following blanks:

A. _____

B. _____

C. _____

D. _____

E. _____

F. _____

- *Click on **TAG 2**, and the following image will appear:*

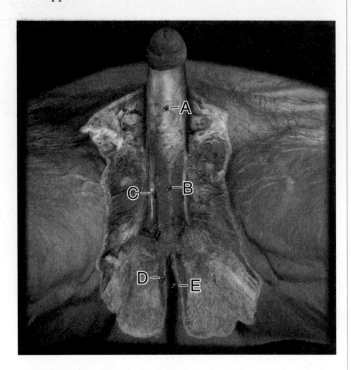

- *Mouse-over the blue pins on the screen to find the information necessary to fill in the following blanks:*

A. _____

B. _____

C. _____

D. _____

E. _____

CHECK POINT:

Perineum—Male, Inferior View

4. Describe the actions of the ischiocavernosus muscle in the male.
5. Describe the actions of the bulbospongiosus muscle in the male. What are the results of these actions?
6. What is the function of the deep fascia of the penis?

- *Click on **TAG 3**, and the following image will appear:*

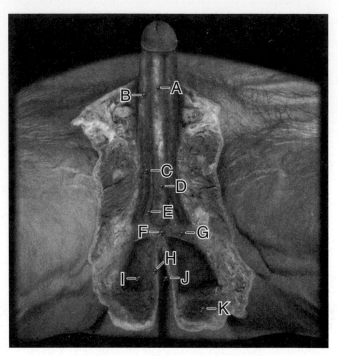

- *Mouse-over the blue pins on the screen to find the information necessary to fill in the following blanks:*

A. _____

B. _____

C. _____

D. _____

E. _____

F. _____

G. _____

H. _____

I. _____

J. _____

K. _____

CHECK POINT:

Perineum—Male, Inferior View

7. Name the single erectile body on the ventral aspect of the penis. This is a continuation of what structure?
8. Name the paired erectile body on the dorsal aspect of the penis. This is a continuation of what structure?
9. What is the function of these erectile tissues?

IN REVIEW

What Have I Learned?

The following questions cover the material that you have just learned—the male perineum. Use the information in the **STRUCTURE INFORMATION** window for these structures to answer the following questions:

1. How does the scrotal temperature compare with body temperature? Why?

2. How is the scrotum divided?

EXERCISE 10.15: Reproductive System—Penis and Scrotum, Anterior View

- *Insert the* Anatomy & Physiology | Revealed® **Digestive, Urinary, Reproductive, and Endocrine Systems** *CD, or, if you are already in the* **Dissection** *section, click the* **CHANGE VIEW** *button at the top of the screen, and skip the next two steps.*

- *From the* **Select system** *menu, select* **Reproductive.**

- *In the* **Home screen,** *select the* **Dissection** *button in the left portion of the screen. You may click either on the* **Dissection** *button or on the word.*

- *In the* **SELECT A VIEW** *window that appears, click on the* **Select topic** *button.*

- *Choose* **Penis and scrotum** *from the menu.*

- **Anterior** *will appear in the* **Select view** *menu.*

- *The* **GO** *button will flash green. Click on it.*

- *Click on* **TAG 1,** *and the following image will appear:*

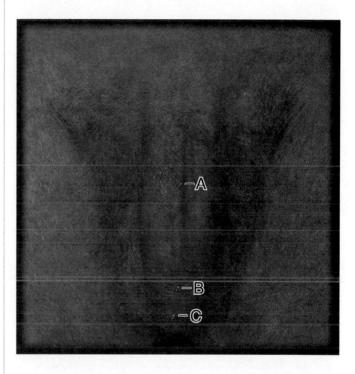

- *Mouse-over the blue pins on the screen to find the information necessary to fill in the following blanks:*

A. _____

B. _____

C. _____

CHECK POINT:

Penis and Scrotum, Anterior View

1. Name the unattached portion of the penis.
2. Name the distal expansion of the penis.
3. Define flaccid. How does the dorsal surface compare when the penis is flaccid and when it is erect?

• *Click on* **TAG 2,** *and the following image will appear:*

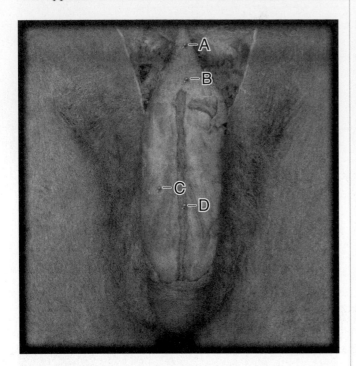

• *Mouse-over the blue pins on the screen to find the information necessary to fill in the following blanks:*

A. _____

B. _____

C. _____

D. _____

CHECK POINT:

Penis and Scrotum, Anterior View

4. Name the large vein on the dorsal surface of the penis that is *not* involved in an erection.
5. Describe the deep fascia of the penis. What is its function?
6. Name the sling of deep fascia that suspends the body of the penis.

• *Click on* **TAG 3,** *and the following image will appear:*

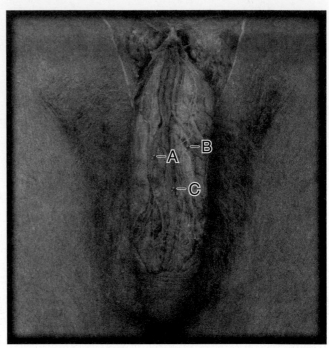

• *Mouse-over the blue pins on the screen to find the information necessary to fill in the following blanks:*

A. _____

B. _____

C. _____

CHECK POINT:

Penis and Scrotum, Anterior View

7. What prevents blood from escaping the erectile tissue?
8. Name the paired arteries on the dorsal side of the penis that distribute to the skin and glans of the penis.
9. Name the nerve that supplies innervation to the body of the penis.

- *Click on **TAG 4**, and the following image will appear:*

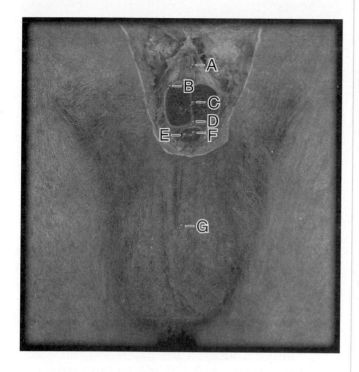

- *Mouse-over the blue pins on the screen to find the information necessary to fill in the following blanks:*

A. _____

B. _____

C. _____

D. _____

E. _____

F. _____

G. _____

CHECK POINT:

Penis and Scrotum, Anterior View

10. Name the distal end cap of the corpora cavernosa.
11. Name the distal end of the corpus spongiosum.
12. Name the opaque, outer fibroelastic capsule of the erectile bodies.

- *Click on **TAG 5**, and the following image will appear:*

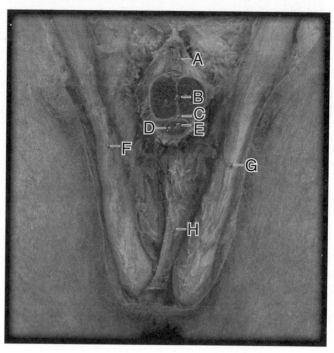

- *Mouse-over the blue pins on the screen to find the information necessary to fill in the following blanks:*

A. _____

B. _____

C. _____

D. _____

E. _____

F. _____

G. _____

H. _____

CHECK POINT:

Penis and Scrotum, Anterior View

13. Name the structures located in the spermatic cord. What muscle is located in the middle fascial layer? What is the function of this muscle?
14. Name the structure that separates the right and left testes.
15. Name the structures that enclose the spermatic cord. What are the three layers that make up these structures?

- *Click on* **TAG 6,** *and the following image will appear:*

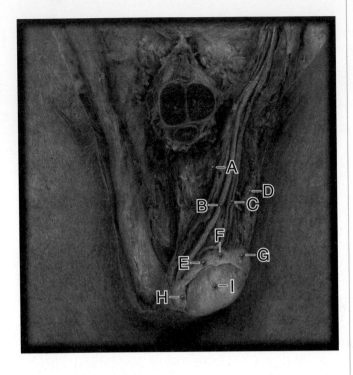

- *Mouse-over the blue pins on the screen to find the information necessary to fill in the following blanks:*

A. _____

B. _____

C. _____

D. _____

E. _____

F. _____

G. _____

H. _____

I. _____

CHECK POINT:

Penis and Scrotum, Anterior View

16. Name the paired male gonads. What are their functions?
17. Name the structure for maturation and storage of spermatozoa. How long can spermatozoa be stored?
18. What are the three parts of the structure in question 17?

IN REVIEW

What Have I Learned?

The following questions cover the material that you have just learned—the penis and scrotum. Use the information in the **STRUCTURE INFORMATION** window for these structures to answer the following questions:

1. What is emission? Name the structure that conducts sperm and testicular fluid during emission.

2. Name the arteries located within the spermatic cord. What is their distribution?

3. Name the network of veins that ascend within the spermatic cord. Where do these drain? (Hint: right is different from left)

The Seminiferous Tubule

EXERCISE 10.16: Reproductive System—
Seminiferous Tubule,
Histology

- *Insert the* Anatomy & Physiology | Revealed® **Digestive, Urinary, Reproductive, and Endocrine Systems** *CD, or, if you are already in the* **Dissection** *section, click the* **CHANGE VIEW** *button at the top of the screen, and skip the next two steps.*

- *From the* **Select system** *menu, select* **Reproductive.**

- *In the* **Home screen,** *select the* **Dissection** *button in the left portion of the screen. You may click either on the* **Dissection** *button or on the word.*

- *In the* **SELECT A VIEW** *window that appears, click on the* **Select topic** *button.*

- *Choose* **Seminiferous tubule** *from the menu.*

- **Histology** *will appear in the* **Select view** *menu.*

- *The* **GO** *button will flash green. Click on it.*

- *Click on* **TAG 1,** *and the following image will appear:*

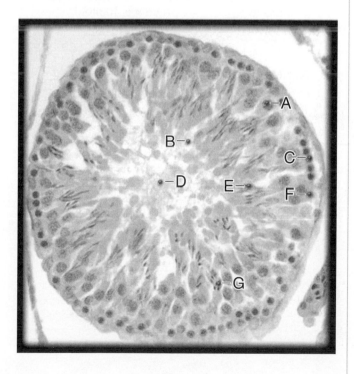

- *Mouse-over the blue pins on the screen to find the information necessary to fill in the following blanks:*

A. _____

B. _____

C. _____

D. _____

E. _____

F. _____

G. _____

CHECK POINT:

Seminiferous Tubule, Histology

1. Name the site of spermatogenesis. Where is this site located? How many are in each location?
2. Name the central tubular cavity of the seminiferous tubules. What is usually contained in this cavity?
3. What are the functions of the sustentacular cells? What other names are they known by?

Animation: Spermatogenesis

- *Insert the* Anatomy & Physiology | Revealed® **Digestive, Urinary, Reproductive, and Endocrine Systems** *CD, or, if you are in the* **Dissection** *section, click the* **ANIMATIONS** *button at the bottom of the screen, and skip the next two steps.*

- *From the* **Select system** *menu, select* **Reproductive.**

- *In the* **Home screen,** *select the* **Animation** *button in the left portion of the screen. You may click either on the* **Animation** *button or on the word.*

- *In the* **Select topic** *menu, select* **Physiology.**

- *From the* **Select animation** *menu, select* **Spermatogenesis.**

- *Click the* **Play** *button, and the animation will run in the* **IMAGE AREA.**

• *After viewing the animation, answer the following questions:*

1. Where are sperm formed?

2. Describe the structure and lining of the seminiferous tubules. What specialized cell types are located there?

3. Name the germ cells from which sperm cells arise.

4. What differences occur in the daughter cells of the spermatogonia? Are they haploid or diploid? How many chromosomes do they each contain?

5. The primary spermatocytes divide by _____ to form _____ _____ spermatocytes, each containing _____ chromosomes. The _____ divide again to form _____.

6. What is spermiogenesis? Where does it occur? What changes occur during this process?

7. Sperm cells are _____ when they leave the seminiferous tubules and testis, and mature _____.

IN REVIEW

What Have I Learned?

The following questions cover the material that you have just learned—the seminiferous tubule. Use the information in the **STRUCTURE IN-FORMATION** window for these structures to answer the following questions:

1. Name the most primitive cell in the male germ line. Diploid or haploid?

2. Name the diploid cells derived from the cells in question 1. These cells give rise to_____, which are (haploid/diploid). These cells in turn give rise to _____, which are (haploid/diploid).

3. Name and describe the structure of the male germ cells. What is their function?

CHAPTER 11

The Endocrine System

Overview: The Endocrine System

The endocrine system is one of two organ systems involved with internal communication. It often works hand-in-hand with the other—the nervous system. An example of this complement between the two systems occurs between the autonomic nervous system and the suprarenal (adrenal) gland of the endocrine system during the fight-or-flight response.

We touched on the function of the endocrine system when we discussed the reproductive system in the last chapter. The body orchestrates the dynamic events of puberty and reproduction with intercellular messengers called hormones. But hormones regulate many other bodily functions, including but not limited to digestion, blood glucose levels, mood, and growth. Your textbook has an excellent discussion of this topic.

We will begin with a discussion of what has been termed the master gland because of its regulation of other endocrine glands and conclude with the endocrine system's involvement with reproduction.

Animation: Hypothalamus and Pituitary Gland

Before beginning your study of the endocrine system, view the correlated *Anatomy & Physiology | Revealed®* animations:

- *Insert the* Anatomy & Physiology | Revealed® **Digestive, Urinary, Reproductive, and Endocrine Systems** *CD.*

- *From the* **Select system** *menu, select* **Endocrine.**

- *In the* **Home screen,** *select the* **Animation** *button in the left portion of the screen. You may click either on the* **Animation** *button or on the word.*

- *In the* **Select topic** *menu, select* **Anatomy.**

- *From the* **Select animation** *menu, select* **Hypothalamus and pituitary gland.**

- *Click the* **Play** *button, and the animation will run in the* **IMAGE AREA.**

- *After viewing the animation, answer the following questions:*

1. The hypothalamus is sometimes referred to as the _____. Why?

2. Where in the brain is the hypothalamus located?

3. Describe the structure of the hypothalamus.

4. What is the infundibulum? What is its function?

5. Where is the pituitary gland located? How is it divided?

6. What is another name for the anterior pituitary? How is it connected to the hypothalamus?

7. What travels along this pathway? What is their function?

8. What is another name for the posterior pituitary? How is it connected to the hypothalamus?

9. What travels along this pathway? How are they transported? What is their destination?

10. Name the two classes of hypothalamic hormones that regulate the anterior pituitary. How do they reach the anterior pituitary? What is their function?

11. How do anterior pituitary hormones arrive at their target tissues?

12. Describe an example of these hormones and their function.

13. Name the hormones produced by the posterior pituitary. What is the source of posterior pituitary hormones?

14. Name two posterior pituitary hormones. How do they arrive at the posterior pituitary?

15. Name the structures that store the posterior pituitary hormones. What causes their release? Where are they released?

16. Name the functions of each posterior pituitary hormone.

Animation: Hormonal Communication

- *Return to the animation menu at the top left of the screen, and from the* **Select topic** *menu select* **Physiology.**

- *From the* **Select animation** *menu, select* **Hormonal communication.**

- *Click the* **Play** *button, and the animation will run in the* **IMAGE AREA.**

- *After viewing the animation, answer the following questions:*

1. In general, how does hormonal communication begin? What reaction then occurs?

2. How are hormones transported to target cells?

3. What occurs when the hormones arrive at their target cells?

4. What then triggers changes in the target cells?

Animation: Intracellulor Receptor Model

- *Return to the animation menu at the top left of the screen, and from the* **Select animation** *menu, select* **Intracellular receptor model.**

- *Click the* **Play** *button, and the animation will run in the* **IMAGE AREA.**

- *After viewing the animation, answer the following questions:*

1. Describe aldosterone, the hormone used in the animation.

2. What does aldosterone bind with in the cytoplasm of the cell?

3. Where does the aldosterone-receptor complex go, and where does it bind?

4. This binding stimulates the synthesis of what molecule? What is the function of this molecule?

5. Where does this mRNA molecule go, and what does it do?

6. What is directed by this binding? What response is produced?

Animation: Receptors and G Proteins

- *Return to the animation menu at the top left of the screen, and from the* **Select animation** *menu, select* **Receptors and G proteins.**

- *Click the* **Play** *button, and the animation will run in the* **IMAGE AREA.**

- *After viewing the animation, answer the following questions:*

1. What is located on the membrane-bound receptor on the outside of the cell?

2. What is a ligand?

3. To what does the portion of the membrane-bound receptor on the inside of the cell bind?

4. What are the three subunits of this protein? What is attached to the alpha subunit?

5. What changes occur in the G protein when the ligand binds to the receptor site? What changes occur to the alpha subunit?

6. What now occurs with the activated alpha subunit? How long can this step be repeated?

7. What occurs when the ligand separates from the receptor site?

8. How is the alpha subunit inactivated?

9. What occurs with the G protein subunits after this inactivation?

The Hypothalamus, Pituitary, and Pineal Glands

EXERCISE 11.1: Endocrine System—Hypothalamus/Pituitary/Pineal, Lateral View

- *Insert the* Anatomy & Physiology | Revealed® **Digestive, Urinary, Reproductive, and Endocrine Systems** *CD, or, if you are in the* **Animation** *section, click the* **DISSECTION** *button at the bottom of the screen, and skip the next two steps.*

- *From the* **Select system** *menu, select* **Endocrine.**

- *In the* **Home screen,** *select the* **Dissection** *button in the left portion of the screen. You may click either on the* **Dissection** *button or on the word.*

- *In the* **SELECT A VIEW** *window that appears, click on the* **Select topic** *button.*

- *Choose* **Hypothalamus/Pituitary/Pineal** *from the menu.*

- **Lateral** *will appear in the* **Select view** *menu.*

- *The* **GO** *button will flash green. Click on it.*

- *Click on* **TAG 3,** *and the following image will appear:*

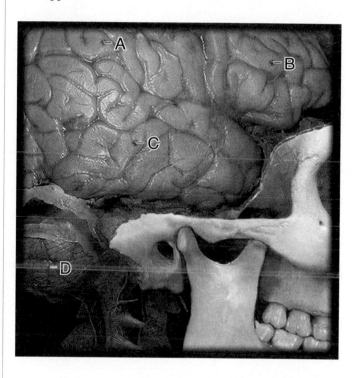

- *Mouse-over the blue pins on the screen to find the information necessary to fill in the following blanks:*

A. _____

B. _____

C. _____

D. _____

- *Click on* **TAG 4,** *and the following image will appear:*

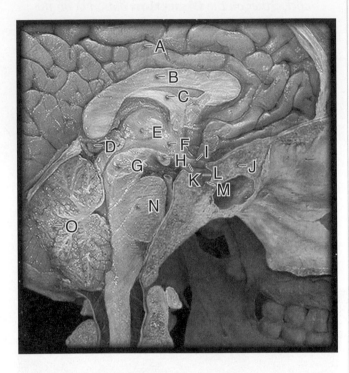

- *Mouse-over the blue pins on the screen to find the information necessary to fill in the following blanks:*

A. _____

B. _____

C. _____

D. _____

E. _____

F. _____

G. _____

H. _____

I. _____

J. _____

K. _____

L. _____

M. _____

N. _____

O. _____

HEADS UP!

Be sure to view the radiographic images for the endocrine system.

- *Click on the* **IMAGING** *button at the bottom of the screen.*

- *Click on the* **Select region** *menu, and select* **Hypothalamus and pituitary.**

- *Click the flashing* **GO** *button.*

- *Click on the* **Select structure** *menu in the* **STRUCTURE LIST** *box and then on the individual structures to highlight them in the* **IMAGE AREA.**

The Pituitary Gland

EXERCISE 11.2   Endocrine System—Anterior Pituitary, Histology

- *Insert the* Anatomy & Physiology | Revealed® **Digestive, Urinary, Reproductive, and Endocrine Systems** *CD, or, if you are already in the* **Dissection** *section, click the* **CHANGE VIEW** *button at the top of the screen, and skip the next two steps.*

- *From the* **Select system** *menu, select* **Endocrine.**

- *In the* **Home screen,** *select the* **Dissection** *button in the left portion of the screen. You may click either on the* **Dissection** *button or on the word.*

- *In the* **SELECT A VIEW** *window that appears, click on the* **Select topic** *button.*

- *Choose* **Anterior pituitary** *from the menu.*

- **Histology** *will appear in the* **Select view** *menu.*

- *The* **GO** *button will flash green. Click on it.*

- *Click on* **TAG 1,** *and the following image will appear:*

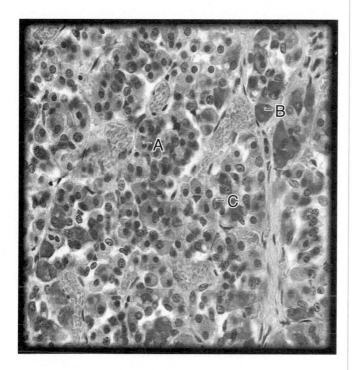

- *Mouse-over the orange pins on the screen to find the information necessary to fill in the following blanks:*

A. _____

B. _____

C. _____

EXERCISE 11.3: Endocrine System—Posterior Pituitary, Histology

- *Insert* Anatomy & Physiology | Revealed® **Digestive, Urinary, Reproductive, and**

Endocrine Systems *CD, or, if you are already in the* **Dissection** *section, click the* **CHANGE VIEW** *button at the top of the screen, and skip the next two steps.*

- *From the* **Select system** *menu, select* **Endocrine.**

- *In the* **Home screen,** *select the* **Dissection** *button in the left portion of the screen. You may click either on the* **Dissection** *button or on the word.*

- *In the* **SELECT A VIEW** *window that appears, click on the* **Select topic** *button.*

- *Choose* **Posterior pituitary** *from the menu.*

- **Histology** *will appear in the* **Select view** *menu.*

- *The* **GO** *button will flash green. Click on it.*

- *Click on* **TAG 1,** *and the following image will appear:*

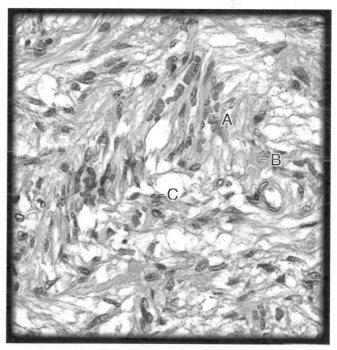

- *Mouse-over the orange pins on the screen to find the information necessary to fill in the following blanks:*

A. _____

B. _____

C. _____

Posterior Pituitary, Histology

1. Name and describe the cells that occupy 25 percent of the volume of the posterior pituitary. What is their function?
2. What are herring bodies? What hormones are associated with them?
3. What is the function of the small blood vessels of the posterior pituitary?

The Thyroid Gland

Animation: Thyroid Gland

- *Insert the* Anatomy & Physiology | Revealed® **Digestive, Urinary, Reproductive, and Endocrine Systems** *CD, or, if you are in the* **Dissection** *section, click the* **ANIMATIONS** *button at the bottom of the screen, and skip the next two steps.*

- *From the* **Select system** *menu, select* **Endocrine.**

- *In the* **Home screen,** *select the* **Animation** *button in the left portion of the screen. You may click either on the* **Animation** *button or on the word.*

- *In the* **Select topic** *menu, select* **Anatomy.**

- *From the* **Select animation** *menu, select* **Thyroid gland.**

- *Click the* **Play** *button, and the animation will run in the* **IMAGE AREA.**

- *After viewing the animation, answer the following questions:*

1. Name the largest endocrine gland. Describe its location and structure.

2. What causes the thyroid's reddish color? Name the blood vessels that distribute to and drain the thyroid gland. What other function do the veins have?

3. The thyroid gland is composed of spherical structures called _____. Describe the wall of these structures. What do they surround?

4. Describe what is located in the lumen of the follicles.

5. What are located in the loose connective tissue between the follicles? Which cells secrete hormones?

6. Name that hormone and its function. What is its antagonist?

7. Name the two different hormones referred to as thyroid hormone.

8. Name the hormone that maintains TH synthesis and secretion. Where is this hormone secreted?

9. What is contained in thyroglobulin?

10. Name the molecules that cross the follicular cell from the blood in the first phase of TH production.

11. What is the destination of these molecules? What are they converted to?

12. What two molecules combine in the lumen of the follicles? What is formed by this combination? How long can it be stored in the follicle lumen?

13. In the second phase of TH production, what becomes of the combined molecules formed in phase one?

14. Name the primary effect of TH. Describe its importance in children.

15. What is hyperthyroidism? What can result from this condition?

16. What is hypothyroidism? What can result from this condition?

EXERCISE 11.4:   Endocrine System—Thyroid Gland, Anterior View and Histology

- *Click the* **DISSECTION** *button at the bottom of the screen.*

- *In the* **SELECT A VIEW** *window that appears, click on the* **Select topic** *button.*

- *Choose* **Thyroid gland** *from the menu.*

- *Click the* **Select view** *menu and select* **Anterior.**

- *The* **GO** *button will flash green. Click on it.*

- *Click on* **TAG 1,** *and the following image will appear:*

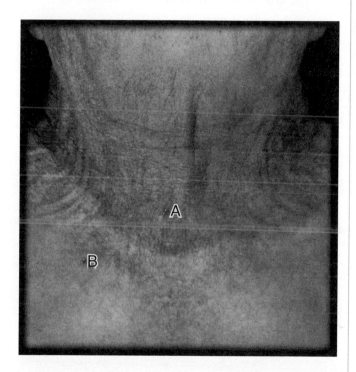

- *Mouse-over the blue pins on the screen to find the information necessary to fill in the following blanks:*

A. _____

B. _____

- *Click on* **TAG 2,** *and the following image will appear:*

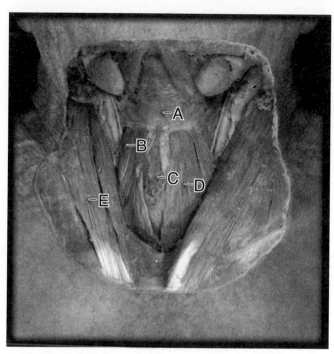

- *Mouse-over the blue pins on the screen to find the information necessary to fill in the following blanks:*

A. _____

B. _____

C. _____

D. _____

E. _____

- *Click on* **TAG 3,** *and the following image will appear:*

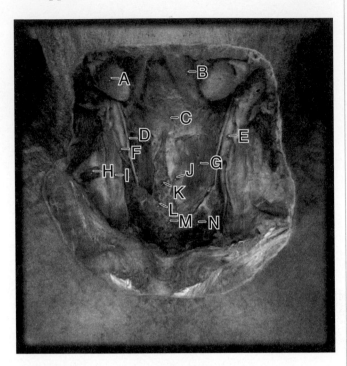

- *Mouse-over the blue pins on the screen to find the information necessary to fill in the following blanks:*

A. _____

B. _____

C. _____

D. _____

E. _____

F. _____

G. _____

H. _____

I. _____

J. _____

K. _____

L. _____

M. _____

N. _____

Thyroid Gland, Anterior View and Histology

1. Describe the structure of the thyroid gland.
2. What is the function of the thyroid gland?
3. What is the name for an enlarged thyroid gland? What is the cause of this condition?

- *Click the* **CHANGE VIEW** *button at the top of the screen.*

- *In the* **SELECT A VIEW** *window,* **Thyroid gland** *appears in the* **Select topic** *menu.*

- *From the* **Select view** *menu, choose* **Histology.**

- *The* **GO** *button will flash green. Click on it.*

- *Click on* **TAG 1,** *and the following image will appear:*

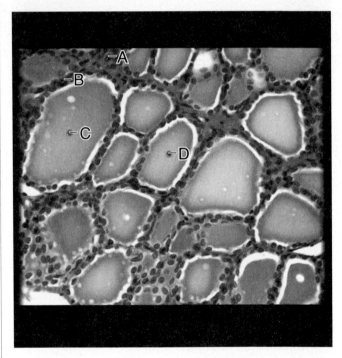

- *Mouse-over the blue pins on the screen to find the information necessary to fill in the following blanks:*

A. _____

B. _____

C. _____

D. _____

CHECK POINT:

Thyroid Gland, Anterior View and Histology

4. Name the thyroid cells that produce calcitonin. What is the function of calcitonin?
5. Describe the structure and function of the
   thyroid follicle.
6. What is follicular colloid? How long of a supply can be stored there?

The Parathyroid Glands

Animation: Parathyroid Glands

- *Insert the* Anatomy & Physiology | Revealed®
  **Digestive, Urinary, Reproductive, and
  Endocrine Systems** *CD, or, if you are in the*
  Dissection *section, click the* **ANIMATIONS**
  *button at the bottom of the screen, and skip the
  next two steps.*

- *From the* **Select system** *menu, select*
  Endocrine.

- *In the* **Home screen,** *select the* **Animation**
  *button in the left portion of the screen. You may
  click either on the* **Animation** *button or on the
  word.*

- *In the* **Select topic** *menu, select* **Anatomy.**

- *From the* **Select animation** *menu, select*
  Parathyroid glands.

- *Click the* **Play** *button, and the animation will
  run in the* **IMAGE AREA.**

- *After viewing the animation, answer the following
  questions:*

1. Describe the location of the parathyroid
   glands. How many are there? What arteries
   supply these glands?

2. Name the two types of parathyroid gland
   cells. What are their functions?

3. What causes the release of PTH?

4. How does PTH raise blood calcium levels?

5. What are the symptoms that may result from
   hyperparathyroidism?

6. What is the most common cause of
   hypoparathyroidism?

7. What are the symptoms of hypoparathyroidism?

EXERCISE 11.5:   Endocrine System—
Parathyroid Gland,
Histology

- *Insert the* Anatomy & Physiology | Revealed®
  **Digestive, Urinary, Reproductive, and
  Endocrine Systems** *CD, or, if you are already
  in the* **Dissection** *section, click the* **CHANGE
  VIEW** *button at the top of the screen, and skip
  the next two steps.*

- *From the* **Select system** *menu, select*
  Endocrine.

- *In the* **Home screen,** *select the* **Dissection**
  *button in the left portion of the screen. You may
  click either on the* **Dissection** *button or on the
  word.*

- *In the* **SELECT A VIEW** *window that appears,
  click on the* **Select topic** *button.*

- *Choose* **Parathyroid gland** *from the menu.*

- **Histology** *will appear in the* **Select view**
  menu.

- *The* **GO** *button will flash green. Click on it.*

- *Click on* **TAG 1,** *and the following image will appear:*

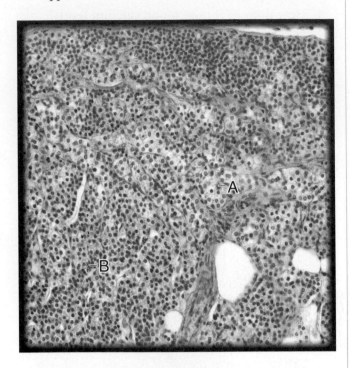

- *Mouse-over the blue pins on the screen to find the information necessary to fill in the following blanks:*

A. _____

B. _____

CHECK POINT:

Parathyroid Gland, Histology

1. Name the most numerous functional cell type in the parathyroid gland. What hormone do they secrete?
2. What is the function of this hormone? What hormone is an antagonist to this one?
3. Name the parathyroid cells that increase in number with age.

The Thymus

EXERCISE 11.6:   Endocrine System—
Thymus—Fetus, Anterior
View

- *Insert the* Anatomy & Physiology | Revealed® **Digestive, Urinary, Reproductive, and Endocrine Systems** *CD, or, if you are already*

in the **Dissection** *section, click the* **CHANGE VIEW** *button at the top of the screen, and skip the next two steps.*

- *From the* **Select system** *menu, select* **Endocrine.**
- *In the* **Home screen,** *select the* **Dissection** *button in the left portion of the screen. You may click either on the* **Dissection** *button or on the word.*
- *In the* **SELECT A VIEW** *window that appears, click on the* **Select topic** *button.*
- *Choose* **Thymus—fetus** *from the menu.*
- **Anterior** *will appear in the* **Select view** *menu.*
- *The* **GO** *button will flash green. Click on it.*
- *Click on* **TAG 2,** *and the following image will appear:*

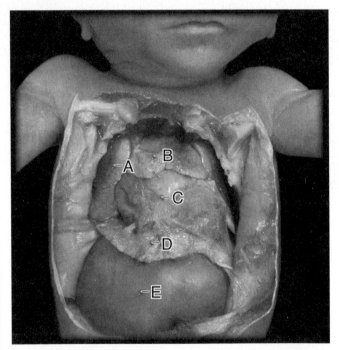

- *Mouse-over the blue pins on the screen to find the information necessary to fill in the following blanks:*

A. _____

B. _____

C. _____

D. _____

E. _____

EXERCISE 11.7: Endocrine System—Thymus—Adult, Anterior View

- *Insert the* Anatomy & Physiology | Revealed® **Digestive, Urinary, Reproductive, and Endocrine Systems** *CD, or, if you are already in the* **Dissection** *section, click the* **CHANGE VIEW** *button at the top of the screen, and skip the next two steps.*

- *From the* **Select system** *menu, select* **Endocrine.**

- *In the* **Home screen,** *select the* **Dissection** *button in the left portion of the screen. You may click either on the* **Dissection** *button or on the word.*

- *In the* **SELECT A VIEW** *window that appears, click on the* **Select topic** *button.*

- *Choose* **Thymus—adult** *from the menu.*

- **Anterior** *will appear in the* **Select view** *menu.*

- *The* **GO** *button will flash green. Click on it.*

- *After clicking on* **TAGs 1** *and* **2** *to orient yourself to this location, click on* **TAG 3,** *and the following image will appear:*

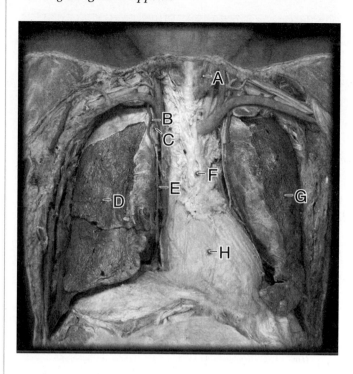

- *Mouse-over the blue pins on the screen to find the information necessary to fill in the following blanks:*

A. _____

B. _____

C. _____

D. _____

E. _____

F. _____

G. _____

H. _____

The Pancreas

Animation: Pancreas

- *Insert the* Anatomy & Physiology | Revealed® **Digestive, Urinary, Reproductive, and Endocrine Systems** *CD, or, if you are in the* **Dissection** *section, click the* **ANIMATIONS** *button at the bottom of the screen, and skip the next two steps.*

- *From the* **Select system** *menu, select* **Endocrine.**

- *In the* **Home screen,** *select the* **Animation** *button in the left portion of the screen. You may click either on the* **Animation** *button or on the word.*

- *In the* **Select topic** *menu, select* **Anatomy.**

- *From the* **Select animation** *menu, select* **Pancreas.**

- *Click the* **Play** *button, and the animation will run in the* **IMAGE AREA.**

- *After viewing the animation, answer the following questions:*

1. The organs of the endocrine system secrete _____ directly into the _____ to _____.

2. Describe the location of the pancreas. Is it endocrine, exocrine, or both?

3. The primary cells of the pancreas are _____. What do these cells secrete, and what is the function of these secretions?

4. Where are the endocrine cells of the pancreas located? What are they called?

5. Name the four types of cells located in the pancreatic islets. What type of substances do these cells release?

6. Which cells are activated by declining blood glucose levels? What hormone do they release? What action does this hormone stimulate?

7. Which cells are activated by increasing blood glucose levels? What hormone do they release? What action does this hormone stimulate?

8. Which cells release somatostatin? What actions does this hormone initiate?

9. Which cells secrete pancreatic polypeptide? What is the function of this secretion?

10. What do these pancreatic hormones together provide for?

EXERCISE 11.8: Endocrine System—Pancreas, Anterior view

- *Click on the* **DISSECTION** *button at the bottom of the screen.*

- *In the* **SELECT A VIEW** *window that appears, click on the* **Select topic** *button.*

- *Choose* **Pancreas** *from the menu.*

- **Anterior** *will appear in the* **Select view** *menu.*

- *The* **GO** *button will flash green. Click on it.*

- *After clicking on* **TAGs 1–3** *to orient yourself to this location, click on* **TAG 4,** *and the following image will appear:*

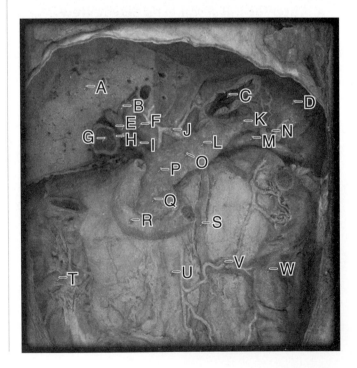

• *Mouse-over the blue pins on the screen to find the information necessary to fill in the following blanks:*

A. _____

B. _____

C. _____

D. _____

E. _____

F. _____

G. _____

H. _____

I. _____

J. _____

K. _____

L. _____

M. _____

N. _____

O. _____

P. _____

Q. _____

R. _____

S. _____

T. _____

U. _____

V. _____

W. _____

CHECK POINT:

Pancreas, Anterior View

1. What is the exocrine function of the pancreas?
2. What is the endocrine function of the pancreas?
3. Name two events that can result in diabetes mellitus.

EXERCISE 11.9:   Endocrine System—Pancreas (Endocrine), Histology

• *Insert the* Anatomy & Physiology | Revealed® **Digestive, Urinary, Reproductive, and Endocrine Systems** *CD, or, if you are already in the* **Dissection** *section, click the* **CHANGE VIEW** *button at the top of the screen, and skip the next two steps.*

• *From the* **Select system** *menu, select* **Endocrine.**

• *In the* **Home screen,** *select the* **Dissection** *button in the left portion of the screen. You may click either on the* **Dissection** *button or on the word.*

• *In the* **SELECT A VIEW** *window that appears, click on the* **Select topic** *button.*

• *Choose* **Pancreas (endocrine)** *from the menu.*

• **Histology** *will appear in the* **Select view** *menu.*

• *The* **GO** *button will flash green. Click on it.*

• *Click on* **TAG 1,** *and the following image will appear:*

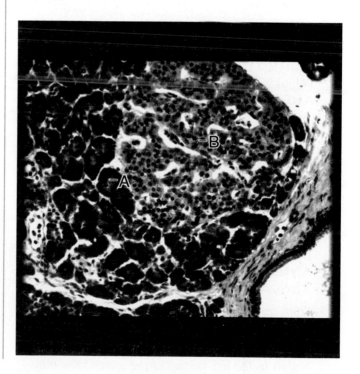

- *Mouse-over the orange pins on the screen to find the information necessary to fill in the following blanks:*

A. _____

B. _____

CHECK POINT:

Pancreas (Endocrine), Histology

1. Name the clusters of cells that form the endocrine portion of the pancreas.
2. Name the hormones produced by these cells.
3. Where are these hormones secreted?

HEADS UP!

View the radiographic images for the endocrine system.

- *Click on the* **IMAGING** *button at the bottom of the screen.*
- *Click on the* **Select region** *menu, and select* **Pancreas.**
- *Click the flashing* **GO** *button.*
- *Click on the* **Select structure** *menu in the* **STRUCTURE LIST** *box, and then on the individual structures to highlight them in the* **IMAGE AREA.**

The Suprarenal (Adrenal) Gland

Animation: Suprarenal (Adrenal) Gland

- *Insert the* Anatomy & Physiology | Revealed® **Digestive, Urinary, Reproductive, and Endocrine Systems** *CD, or, if you are in the* **Dissection** *or image section, click the* **ANIMATIONS** *button at the bottom of the screen, and skip the next two steps.*
- *From the* **Select system** *menu, select* **Endocrine.**
- *In the* **Home screen,** *select the* **Animation** *button in the left portion of the screen. You may click either on the* **Animation** *button or on the word.*
- *In the* **Select topic** *menu, select* **Anatomy.**
- *From the* **Select animation** *menu, select* **Suprarenal (adrenal) glands.**

- *Click the* **Play** *button, and the animation will run in the* **IMAGE AREA.**
- *After viewing the animation, answer the following questions:*

1. The organs of the endocrine system secrete _____ directly into the _____ to _____.

2. Describe the location of the suprarenal glands. Where do they receive their blood supply? Name the single vein that drains them.

3. The suprarenal glands are composed of what two layers?

4. What are corticosteroids? Where are they synthesized?

5. What are mineralocorticoids? Where are they synthesized?

6. What is the principal mineralocorticoid? What is its function?

7. What are glucocorticoids? Where are they synthesized?

8. What are the two most common glucocorticoids?

9. What are the adrenal sex hormones? Where are they synthesized?

10. Name the inner core of the suprarenal glands. What cells are located in large numbers there?

11. What hormones are produced in this inner core area? What stimulates their release? What is their function?

EXERCISE 11.10: Endocrine System— Suprarenal (Adrenal) Gland, Anterior View and Histology

- *Click the* **DISSECTION** *button at the bottom of the screen.*
- *In the* **SELECT A VIEW** *window that appears, click on the* **Select topic** *button.*

- *Choose* **Suprarenal (adrenal) gland** *from the menu.*
- *Click the* **Select view** *menu, and select* **Anterior.**
- *The* **GO** *button will flash green. Click on it.*
- *Click on* **TAG 1,** *and the following image will appear:*

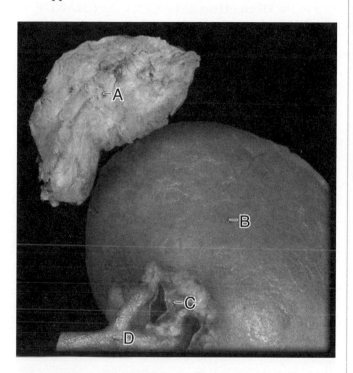

- *Mouse-over the blue pins on the screen to find the information necessary to fill in the following blanks:*

A. _____

B. _____

C. _____

D. _____

CHECK POINT:

Suprarenal (Adrenal) Gland, Anterior View and Histology

1. Name the endocrine gland located superior to the kidney.
2. Name the two parts of this gland.
3. What are the functions of these two parts?

- *Click on* **TAG 2,** *and the following image will appear:*

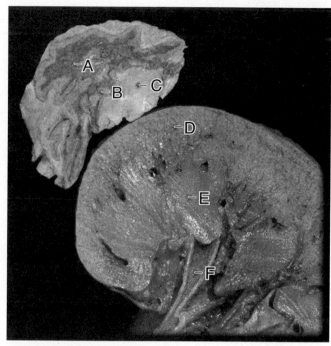

- *Mouse-over the blue pins on the screen to find the information necessary to fill in the following blanks:*

A. _____

B. _____

C. _____

D. _____

E. _____

F._____

CHECK POINT:

Suprarenal (Adrenal) Gland, Anterior View and Histology

4. Name the connective tissue that surrounds the suprarenal gland.
5. Name the lipid-rich outer part of the suprarenal gland. What is the function of each of the three regions of this part?
6. Name the reddish-brown core of the suprarenal gland. What is its function?

- *Click the* **CHANGE VIEW** *button at the top of the screen.*
- *In the* **SELECT A VIEW** *window,* **Suprarenal (adrenal) gland** *appears in the* **Select topic** *menu.*

- Click the **Select view** menu and choose **Histology.**
- The **GO** button will flash green. Click on it.
- Click on **TAG 1,** and the following image will appear:

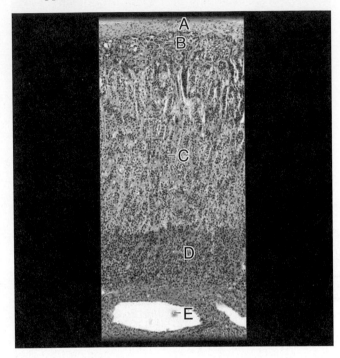

- Mouse-over the orange pins on the screen to find the information necessary to fill in the following blanks:

A. _____

B. _____

C. _____

D. _____

E. _____

CHECK POINT:

Suprarenal (Adrenal) Gland, Anterior View and Histology

7. Name the five layers of the suprarenal gland from superficial to deep.
8. Name the thick fibroblastic outer coat of the suprarenal gland. What is its function?
9. Name the layer of the suprarenal gland responsible for sex hormone production.

The Ovary

EXERCISE 11.11: Endocrine System—Ovary, Superior View

- Insert the Anatomy & Physiology | Revealed® **Digestive, Urinary, Reproductive, and Endocrine Systems** CD, or, if you are already in the **Dissection** section, click the **CHANGE VIEW** button at the top of the screen, and skip the next two steps.
- From the **Select system** menu, select **Endocrine.**
- In the **Home screen,** select the **Dissection** button in the left portion of the screen. You may click either on the **Dissection** button or on the word.
- In the **SELECT A VIEW** window that appears, click on the **Select topic** button.
- Choose **Ovary** from the menu.
- **Superior** will appear in the **Select view** menu.
- The **GO** button will flash green. Click on it.
- After clicking on **TAG 1** to orient yourself to this location, click on **TAG 2,** and the following image will appear:

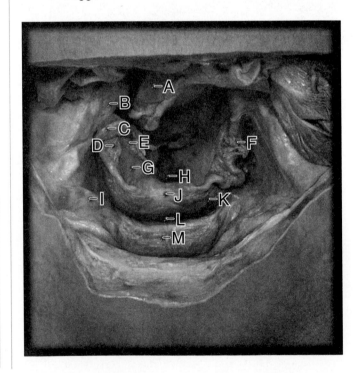

- *Mouse-over the blue pins on the screen to find the information necessary to fill in the following blanks:*

A. _____

B. _____

C. _____

D. _____

E. _____

F. _____

G. _____

H. _____

I. _____

J. _____

K. _____

L. _____

M. _____

CHECK POINT:

Ovary, Superior View

1. Describe the structure of the female gonads.
2. Name the hormones produced by the female gonads.
3. What change occurs in these structures after menopause?

The Testis

EXERCISE 11.12: Endocrine System—Testis, Anterior View

- *Insert the Anatomy & Physiology | Revealed® **Digestive, Urinary, Reproductive, and Endocrine Systems** CD, or, if you are already in the **Dissection** section, click the **CHANGE VIEW** button at the top of the screen, and skip the next two steps.*

- *From the **Select system** menu, select **Endocrine.***

- *In the **Home screen**, select the **Dissection** button in the left portion of the screen. You may click either on the **Dissection** button or on the word.*

- *In the **SELECT A VIEW** window that appears, click on the **Select topic** button.*

- *Choose **Testis** from the menu.*

- **Anterior** *will appear in the **Select view** menu.*

- *The **GO** button will flash green. Click on it.*

- *After clicking on **TAGs 1–3** to orient yourself to this location, click on **TAG 4**, and the following image will appear:*

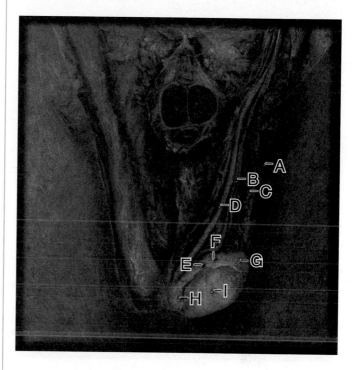

- *Mouse-over the blue pins on the screen to find the information necessary to fill in the following blanks:*

A. _____

B. _____

C. _____

D. _____

E. _____

F. _____

G. _____

H. _____

I. _____

CHECK POINT:

Testis, Anterior View

1. Describe the structure of the male gonads.
2. Name the hormones produced by the male gonads.
3. What is the function of testosterone?

- *Click on* **TAG 5,** *and the following image will appear:*

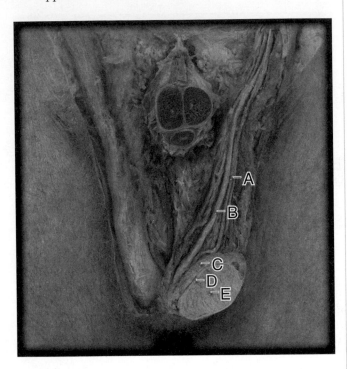

- *Mouse-over the blue pins on the screen to find the information necessary to fill in the following blanks:*

A. _____

B. _____

C. _____

D. _____

E. _____

The Seminiferous Tubule

EXERCISE 11.13: Endocrine System—Seminiferous Tubule, Histology

- *Insert the* Anatomy & Physiology | Revealed® **Digestive, Urinary, Reproductive, and Endocrine Systems** *CD, or, if you are already in the* **Dissection** *section, click the* **CHANGE VIEW** *button at the top of the screen, and skip the next two steps.*

- *From the* **Select system** *menu, select* **Endocrine.**

- *In the* **Home screen,** *select the* **Dissection** *button in the left portion of the screen. You may click either on the* **Dissection** *button or on the word.*

- *In the* **SELECT A VIEW** *window that appears, click on the* **Select topic** *button.*

- *Choose* **Seminiferous tubule** *from the menu.*

- **Histology** *will appear in the* **Select view** *menu.*

- *The* **GO** *button will flash green. Click on it.*

- *Click on* **TAG 1,** *and the following image will appear:*

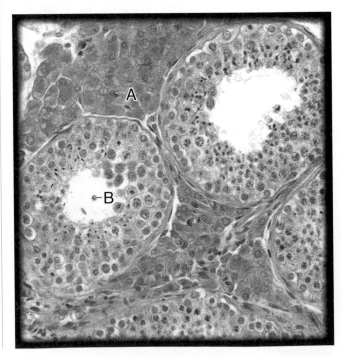

• *Mouse-over the blue pins on the screen to find the information necessary to fill in the following blanks:*

A. _____

B. _____

Seminiferous Tubule, Histology

1. Name the structures responsible for testosterone production.
2. What is another name for these cells?
3. Name the structures that these cells surround. What correlation do they have with these structures?

IN REVIEW

What Have I Learned?

The following questions cover the material that you have just learned—the endocrine system. Use the information in the **STRUCTURE INFORMATION** window for these structures to answer the following questions:

1. The pituitary gland rests in what structure of which bone?

2. Name the fossa that surrounds the pituitary gland.

3. Describe the structure and function of the mammillary body.

4. Name the cell that represents a chromophil depleted of hormone.

5. List the hormones released by the anterior pituitary gland.

6. List the hormones released by the posterior pituitary gland.

7. Name the function of the following hormones: T_3 and T_4 and calcitonin. (These may require a little research on your part.)

8. Which division of the autonomic nervous system is responsible for the fight-or-flight response? What two hormones regulate this response? What is the origin of these hormones?

9. Name two disorders of suprarenal cortex hormone secretion.

10. Name the layer of the suprarenal gland responsible for mineralocorticoid production.

11. Name the layer of the suprarenal gland responsible for glucocorticoid production.

12. In the following chart, list the hormones produced by the suprarenal gland, where they are produced, and their function.

HORMONE	WHERE PRODUCED	FUNCTION
_____	_____	_____
_____	_____	_____